Implementing Discrete Mathematics

Combinatorics and Graph Theory with Mathematica®

Implementing Discrete Mathematics

Combinatorics and Graph Theory with Mathematica®

With programs by Steven Skiena and Anil Bhansali

Steven Skiena
State University of New York, Stony Brook

ADDISON-WESLEY PUBLISHING COMPANY
The Advanced Book Program
Redwood City, California • Menlo Park, California • Reading, Massachusetts
New York • Don Mills, Ontario • Wokingham, United Kingdom • Amsterdam
Bonn • Sydney • Singapore • Tokyo • Madrid • San Juan

Mathematica is not associated with Mathematica, Inc., Mathematica Policy Research, Inc. or MathTech, Inc.

Publisher: *Allan M. Wylde*
Production Manager: *Jan V. Benes*
Marketing Manager: *Laura Likely*

Library of Congress Cataloging-in-Publication Data

Skiena, Steven S.
Implementing discrete Mathematics: combinatorics and graph theory with Mathematica/Steven Skiena.
p. cm.
"The Advanced Book Program."
Includes bibliographical references.
1. Combinatorial analysis—Data processing. 2. Graph theory—Data processing. 3. Mathematica (Computer program) I. Title.
QA164.S56 1990 511′.6′0285—dc 20 90-695
ISBN 0-201-50943-1

This book was prepared by the author, using the TEX typesetting language .

ISBN 0-201-50943-1

3 4 5 6 7 8 9 10-MU-95 94 93 92 91

Preface

The Reason for This Book

To me, the most exciting aspect of discrete mathematics is studying a particular mathematical structure and discovering that it has beautiful and interesting properties. Structures often turn out to have practical importance because of these unique properties. More often they are just interesting to think about. The appeal of discrete mathematics is that you can draw these objects on a blackboard and get a feel for them in way that you cannot in other, more abstract areas of mathematics.

Unfortunately, only very small structures can be built on a blackboard, and we are always interested in larger examples than we can draw. Computers can be used to play with larger ones, but at a cost, for time must be invested to write programs to support this. This barrier complicates doodling with ideas, formulating wild conjectures and identifying counter-examples to them in the casual way that most theorems are understood and discovered. The goal of this book is to facilitate an appreciation for combinatorics and graph theory by collecting a wide variety of operations on interesting structures in one place for easy reference and study. The goal of this package is to facilitate experimentation by computer, to enable you to answer questions not mentioned in the book. It is both a reference and a laboratory in discrete mathematics, because it is both a book and a computer program.

What's between the Covers

This book concentrates on two distinct areas in discrete mathematics. The first section deals with combinatorics, loosely defined as the study of counting. We provide functions for generating combinatorial objects such as permutations, partitions, and Young tableaux, as well as for studying various aspects of these structures.

The second section considers graph theory, which can be defined equally loosely as the study of binary relations. We consider a wide variety of graphs, provide functions to create them, and functions to show what special properties they have. Although graphs are combinatorial structures, understanding them requires pictures or embeddings. Thus we provide functions to create a variety of graph embeddings, so the same structure can be viewed in several different ways. Algorithmic graph theory is an important interface between mathematics and computer science, and so we study a variety of polynomial and exponential time problems.

This book is designed as a tool for manipulating discrete structures. We collect algorithms and problems from a wide variety of sources, but unfortunately do not have the time or space to fully develop the theory behind them. Thus you will find no formal proofs in this book, but hopefully enough discussion to understand

and appreciate the algorithms. Further, we provide extensive references as pointers to the appropriate results. Since the body of the text contains over two hundred functions written in *Mathematica*, it is also an excellent guide for writing your own *Mathematica* programs. The appendix includes a brief guide to *Mathematica* for the uninitiated to help in this regard.

This book is also a complete reference manual for using these functions to explore discrete mathematics. As you read the book I urge you to play with the package, which is available via anonymous FTP from cs.sunysb.edu. For Apple Macintosh and MS-DOS systems, disks can be ordered at a nominal fee from Wolfram Research, Inc. The package is also available through the *Mathematica* users group and various electronic bulletin boards. Starting with Version 2.0 it is included with the standard distribution of *Mathematica*. Documentation for all useful functions appears in the *Glossary of Functions* at the end of the book, and cross-references to examples using a particular function in an interesting way appear in the index.

Finally, we have included exercises at the end of each chapter which make use of the functions in this book. Some are programming exercises, with the rest suggesting some experiment to conduct using these routines. Several "research problems" have been included as well, which are research in the sense that they sound interesting and we do not know the answer. We cannot promise that all are truly open and of interest to anyone except the author.

■ Why *Mathematica*?

A fundamental decision in creating any program is selecting the right programming language. There are a wide variety of alternatives, with inherent tradeoffs in efficiency, expressibility, and portability. English style pseudo-code is traditionally used for exposition, but the resulting algorithms cannot be executed and usually do not contain enough detail to make them easy to code. Standard programming languages such as Pascal and C are possibilities, but the resulting package is doomed to fail as a workbench, since there is no standard environment/interface portable across a wide range of machines. Even if we shoot for Unix as a vanilla environment, we do not get to take advantage of graphics. Nobody should be excluded from using this package if they have access to a powerful enough machine. What we want is a standard environment which every person who is interested in discrete mathematics will have access to. If what Stephen Wolfram has promised me is true, *Mathematica* will be such a system.

Building a discrete mathematics package in *Mathematica* has several advantages. Arbitrary precision arithmetic means we are free of the burdens of computer word length. Where appropriate we have access to portable PostScript graphics, bringing to view graphs and their embeddings and partitions and their Ferrers diagrams.

Working with symbolic formulae makes convenient techniques such as generating functions and chromatic polynomials. The freedom of a high-level language with so much mathematics already under the hood liberates us to explore a much larger fraction of what is known about discrete mathematics.

Of course, every silver lining has its cloud. The chief drawback to using such a high-level language as *Mathematica* is that we lose tight control over the time complexity of our algorithms. The model of computation which it presents (the Wolf-RAM?) is dramatically and mysteriously different from the traditional random access machine. This distorts the way we must view the world, for traditional algorithms are often doomed to be slow or difficult to express with the available operations and structures. Since the goal of using a workbench is to work with as large examples as possible, we cannot completely neglect efficiency.

The goal of this book is to make available as wide a selection of functions as possible, to permit free exploration of many aspects of discrete mathematics. Thus we use the simplest reasonable algorithm we can find, where simplest is measured in length. We will give up a polynomial time factor if it helps make the code concise, particularly since a *Mathematica* rule of thumb is that the shortest programs are fastest. To get decent performance, we use a wide variety of control structures. We keep in mind the distinction between polynomial and exponential time, so brute-force search has been used only when no alternative is known.

All functions have been tested with Version 1.2 of *Mathematica*. Some of the functions assume language constructs not available in earlier versions. The timings have been done on a Silicon Graphics IRIS 4D/120 running Version 1.2 of *Mathematica*.

■ Acknowledgements and Caveat

I would like to acknowledge the support of Apple Computer, for providing a stimulating sanctuary where version 0.1 of *The Act of Counting* was written in the Summer of 1988. I thank Larry Tesler and Al Alcorn for providing the freedom to work on this project and Ivan Sutherland for providing so little supervision. Allan Wylde of Addison-Wesley provided encouragement and a means to communicate this to the world. I appreciate his patience and integrity. Addison-Wesley, Wolfram Research, and a Stony Brook Research Development Grant provided additional support. My Ph.D advisor, Herbert Edelsbrunner, helped me escape from Illinois, which permitted me to focus my interests in this direction. Mom and Dad and Len and Rob provided moral and emotional support.

Anil Bhansali made a significant contribution to this book by writing many of the functions within during the Summer and Fall of 1989. In addition, Anil managed the testing of the code and performed sundry tasks related to the mechanical generation of the manuscript. Stephen Wolfram's continuing interest in the project over a two year period had a lot to do with its becoming reality. Fred Buckley and Nora Hartsfield reviewed an earlier version of the manuscript for Addison-Wesley. Matthew Markert, Marko Petkovšek and Ilan Vardi provided extremely detailed and helpful critiques, and I thank them for their care and dedication. Jürgen Koslowski and Rick Wilson experimented with the alpha version of the software and found some of the bugs. A variety of other Stony Brook (Philip Hsu, Phil Lewis, Yaw-Ling Lin, Gene Stark, Brian Tria, Alan Tucker, Shipei Zhang), Addison-Wesley (Jan Benes, Laura Likely, Karl Matsumoto), and WRI (Dave Ballman, John Bonadies, Martin Buchholz, Joe Grohens, Igor Rivin, Monte Seyer, Lisa Shipley, Cameron Smith) people all helped out in one way or another.

Since we cannot promise efficiency, we have strived for accuracy instead. We hope that every function works as promised so your explorations will be in discrete mathematics instead of debugging. Of course it doesn't, but what is broken should be fixed in later editions. It is traditional for the author to magnanimously accept the blame for whatever deficiencies remain. We don't. Any errors, deficiencies, or problems in this book are somebody else's fault, but we would appreciate knowing about them so we can determine who to blame.

Steven S. Skiena
Department of Computer Science
State University of New York
Stony Brook, NY 11794
skiena@sbcs.sunysb.edu
April 1990

Contents

1. Permutations and Combinations 1
   1.1 Permutations 3
   1.2 Permutation Groups 17
   1.3 Inversions and Inversion Vectors 27
   1.4 Special Classes of Permutations 32
   1.5 Combinations 40
   1.6 Exercises and Research Problems 47

2. Partitions, Compositions, and Young Tableaux 49
   2.1 Partitions 51
   2.2 Compositions 60
   2.3 Young Tableaux 63
   2.4 Exercises and Research Problems 77

3. Representing Graphs 79
   3.1 Data Structures for Graphs 81
   3.2 Elementary Graph Operations 89
   3.3 Graph Embeddings 98
   3.4 Storage Formats 119
   3.5 Exercises and Research Problems 124

4. Generating Graphs 127
   4.1 Building Graphs from Other Graphs 129
   4.2 Regular Structures 140
   4.3 Trees 151
   4.4 Random Graphs 154
   4.5 Relations and Functional Graphs 161
   4.6 Exercises and Research Problems 166

5. Properties of Graphs 169
5.1 Connectivity 171
5.2 Graph Isomorphism 181
5.3 Cycles in Graphs 188
5.4 Partial Orders 203
5.5 Graph Coloring 210
5.6 Cliques, Vertex Covers, and Independent Sets 217
5.7 Exercises and Research Problems 220

6. Algorithmic Graph Theory 223
6.1 Shortest Paths 225
6.2 Minimum Spanning Trees 232
6.3 Network Flow 237
6.4 Matching 240
6.5 Planar Graphs 247
6.6 Exercises and Research Problems 252

Appendix 255
Mathematical Preliminaries 257
Mathematica Preliminaries 259
Glossary of Functions 270
References 301
Index 318

Chapter 1.

Permutations and Combinations

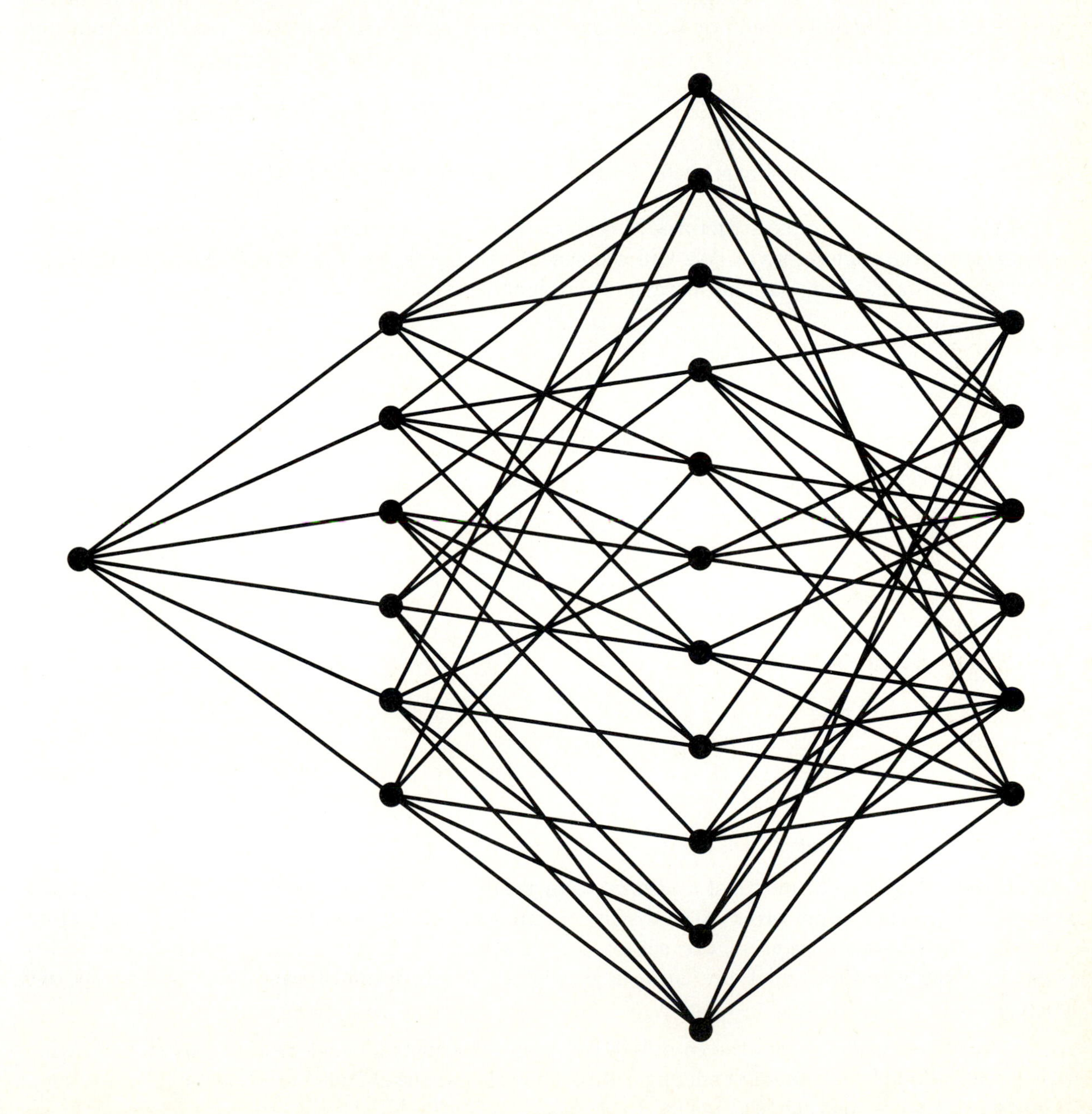

Combinatorics is the study of counting, and over the years mathematicians have counted a variety of different things. Perhaps the most fundamental combinatorial object is the *permutation.* We define a permutation as an ordering of the integers 1 to n without repetition. A permutation of items defines a particular arrangement of them, and so permutations can be viewed as either objects (the actual arrangement) or operations (the rearrangement necessary to move from one arrangement to another).

Closely related to permutations are *combinations*, distinct subsets of a set whose order doesn't matter. In this chapter, we provide algorithms for generating permutations and combinations in a variety of different ways, each of which has distinct useful properties.

These relatively simple structures and algorithms also provide a proper introduction to the *Mathematica* programming style. We will see that writing efficient functions in *Mathematica* is a different game than in conventional programming languages.

About the illustration overleaf:

An exchange of any two elements of a permutation yields a different permutation, and by applying transpositions in the proper order all $n!$ distinct permutations can be constructed. The illustration shows the *transposition graph* of the set of permutations on four elements, where each vertex represents a permutation and there is an edge whenever two permutations differ by exactly one transposition.

Using tools we will develop, the complete command to construct and display this transposition graph is: `ShowGraph[ RankedEmbedding[ MakeGraph[ Permutations[{1,2,3,4}], (Count[#1-#2,0]==(Length[#1]-2))& ], {1} ] ];`.

1.1 Permutations

Permutations can viewed as either objects or rearrangement operations. When thought of as an operation, a permutation specifies where each of n objects in distinct positions should go. Since this is well defined only if the permutation is on the set 1 to n, we define a *permutation* to be an ordering of the integers 1 to n without repetition.

```
PermutationQ[p_List] := (Sort[p] == Range[Length[p]])
```

Testing a Permutation

Mathematica array notation provides a slick way to permute a set of objects according to a particular permutation, since elements of one list can be used as indices to select elements of another. The same technique can be used to swap two values without an explicit intermediate variable or to select an arbitrary subset of a set.

```
Permute[l_List,p_?PermutationQ] := l [[ p ]]
```

Permuting a List

Our restriction limiting permutations to the integers from 1 to n is not significant, since we can easily permute a list of whatever we are interested in.

```
In[1]:= Permute[{a,b,c,d},{1,3,2,4}]
Out[1]= {a, c, b, d}
```

Use of the `PermutationQ` predicate prevents `Permute` from matching arguments that are not legitimate permutations.

```
In[2]:= Permute[{a,b,c,d},{1,2,2,4}]
Out[2]= Permute[{a, b, c, d}, {1, 2, 2, 4}]
```

The rest of this section will be devoted to the problem of constructing permutations. There are a variety of interesting algorithms which provide an introduction to building combinatorial objects with *Mathematica*.

1.1.1 Lexicographically Ordered Permutations

The operation of constructing all permutations is common enough that the built-in *Mathematica* function `Permutations` generates all $n!$ permutations of a list. A number of interesting algorithms for constructing permutations are surveyed in [Sed77].

The most straightforward way to construct all permutations of the n items of a list l is to successively pull each element out of l and prepend it to each permutation of the $n-1$ other items. This recurrence bottoms out on lists of one or two elements, the permutations of which are easily enumerated.

```
LexicographicPermutations[{l_}] := {{l}}

LexicographicPermutations[{a_,b_}] := {{a,b},{b,a}}

LexicographicPermutations[l_List] :=
        Block[{i,n=Length[l]},
                Apply[
                        Join,
                        Table[
                                Map[
                                        (Prepend[#,l[[i]]])&,
                                        LexicographicPermutations[
                                                Complement[l,{l[[i]]}]
                                        ]
                                ],
                                {i,n}
                        ]
                ]
        ]
```

Constructing Lexicographically Ordered Permutations

The permutations constructed by this algorithm have the property that they are *lexicographically* or alphabetically ordered. If the initial list of items is in alphabetical order, the resultant permutations will also appear in alphabetical order.

As you can see, the permutations are generated in lexicographic order. The built-in generator `Permutations` also constructs them in lexicographic order.

```
In[3]:= LexicographicPermutations[{a,b,c,d}]
Out[3]= {{a, b, c, d}, {a, b, d, c}, {a, c, b, d},
   {a, c, d, b}, {a, d, b, c}, {a, d, c, b}, {b, a, c, d},
   {b, a, d, c}, {b, c, a, d}, {b, c, d, a}, {b, d, a, c},
   {b, d, c, a}, {c, a, b, d}, {c, a, d, b}, {c, b, a, d},
   {c, b, d, a}, {c, d, a, b}, {c, d, b, a}, {d, a, b, c},
   {d, a, c, b}, {d, b, a, c}, {d, b, c, a}, {d, c, a, b},
   {d, c, b, a}}
```

Here we compute the ratio of the execution times of two permutation generating functions. The built-in *Mathematica* `Permutations` is a good 25 times faster, which is not terribly surprising, even if both functions were to use the same underlying algorithm. `Permutations` is compiled instead of interpreted, and uses lower-level data structures, which together accounts for most of the difference in speed.

```
In[4]:= Table[ First[ Timing[
LexicographicPermutations[Range[i]]; ] / Timing[
Permutations[Range[i]]; ] ], {i,5,7}]

Out[4]= {26., 22.75, 25.3621}
```

1.1.2 Ranking and Unranking Permutations

Since there are $n!$ distinct permutations of size n, any ordering of permutations defines a bijection with the integers from 1 to $n!$. Such a bijection can be used to give several useful operations. For example, if we are seeking an instance of a permutation with a certain property, where such instances are plentiful, it will be more efficient to repeatedly construct and test the ith permutation, starting from $i = 0$, than to build all $n!$ permutations in advance and search.

A *ranking* function for an object determines its position in the total order. For a permutation in lexicographical order, its rank may be computed by observing that all permutations sharing the same first element k are ranked $(k-1)(n-1)!$ to $k(n-1)! - 1$ in the total order, and after deleting the first element we may recur on a permutation of length $n-1$ to determine the exact rank.

An *unranking* algorithm constructs the ith object for a given i. The previous analysis shows that i mod $(n-1)!$ gives the index of the first element of the permutation, and by induction on the residue we can determine the rest.

```
RankPermutation[{1}] = 0

RankPermutation[p_?PermutationQ] := (p[[1]]-1) (Length[Rest[p]]!) +
        RankPermutation[ Map[(If[#>p[[1]], #-1, #])&, Rest[p]] ]

NthPermutation[n1_Integer,l_List] :=
        Block[{k, n=n1, s=l, i},
                Table[
                        n = Mod[n,(i+1)!];
                        k = s [[Quotient[n,i!]+1]];
                        s = Complement[s,{k}];
                        k,
                        {i,Length[l]-1,0,-1}
                ]
        ]
```

```
NextPermutation[p_?PermutationQ] :=
        NthPermutation[ RankPermutation[p]+1, Sort[p] ]
```

Constructing the Next Permutation

Applying the unranking algorithm $n!$ times gives a different but slower method for constructing all permutations in lexicographic order.

```
In[5]:= Table[ NthPermutation[n,Range[4]], {n,0,23}]
Out[5]= {{1, 2, 3, 4}, {1, 2, 4, 3}, {1, 3, 2, 4},
  {1, 3, 4, 2}, {1, 4, 2, 3}, {1, 4, 3, 2}, {2, 1, 3, 4},
  {2, 1, 4, 3}, {2, 3, 1, 4}, {2, 3, 4, 1}, {2, 4, 1, 3},
  {2, 4, 3, 1}, {3, 1, 2, 4}, {3, 1, 4, 2}, {3, 2, 1, 4},
  {3, 2, 4, 1}, {3, 4, 1, 2}, {3, 4, 2, 1}, {4, 1, 2, 3},
  {4, 1, 3, 2}, {4, 2, 1, 3}, {4, 2, 3, 1}, {4, 3, 1, 2},
  {4, 3, 2, 1}}
```

The ranking function illustrates that `Permutations` uses lexicographic sequencing.

```
In[6]:= Map[RankPermutation, Permutations[{1,2,3,4}]]
Out[6]= {0, 1, 2, 3, 4, 5, 6, 7, 8, 9, 10, 11, 12, 13,
  14, 15, 16, 17, 18, 19, 20, 21, 22, 23}
```

The sequencing order wraps around, so the permutation following the reverse order is the total order.

```
In[7]:= NextPermutation[ Reverse[Range[8]] ]
Out[7]= {1, 2, 3, 4, 5, 6, 7, 8}
```

■ 1.1.3 Random Permutations

With $n!$ distinct permutations of n items, it is impractical to test a conjecture on all permutations for any reasonable value of n. However, experiments on several large, *random* permutations can give confidence that any results we get are not an artifact of how the examples were selected, as they might be if the permutations were all near each other in lexicographic order. For this reason, functions are included for generating random instances of most of the combinatorial structures discussed in this book. Many start by constructing a random permutation. There are entire books consisting of nothing but random permutations [MO63], which we will make obsolete with a one-line function.

There are two common algorithms for constructing random permutations of length n. The first constructs a vector of random real numbers and uses them as keys to records containing the integers 1 to n. Assuming no two random numbers are equal, they describe a unique permutation, which can be obtained by sorting the records and then striping off the random numbers. Under the conventional RAM model of computation, this algorithm takes $O(n \log n)$ time for the sorting.

A different algorithm starts with an arbitrary permutation and exchanges the ith element with a randomly selected one from the first i elements, for each i from n to 1. The nth element is therefore equally likely to be anything from 1 to n, and the randomness of the permutation follows by induction. Since there are only $n-1$ iterations, this is an $O(n)$ algorithm and thus destined to be more efficient for large enough n on a conventional machine than the previous algorithm.

```
RandomPermutation1[n_Integer?Positive] :=
        Map[ Last, Sort[ Map[({Random[],#})&,Range[n]] ] ]

RandomPermutation2[n_Integer?Positive] :=
        Block[{p = Range[n],i,x},
                Do [
                        x = Random[Integer,{1,i}];
                        {p[[i]],p[[x]]} = {p[[x]],p[[i]]},
                        {i,n,2,-1}
                ];
                p
        ]

RandomPermutation[n_Integer?Positive] := RandomPermutation1[n]
```

Constructing a Random Permutation

With $3! = 6$ distinct permutations of three elements, within twenty random permutations we are likely to see all of them. Observe that it is unlikely for the first six permutations to all be distinct.

```
In[8]:= Table[ RandomPermutation[3], {20}]
Out[8]= {{1, 2, 3}, {2, 1, 3}, {1, 2, 3}, {3, 1, 2},
  {1, 3, 2}, {2, 3, 1}, {3, 1, 2}, {3, 2, 1}, {2, 3, 1},
  {3, 2, 1}, {1, 3, 2}, {2, 1, 3}, {2, 1, 3}, {2, 3, 1},
  {1, 3, 2}, {3, 1, 2}, {3, 1, 2}, {2, 3, 1}, {1, 3, 2},
  {3, 1, 2}}
```

Since each of the 30! permutations are equally likely to be constructed by either algorithm, we can be mighty confident that these two permutations will not be identical.

```
In[9]:= RandomPermutation1[30] === RandomPermutation2[30]
Out[9]= False
```

Theory aside, which of these functions *is* faster in *Mathematica*? We are in for a surprise, because the model of computation *Mathematica* presents is not truly a random access machine. When modifying a list structure, it often rewrites the structure in its entirety. Therefore, each exchange of elements in the permutation becomes a *linear*, not constant time, operation meaning the time complexity of `RandomPermutation2` is in fact $O(n^2)$. It is important to know that such hidden dangers can lurk in *Mathematica* programs.

Here the execution time of RandomPermutation1 in seconds is plotted as a function of the number of elements requested. The time complexity of $O(n \log n)$ appears very close to linear, as should be expected for relatively small n.

```
In[10]:= ListPlot[ Table[{i, Timing[
RandomPermutation1[i]; ][[1,1]]}, {i,50,1000,50}] ];
```

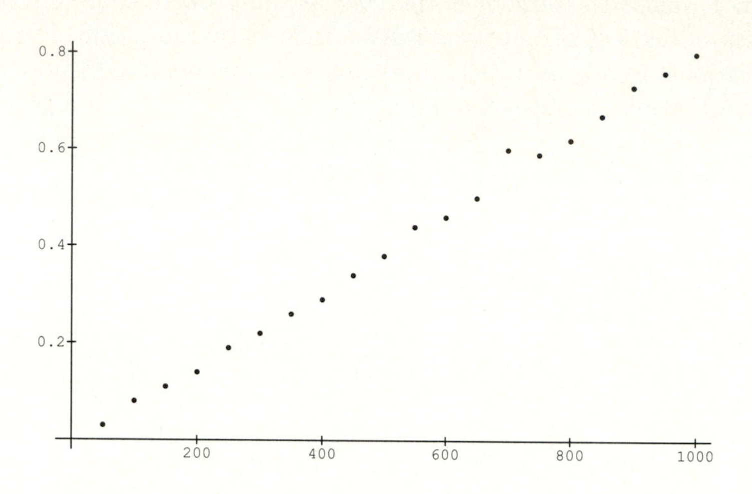

Pictures do not lie, and this graph makes it clear that the time complexity of the erstwhile linear algorithm is in fact quadratic. Comparing the times shows that there are essentially no values of n for which RandomPermutation2 is faster then RandomPermutation1, so the latter will be used in practice for constructing random permutation.

```
In[11]:= ListPlot[ Table[{i, Timing[
RandomPermutation2[i]; ][[1,1]]}, {i,50,1000,50}] ];
```

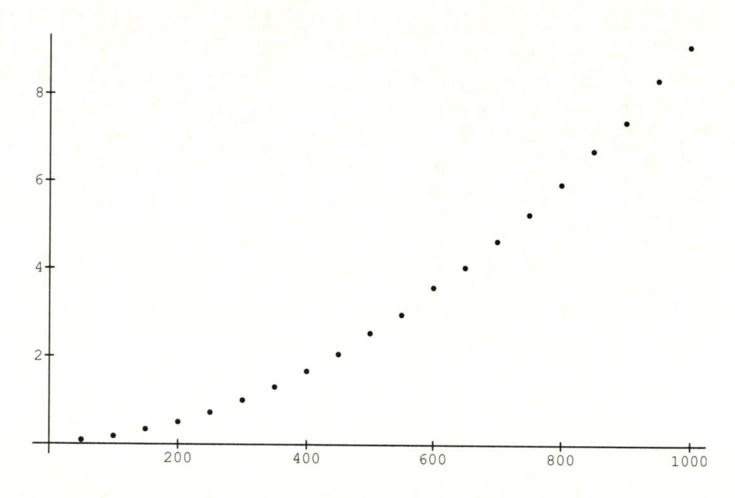

In fairness to the creators of *Mathematica*, it would be difficult to implement lists as arrays while providing the same degree of flexibility provided by *Mathematica*'s transformation rules. Still, the fact that list modification operations are expensive must have a big effect on the algorithms we implement. Most combinatorial algorithms make extensive use of arrays and pointers, and so must be reformulated.

The correct model of computation presented by *Mathematica* is perhaps closer to an *applicative machine*, where all memory except for a constant number of registers is write-once. The problem of finding an $O(n)$ algorithm for generating random permutations on an applicative machine is posed in [PMN88].

Care must be taken that random objects are indeed selected with equal probability from the set of possibilities. For many types of structures, this is a non-trivial problem. Nijenhuis and Wilf [NW78, Wil89] consider this problem in depth, and their books are the source of many of the algorithms we implement.

The best way to verify that the *implementation* gives truly random objects is to conduct experiments. Here we see that all six permutations of three elements are generated with roughly equal probability.

```
In[12]:= Distribution[Table[RandomPermutation[3],{300}]]
Out[12]= {49, 53, 34, 45, 65, 54}
```

■ 1.1.4 Permutations from Transpositions

Beyond lexicographic order, another way to sequence permutations minimizes the amount of change between adjacent permutations. Since two distinct permutations contain the same elements in a different order, the minimum change between two distinct permutations consists of an exchange of two elements. In fact, it is possible to arrange the permutations such that every two adjacent permutations differ from each other by exactly one exchange or *transposition*.

To best illustrate the transposition structure of permutations, we introduce some of the graph theoretic tools we will develop in the second half of this book. A *graph* is a collection of pairs of elements from some set. If the elements of the set are represented by points or *vertices* and the pairs by lines or *edges* connecting the points, we can get a visual representation of the underlying structure. We may construct a graph where the vertices correspond to permutations, with an edge between permutations that differ by exactly one transposition.

The transposition graph illustrates that each permutation is one transposition removed from $\binom{n}{2}$ other permutations. Further, tracing all the paths in the graph between two vertices shows that any particular pair of permutations differ from each other by either an even or odd number of transpositions, independent of the order in which the transpositions are applied.

```
In[13]:= ShowGraph[ r = RankedEmbedding[ MakeGraph[
Permutations[{1,2,3,4}], (Count[#1-#2,0]==(Length[#1]-2))&
], {1} ] ];
```

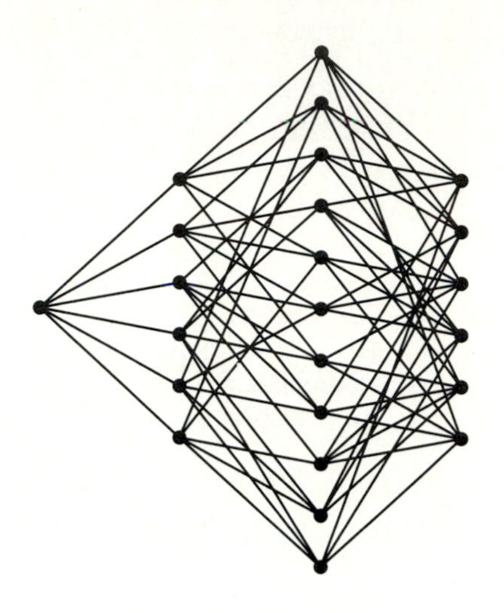

All cycles in the transposition graph are of even length, meaning that the graph is *bipartite*. This accounts for the even-odd transposition property.

```
In[14]:= BipartiteQ[r]
Out[14]= True
```

A *Hamiltonian cycle* of a graph is a tour which visits each vertex of the graph without repetition. The Hamiltonian cycle of a transposition graph corresponds to an ordering where each successive permutation differs by exactly one transposition. The transposition graph of a multiset always contains a Hamiltonian cycle [Cha73].

```
In[15]:= Permutations[{1,2,3,4}] [[ HamiltonianCycle[r] ]]
Out[15]= {{1, 2, 3, 4}, {1, 2, 4, 3}, {1, 3, 4, 2},
  {1, 3, 2, 4}, {1, 4, 2, 3}, {1, 4, 3, 2}, {2, 4, 3, 1},
  {2, 1, 3, 4}, {2, 1, 4, 3}, {2, 3, 4, 1}, {2, 3, 1, 4},
  {2, 4, 1, 3}, {3, 4, 1, 2}, {3, 1, 4, 2}, {3, 1, 2, 4},
  {3, 2, 1, 4}, {3, 2, 4, 1}, {3, 4, 2, 1}, {4, 3, 2, 1},
  {4, 1, 2, 3}, {4, 1, 3, 2}, {4, 3, 1, 2}, {4, 2, 1, 3},
  {4, 2, 3, 1}, {1, 2, 3, 4}}
```

Since a minimum amount of work has to be done to construct each subsequent permutation from its predecessor, minimum change order is the most efficient way to construct permutations. Sedgewick's study [Sed77] revealed the following algorithm [Hea63] to be generally the fastest permutation generation algorithm. The pattern of transpositions is recursively defined, and each transposition almost always includes the first element.

```
MinimumChangePermutations[l_List] :=
        Block[{i=1,c,p=l,n=Length[l],k},
                c = Table[1,{n}];
                Join[
                        {l},
                        Table[
                                While [ c[[i]] >= i, c[[i]] = 1; i++];
                                If[OddQ[i], k=1, k=c[[i]] ];
                                {p[[i]],p[[k]]} = {p[[k]],p[[i]]};
                                c[[i]]++;
                                i = 2;
                                p,
                                {n!-1}
                        ]
                ]
        ]
```

Sequencing Permutations in Minimum Change Order

Although successive permutations differ by exactly one transposition, the last permutation is more than one transposition removed from the first, so this ordering is not cyclic. The Hamiltonian cycle construction above and the Johnson-Trotter algorithm [Joh63, Tro62] both define cyclic minimum change orderings.

```
In[16]:= MinimumChangePermutations[{1,2,3,4}]

Out[16]= {{1, 2, 3, 4}, {2, 1, 3, 4}, {3, 1, 2, 4},
  {1, 3, 2, 4}, {2, 3, 1, 4}, {3, 2, 1, 4}, {4, 2, 1, 3},
  {2, 4, 1, 3}, {1, 4, 2, 3}, {4, 1, 2, 3}, {2, 1, 4, 3},
  {1, 2, 4, 3}, {1, 3, 4, 2}, {3, 1, 4, 2}, {4, 1, 3, 2},
  {1, 4, 3, 2}, {3, 4, 1, 2}, {4, 3, 1, 2}, {4, 3, 2, 1},
  {3, 4, 2, 1}, {2, 4, 3, 1}, {4, 2, 3, 1}, {3, 2, 4, 1},
  {2, 3, 4, 1}}
```

The permutations of $n \geq 4$ elements can be sequenced in *maximum change order* where neighboring permutations differ in all positions. Such a sequence can be constructed by finding a Hamiltonian cycle on the appropriate adjacency graph [Wil89]. This graph is disconnected for $n = 3$.

```
In[17]:= Permutations[{1,2,3,4}] [[ HamiltonianCycle[
MakeGraph[Permutations[{1,2,3,4}], (Count[#1-#2,0] == 0)&]
] ]]

Out[17]= {{1, 2, 3, 4}, {2, 1, 4, 3}, {1, 3, 2, 4},
  {2, 4, 1, 3}, {1, 3, 4, 2}, {2, 1, 3, 4}, {1, 2, 4, 3},
  {2, 3, 1, 4}, {1, 4, 2, 3}, {2, 3, 4, 1}, {1, 4, 3, 2},
  {3, 1, 2, 4}, {2, 4, 3, 1}, {4, 1, 2, 3}, {3, 2, 1, 4},
  {4, 1, 3, 2}, {3, 2, 4, 1}, {4, 3, 1, 2}, {3, 4, 2, 1},
  {4, 2, 1, 3}, {3, 1, 4, 2}, {4, 2, 3, 1}, {3, 4, 1, 2},
  {4, 3, 2, 1}, {1, 2, 3, 4}}
```

1.1.5 Backtracking and Distinct Permutations

Generating distinct permutations becomes more difficult for a *multiset*, a set of elements which are not necessarily distinct. In a multiset of k distinct elements of multiplicity n_i, $1 \leq i \leq k$, the number of distinct permutations is $(\sum_{i=1}^{k} n_i)!/(\prod_{i=1}^{k} n_i!)$, a quantity which is known as the *multinomial coefficient* of $n_1, ..., n_k$.

Permutations does not correctly handle the situation when the list defines a *multiset*, meaning the elements are not unique. Obviously, there is only one permutation of five a's.

```
In[18]:= Length[ Permutations[{a,a,a,a,a}] ]
Out[18]= 120
```

A simple but wasteful solution uses Permutations to construct all configurations and then Union to eliminate the duplicates.

```
In[19]:= Union[ Permutations[{a,a,a,a,a}] ]
Out[19]= {{a, a, a, a, a}}
```

However, there is a much better way to do exhaustive search. When the solution space to a problem consists of ordered configurations of elements, each prefix to a solution represents a *partial solution*. If we can demonstrate that a given prefix does not lead to to any solutions, there is no reason to expand the prefix and look further. *Backtracking* lets us start with the smallest possible configuration, and keep on adding elements to it until either we have achieved a final solution or we can prove that there are no solutions with this prefix. In the first case, we return the answer; in the second we remove the last element from the configuration and then replace it with the next possibility. Backtracking can greatly reduce the amount of work in an exhaustive search, or in this case, an exhaustive construction.

```
Backtrack[space_List,partialQ_,solutionQ_,flag_:One] :=
        Block[{n=Length[space],all={},done,index,v=2,solution},
                index=Prepend[ Table[0,{n-1}],1];
                While[v > 0,
                        done = False;
                        While[!done && (index[[v]] < Length[space[[v]]]),
                                index[[v]]++;
                                done = Apply[partialQ,{Solution[space,index,v]}];
                        ];
                        If [done, v++, index[[v--]]=0 ];
                        If [v > n,
                                solution = Solution[space,index,n];
                                If [Apply[solutionQ,{solution}],
                                        If [SameQ[flag,All],
                                                AppendTo[all,solution],
                                                all = solution; v=0
                                        ]
```

```
                    ];
                    v--
                ]
            ];
            all
    ]

Solution[space_List,index_List,count_Integer] :=
    Block[{i}, Table[space[[ i,index[[i]] ]], {i,count}] ]
```

Generalized Backtracking

`Backtrack` takes four arguments. The first describes the *state space*, which are the possible elements which can be in each position. The second is a predicate which is applied in testing whether a partial solution can be extended, while the third argument is a predicate to test whether a configuration is the final solution. Finally, an optional flag specifies whether the search should stop on the first solution or continue the search to find all of them.

The key to `Backtrack` is the index array, which keeps track of the state of the partial solution. Since index always increases lexicographically, the search will terminate without finding duplicates. The partial solution at any given point can be obtained by copying the indexed elements of space.

```
DistinctPermutations[s_List] :=
    Block[{freq,alph=Union[s],n=Length[s]},
        freq = Map[ (Count[s,#])&, alph];
        Map[
            (alph[[#]])&,
            Backtrack[
                Table[Range[Length[alph]],{n}],
                (Count[#,Last[#]] <= freq[[Last[#]]])&,
                (Count[#,Last[#]] <= freq[[Last[#]]])&,
                All
            ]
        ]
    ]
```

Constructing the Permutations of a Multiset

For the problem of constructing distinct permutations, the state space consists of the entire alphabet for each of the n positions, because every symbol appears in every position in some permutation. The predicate to test the legality of both partial and complete solutions is whether the last symbol occurs in the partial solution at most as many times as in the multiset. If so, it can be extended to distinct permutations; otherwise, it represents a dead end and the last element is deleted. Since we are interested in all distinct permutations, we do not stop the search on finding the first solution.

The permutations of this multiset are returned in lexicographic order, so it is apparent that each is constructed only once. There are $\binom{6}{3}$ permutations of this multiset, since there are that many ways to select the position of the ones.

```
In[20]:= DistinctPermutations[{1,1,1,2,2,2}]
Out[20]= {{1, 1, 1, 2, 2, 2}, {1, 1, 2, 1, 2, 2},
  {1, 1, 2, 2, 1, 2}, {1, 1, 2, 2, 2, 1},
  {1, 2, 1, 1, 2, 2}, {1, 2, 1, 2, 1, 2},
  {1, 2, 1, 2, 2, 1}, {1, 2, 2, 1, 1, 2},
  {1, 2, 2, 1, 2, 1}, {1, 2, 2, 2, 1, 1},
  {2, 1, 1, 1, 2, 2}, {2, 1, 1, 2, 1, 2},
  {2, 1, 1, 2, 2, 1}, {2, 1, 2, 1, 1, 2},
  {2, 1, 2, 1, 2, 1}, {2, 1, 2, 2, 1, 1},
  {2, 2, 1, 1, 1, 2}, {2, 2, 1, 1, 2, 1},
  {2, 2, 1, 2, 1, 1}, {2, 2, 2, 1, 1, 1}}
```

The built-in *Mathematica* function `Multinomial` computes multinomial coefficients and thus can be used to determine the number of permutations of a given multiset.

```
In[21]:= Multinomial[3,3]
Out[21]= 20
```

With a set of distinct elements, all $n!$ permutations are constructed.

```
In[22]:= DistinctPermutations[{A,B,C}]
Out[22]= {{A, B, C}, {A, C, B}, {B, A, C}, {B, C, A},
  {C, A, B}, {C, B, A}}
```

■ 1.1.6 Sorting and Searching

Each permutation of distinct elements defines a total order. One such permutation is usually defined as *the* total order, and sorting is the operation which takes you there. Foremost among the reasons to sort a set of items is to permit fast membership queries over it; hence the problems of sorting and searching are linked together.

Knuth [Knu73b] gives a thorough treatment of sorting and searching, and we limit our discussion to one algorithm for each problem. *Selection sort* makes n passes over the set, in each pass selecting the smallest element and then deleting it from the set. The elements will, as a result, be selected in increasing order. Selection sort is a quadratic time algorithm to solve a problem for which numerous $O(n \log n)$ algorithms are known, and is given here only as an illustration. The built-in *Mathematica* `Sort` function is much faster. An improved implementation of selection sort will appear in Section 1.4.4.

```
MinOp[l_List,f_] :=
        Block[{min=First[l]},
                Scan[ (If[ Apply[f,{#,min}], min = #])&, l];
                Return[min];
        ]

SelectionSort[l_List,f_] :=
        Block[{where,item,unsorted=l},
                Table[
                        item = MinOp[unsorted, f];
                        {where} = First[ Position[unsorted,item] ];
                        unsorted = Drop[unsorted,{where,where}];
                        item,
                        {Length[l]}
                ]
        ]
```

Selection Sorting

Sort can be used to obtain the total order defined by any pairwise comparison operator. Here we sort all subsets of four elements in order of non-decreasing length. Subsets will be discussed in Section 1.5.

```
In[23]:= Sort[ Subsets[{1,2,3,4}],
(Apply[Plus,#1]<=Apply[Plus,#2])& ]

Out[23]= {{}, {1}, {2}, {1, 2}, {3}, {1, 3}, {4}, {2, 3},
  {1, 4}, {1, 2, 3}, {2, 4}, {3, 4}, {1, 2, 4},
  {1, 3, 4}, {2, 3, 4}, {1, 2, 3, 4}}
```

The Sort function is fastest when it is given with only one argument. Explicitly specifying Less is unnecessary. With or without a second argument, this home-grown sorting function will be left in the dust by the $O(n \log n)$ algorithm used by Sort.

```
In[24]:= Table[ p = RandomPermutation[i]; First[
Timing[SelectionSort[p,Less];] / Timing[Sort[p,Less];] ],
{i,25,150,25}]

Out[24]= {25., 20.75, 34.8, 38.5, 47., 39.}
```

The most efficient way to locate an item in a sorted table is via *binary search.* This is the algorithm of choice when looking somebody up in a telephone book, assuming we hope to do something else with the rest of our day. By testing in the middle of a sorted interval, we eliminate half of the keys with each comparison. This implementation permits the key extraction function to be specified, so the binary search can be used in several contexts. It returns the index of the key of the item if it exists or else where it would be if it existed.

```
BinarySearch[l_List,k_Integer] := BinarySearch[l,k,1,Length[l],Identity]
BinarySearch[l_List,k_Integer,f_] := BinarySearch[l,k,1,Length[l],f]

BinarySearch[l_List,k_Integer,low_Integer,high_Integer,f_] :=
```

```
Block[{mid = Floor[ (low + high)/2 ]},
        If [low > high, Return[low - 1/2]];
        If [f[ l[[mid]] ] == k, Return[mid]];
        If [f[ l[[mid]] ] > k,
                BinarySearch[l,k,1,mid-1,f],
                BinarySearch[l,k,mid+1,high,f]
        ]
]
```

Binary Search

Clearly 40 is one of the first 30 even numbers.

In[25]:= **BinarySearch[Table[2i,{i,30}],40]**

Out[25]= 20

Just as clearly, 41 isn't. Returning a fractional pointer is more useful than an error code, since taking the `Floor` gives the index of the largest element less than the key and `Ceiling` gives the index of the smallest element greater than the key.

In[26]:= **BinarySearch[Table[2i,{i,30}],41]**

Out[26]= $\frac{41}{2}$

Strictly speaking, we use *interpolation search* [LBB81] when looking somebody up in a telephone book, since we exploit our understanding of letter distribution in English. Interpolation search can outperform binary search if the distribution of keys is as we expect.

1.2 Permutation Groups

A mathematical *group* is a set of objects that is closed under an associative multiplication operator, that contains an identity element, and for which every element has an inverse. The complete set of permutations defines a group, and other *permutation groups* are defined by appropriate subsets of permutations.

Computers have been widely used in group theory [Can69], with programs such as *CAYLEY* [Can90, Hol88].

1.2.1 Multiplying Permutations

For binary operators over a closed set of elements, it is useful to define a *multiplication table* showing how every pair of elements relates to each other. For permutation groups, `Permute` serves as the multiplication operator, and the permutation `Range[n]` as the identity element.

```
MultiplicationTable[elems_List,op_] :=
        Block[{i,j,n=Length[elems],p},
                Table[
                        p = Position[elems, Apply[op,{elems[[i]],elems[[j]]}]];
                        If [p === {}, 0, p[[1,1]]],
                        {i,n},{j,n}
                ]
        ]
```

Constructing a Multiplication Table

The complete set of permutations on n elements, or *symmetric group* S_n, is closed under composition. The multiplication table shows that the first permutation is the identity element, each permutation has an inverse, and that the set of permutations is closed under multiplication. Since the table is not symmetric around the diagonal, S_n is not a commutative or *abelian* group.

```
In[27]:= TableForm[ MultiplicationTable[
Permutations[{1,2,3}], Permute ] ]

                   1  2  3  4  5  6
                   2  1  5  6  3  4
                   3  4  1  2  6  5
                   4  3  6  5  1  2
                   5  6  2  1  4  3
Out[27]//TableForm= 6  5  4  3  2  1
```

1.2.2 The Inverse of a Permutation

The identity permutation maps each element to itself. For any permutation p, its *inverse* maps it to the identity permutation. The inverse p' of a permutation p is defined so that ${p'}_j = k$ if and only if $p_k = j$.

```
InversePermutation[p_?PermutationQ] :=
        Block[{inverse=p, i},
                Do[ inverse[[ p[[i]] ]] = i, {i,Length[p]} ];
                inverse
        ]
```

The Inverse of a Permutation

A *fixed-point* of a permutation p is an element in the same position in p as in the inverse of p. Thus the only fixed point in this permutation is 7.

```
In[28]:= InversePermutation[{4,8,5,2,1,3,7,6}]
Out[28]= {5, 4, 6, 1, 3, 8, 7, 2}
```

■ 1.2.3 Equivalence Relations and Classes

A binary relation on a set S is an *equivalence relation* if it is reflexive, symmetric, and transitive. Equivalence relations have several nice properties, particularly that the elements can be partitioned into *equivalence classes* such that every pair of elements within the same equivalence class are related and no pair of elements from different classes are related.

```
EquivalenceRelationQ[r_?SquareMatrixQ] :=
        ReflexiveQ[r] && SymmetricQ[r] && TransitiveQ[r]
EquivalenceRelationQ[g_Graph] := EquivalenceRelationQ[ Edges[g] ]

SquareMatrixQ[{}] = True
SquareMatrixQ[r_] := MatrixQ[r] && (Length[r] == Length[r[[1]]])

ReflexiveQ[r_?SquareMatrixQ] :=
        Block[{i}, Apply[And, Table[(r[[i,i]]!=0),{i,Length[r]}] ] ]

TransitiveQ[r_?SquareMatrixQ] := TransitiveQ[ Graph[v,RandomVertices[Length[r]]] ]
TransitiveQ[r_Graph] := IdenticalQ[r,TransitiveClosure[r]]

SymmetricQ[r_?SquareMatrixQ] := (r === Transpose[r])

EquivalenceClasses[r_List?EquivalenceRelationQ] :=
        ConnectedComponents[ Graph[r,RandomVertices[Length[r]]] ]
EquivalenceClasses[g_Graph?EquivalenceRelationQ] := ConnectedComponents[g]
```

Identifying Equivalence Relations and Classes

For any permutation group $G = \{P_m, 1 \leq m \leq k\}$ on the elements $p_1, ..., p_n$, we may define the *equivalence relation induced by G* such that p_i is related to p_j,

$1 \leq i, j \leq n$ if and only if there exists a permutation P_m such that the ith element of P_m is p_j.

This *sameness* relation provides a method for testing whether a set of permutations represents a group, since it is readily verified [Rob84] that G is a permutation group if and only if the sameness relation on G is an equivalence relation.

```
PermutationGroupQ[perms_List] := EquivalenceRelationQ[SamenessRelation[perms]]

SamenessRelation[perms_List] :=
        Block[{positions = Transpose[perms], i, j, n=Length[First[perms]]},
                Table[
                        If[ MemberQ[positions[[i]],j], 1, 0],
                        {i,n}, {j,n}
                ]
        ] /; perms != {}
```

Identifying Permutation Groups

Star[n] is the graph of $n-1$ vertices of degree 1 and 1 vertex of degree $n-1$. An *automorphism* is a permutation of the vertices so that the graph is isomorphic to itself. The automorphisms of a graph always describe a group.

```
In[29]:= star = Automorphisms[Star[5]]
Out[29]= {{1, 2, 3, 4, 5}, {1, 2, 4, 3, 5},
  {1, 3, 2, 4, 5}, {1, 3, 4, 2, 5}, {1, 4, 2, 3, 5},
  {1, 4, 3, 2, 5}, {2, 1, 3, 4, 5}, {2, 1, 4, 3, 5},
  {2, 3, 1, 4, 5}, {2, 3, 4, 1, 5}, {2, 4, 1, 3, 5},
  {2, 4, 3, 1, 5}, {3, 1, 2, 4, 5}, {3, 1, 4, 2, 5},
  {3, 2, 1, 4, 5}, {3, 2, 4, 1, 5}, {3, 4, 1, 2, 5},
  {3, 4, 2, 1, 5}, {4, 1, 2, 3, 5}, {4, 1, 3, 2, 5},
  {4, 2, 1, 3, 5}, {4, 2, 3, 1, 5}, {4, 3, 1, 2, 5},
  {4, 3, 2, 1, 5}}
```

The sameness relation clearly shows that the four vertices of degree 1 form an equivalence class. The 24 automorphisms listed above represent all the permutations of these four vertices.

```
In[30]:= TableForm[ relation=SamenessRelation[star] ]
                       1  1  1  1  0
                       1  1  1  1  0
                       1  1  1  1  0
                       1  1  1  1  0
Out[30]//TableForm=    0  0  0  0  1
```

The two equivalence classes correspond to the vertices of different degree.

```
In[31]:= EquivalenceClasses[relation]
Out[31]= {{1, 2, 3, 4}, {5}}
```

Not all subsets of the symmetric group are permutation groups. Two permutations form a group only if one is the identity element and the other is an *involution*, a permutation which is its own inverse.

```
In[32]:= PermutationGroupQ[{{1,2,3,4},{4,2,3,1}}]
Out[32]= True
```

Instead of just listing all the permutations in a group, a more concise representation of a group is in terms of a minimum set of permutations which *generate* all the permutations in the group. For example, $\{1,2,3,4\}$, $\{4,2,1,3\}$ and $\{2,3,1,4\}$ together generate the symmetric group, since by repeated multiplication of these we can construct all permutations. Furst, Hopcroft, and Luks [FHL80] give a polynomial-time algorithm for testing group membership with such a representation.

■ 1.2.4 The Cycle Structure of Permutations

Suppose multiplying by a particular permutation takes element p_i to p_j and p_j to p_k. Eventually, we always get back to the first element p_i, forming a cycle. It turns out that any permutation can be considered as a composition of disjoint cycles, where each cycle specifies a circular shift of the elements to be permuted. For example, the permutation $p = \{5,3,4,2,6,1\}$ has a cycle structure of $\{\{5,6,1\},\{3,4,2\}\}$. Observe that multiplying a permutation by p takes the first element to the fifth position, the fifth element to the sixth position, and the sixth element to the first position, closing the first cycle. This suggests an alternate represention for permutations in terms of their *cycle structure.*

There is a lot of freedom associated with representing permutations by their cycle structure. Since the cycles are disjoint, they can be specified in any order. Further, since each cycle specifies a circular shift of the given elements, there is no distinguished starting point, and any rotation of a cycle describes the same cycle. Thus the cycles $\{\{2,3,4\},\{6,1,5\}\}$ also define the permutation p.

The cycle representation of a permutation p can be constructed by starting with p_1 and following the chain of successors until it gets back to p_1, finishing the cycle. We can then trace the cycle associated with the smallest element which isn't in this first cycle, and repeat until all elements are covered.

```
ToCycles[p1_?PermutationQ] :=
        Block[{p=p1,m,n,cycle,i},
                Select[
                        Table[
                                m = n = p[[i]];
                                cycle = {};
                                While[ p[[n]] != 0,
```

```
                                        AppendTo[cycle,m=n];
                                        n = p[[n]];
                                        p[[m]] = 0
                                ];
                                cycle,
                                {i,Length[p]}
                        ],
                        (# =!= {})&
                ]
        ]
```

Extracting the Cycle Structure of a Permutation

From the cycle structure, we can recover the original permutation by stepping through each element of each cycle.

```
FromCycles[cyc_List] :=
        Block[{p=Table[0,{Length[Flatten[cyc]]}], pos},
                Scan[
                        (pos = Last[#];
                         Scan[ Function[c, pos = p[[pos]] = c], #])&,
                        cyc
                ];
                p
        ]
```

Reconstructing a Permutation from its Cycle Structure

The identity permutation consists of n singleton cycles or fixed-points.

```
In[33]:= ToCycles[{1,2,3,4,5,6,7,8,9,10}]
Out[33]= {{1}, {2}, {3}, {4}, {5}, {6}, {7}, {8}, {9},
  {10}}
```

There are $(n-1)!$ permutations on n elements which define just one cycle, since after placing the smallest element at the head of the cycle, any permutation of the rest of the elements finishes the job.

```
In[34]:= Select[ Permutations[{1,2,3,4}],
(Length[ToCycles[#]] == 1)&]
Out[34]= {{2, 3, 4, 1}, {2, 4, 1, 3}, {3, 1, 4, 2},
  {3, 4, 2, 1}, {4, 1, 2, 3}, {4, 3, 1, 2}}
```

The reverse of the identity permutation contains only cycles of length two, with one singleton cycle if n is odd. Cycles of length two are transpositions.

```
In[35]:= ToCycles[ Reverse[Range[10]] ]
Out[35]= {{10, 1}, {9, 2}, {8, 3}, {7, 4}, {6, 5}}
```

Any permutation p with a maximum cycle length of two is an involution, meaning that multiplying p by itself gives the identity.

```
In[36]:= Permute[ Reverse[Range[10]], Reverse[Range[10]] ]
Out[36]= {1, 2, 3, 4, 5, 6, 7, 8, 9, 10}
```

Since each cycle structure defines a unique permutation, converting to and from the cycle structure is an identity operation.

```
In[37]:= Apply[And, Table[ p=RandomPermutation[20];
p===FromCycles[ToCycles[p]], {50}] ]

Out[37]= True
```

Hiding Cycles

The parentheses in the cyclic representation of a permutation turn out to be unnecessary, since the cycles can be unambiguously represented without them. This gives a bijection between permutations and cyclic representations. Our canonical representation rotates each cycle so the minimum element is first and then sorts the cycles in decreasing order of the smallest element. This construction is not unique, and [SW86] present an alternate way to hide cycles.

```
HideCycles[c_List] :=
        Flatten[
                Sort[
                        Map[(RotateLeft[#,Position[#,Min[#]] [[1,1]] - 1])&, c],
                        (#1[[1]] > #2[[1]])&
                ]
        ]
```

Hiding the Cycle Structure

The cycle structure can be revealed again by scanning from left to right and starting a new cycle before each left-to-right minimum.

```
RevealCycles[p_?PermutationQ] :=
        Block[{start=end=1, cycles={}},
                While [end <= Length[p],
                        If [p[[start]] > p[[end]],
                                AppendTo[ cycles, Take[p,{start,end-1}] ];
                                start = end,
                                end++
                        ]
                ];
                Append[cycles,Take[p,{start,end-1}]]
        ]
```

Revealing the Hidden Cycle Structure

This permutation contains three cycles, one of each size.

```
In[38]:= ToCycles[{6,2,1,5,4,3}]

Out[38]= {{6, 3, 1}, {2}, {5, 4}}
```

By ordering the cycles by their last elements, the parentheses become unnecessary. Observe that the permutation we get is *not* the one we started with.

```
In[39]:= HideCycles[%]

Out[39]= {4, 5, 2, 1, 6, 3}
```

Although HideCycles permuted the order of the cycles, the permutation is correctly defined.

```
In[40]:= RevealCycles[ HideCycles[ ToCycles[{6,2,1,5,4,3}]
] ]

Out[40]= {{4, 5}, {2}, {1, 6, 3}}
```

None of the permutations on five elements is identical to its hidden cycle representation.

```
In[41]:= Apply[Or, Map[(# === HideCycles[ToCycles[#]])&,
Permutations[{1,2,3,4,5}]] ]

Out[41]= False
```

Counting Cycles

To determine the number of permutations $\left[{n \atop k}\right]$ on n elements with exactly k cycles, we can formulate a recurrence based on the cycle structures of all permutations on $n-1$ elements. Either the nth element forms a singleton cycle or it doesn't. If it does, there are $\left[{n-1 \atop k-1}\right]$ ways to arrange the rest of the elements to form $k-1$ cycles. If not, the nth element can be inserted in every possible position of every cycle of the $\left[{n-1 \atop k}\right]$ ways to make k cycles out of $n-1$ elements. This recurrence defines the *Stirling numbers of the first kind.*

```
NumberOfPermutationsByCycles[n_Integer,m_Integer] := (-1)^(n-m) StirlingS1[n,m]

StirlingFirst[n_Integer,0] := If [n == 0, 1, 0]
StirlingFirst[0,m_Integer] := If [m == 0, 1, 0]

StirlingFirst[n_Integer,m_Integer] := StirlingFirst[n,m] =
        (n-1) StirlingFirst[n-1,m] + StirlingFirst[n-1, m-1]

StirlingSecond[n_Integer,0] := If [n == 0, 1, 0]
StirlingSecond[0,m_Integer] := If [m == 0, 1, 0]

StirlingSecond[n_Integer,m_Integer] := StirlingSecond[n,m] =
        m StirlingSecond[n-1,m] + StirlingSecond[n-1,m-1]
```

Counting Permutations by Cycles

To avoid recomputing partial results, the algorithmic technique of *dynamic programming* has been used to implement both recurrences. In *Mathematica*, this is implemented by explicitly defining StirlingFirst[m,n] for each value of m and n, thus replacing future recursive calls by a pattern-match operation.

Our definition of the Stirling numbers of the first kind follows Knuth [Knu73a, GKP89], and differs from the built-in *Mathematica* function `StirlingS1`. The latter alternates positive and negative depending upon the parity of the sum of the arguments.

```
In[42]:= {StirlingFirst[6,3], StirlingS1[6,3]}
Out[42]= {225, -225}
```

These are the eleven permutations of four elements with exactly two cycles.

```
In[43]:= Select[ Map[ToCycles,Permutations[{1,2,3,4}]],
(Length[#]==2)&]
Out[43]= {{{1}, {3, 4, 2}}, {{1}, {4, 3, 2}},
  {{2, 1}, {4, 3}}, {{2, 3, 1}, {4}}, {{2, 4, 1}, {3}},
  {{3, 2, 1}, {4}}, {{3, 4, 1}, {2}}, {{3, 1}, {4, 2}},
  {{4, 2, 1}, {3}}, {{4, 3, 1}, {2}}, {{4, 1}, {3, 2}}}
```

This total is consistant with the enumeration formula.

```
In[44]:= NumberOfPermutationsByCycles[4,2]
Out[44]= 11
```

The *Stirling numbers of the second kind* $\left\{ n \atop k \right\}$ count the number of ways to partition n items into k sets. By a similar argument, they are defined by the recurrence $\left\{ n \atop k \right\} = k\left\{ n-1 \atop k \right\} + \left\{ n-1 \atop k \right\}$. The built-in *Mathematica* function `StirlingS2` computes them appropriately.

```
In[45]:= {StirlingSecond[6,3], StirlingS2[6,3]}
Out[45]= {90, 90}
```

1.2.5 Signatures

As discussed in Section 1.1.4, permutations can be constructed as a sequence of *transpositions*, or swaps of two elements, starting from the identity permutation. The parity of the size of any such sequence of transpositions is invariant for any particular permutation. The *signature* or *sign* of a permutation p is $+1$ if the number of transpositions in the sequence is even and -1 if it is odd.

This signature of a permutation can be computed from its cycle decomposition. The effect of multiplying a permutation by cycle of length n can be obtained by performing $n-1$ transpositions, so the total length of all the cycles minus the number of cycles determines the parity of the signature.

```
SignaturePermutation[p_?PermutationQ] := (-1) ^ (Length[p]-Length[ToCycles[p]])
```

Finding the Signature of a Permutation

Calculating the signature happens to be one of the combinatorial functions built into *Mathematica*. As expected, the built-in version is faster, but not dramatically so.

```
In[46]:= Table[ p=RandomPermutation[i]; First[
Timing[SignaturePermutation[p];] / Timing[Signature[p];]
], {i,50,500,50}]

Out[46]= {15., 5.33333, 4., 3.45455, 3.08333, 2.90196,

  2.61429, 2.83871, 2.38136, 2.17687}
```

One transposition is sufficient to move this to the identity permutation, meaning an odd number of transpositions no matter how they are selected.

```
In[47]:= SignaturePermutation[{1,3,2,4,5,6,7,8}]

Out[47]= -1
```

All permutations have the same sign as their inverse.

```
In[48]:= {SignaturePermutation[p],
SignaturePermutation[InversePermutation[p]]}

Out[48]= {1, 1}
```

The set of all even permutations forms a group. This proves that the product of two even permutations is even, as is the product of two odd permutations.

```
In[49]:= PermutationGroupQ[ Select[
Permutations[{1,2,3,4}], (SignaturePermutation[#]==1)&] ]

Out[49]= True
```

1.2.6 Polya's Theory of Counting

Polya's theory of counting [Pol37] provides a tool for enumerating the number of distinct ways of a structure can be built when symmetry is taken into account. The classic problem in Polya theory is counting how many different ways necklaces can be made out of k beads, when there are m different types or colors of beads to choose from. The symmetry in the necklaces is reflected by the automorphism group of a simple cycle. Polya's theory is developed in most combinatorics texts [Rob84, Tuc84].

```
Polya[g_List,m_] := Apply[ Plus, Map[(m^Length[ToCycles[#]])&,g] ] / Length[g]
```

Counting Colorings with Polya's Theorem

When two necklaces are considered the same if they can be obtained only by rotating the beads (as opposed to turning the necklace over), the symmetries are defined by k permutations, each of which is a cyclic shift of the identity permutation. When a variable is specified for the number of colors, a polynomial results.

```
In[50]:= Polya[Table[RotateRight[Range[8],i], {i,8}], m]

                    2    4    8
         4 m + 2 m  + m  + m
Out[50]= ---------------------
                   8
```

The number of colorings for unrestricted necklaces is determined by the automorphism group of a cycle. There are approximately half as many unrestricted necklaces as restricted ones, because few necklaces are equivalent to their reverse.

```
In[51]:= Polya[Automorphisms[Cycle[8]], m]

                2      4      5    8
         4 m + 2 m  + 5 m  + 4 m  + m
Out[51]= ----------------------------
                      16
```

The resultant polynomials can be manipulated by any of the functions for symbolic computation available in *Mathematica*.

```
In[52]:= Factor[%]

                                 2      3    4    5    6
         m (1 + m) (4 - 2 m + 2 m  + 3 m  + m  - m  + m )
Out[52]= ------------------------------------------------
                               16
```

1.3 Inversions and Inversion Vectors

A pair of elements p_i and p_j represent an *inversion* in a permutation p if $i > j$ and $p_i < p_j$. Inversions are pairs which are out of order, and so they play a prominent role in the analysis of sorting algorithms.

1.3.1 Inversion Vectors

The ith element of the *inversion vector* v of permutation is the number of elements greater than i to the left of i. Marshall Hall [Tho56] demonstrated that there is a nice bijection between permutations and inversion vectors.

```
ToInversionVector[p_?PermutationQ] :=
        Block[{i,inverse=InversePermutation[p]},
                Table[
                        Length[ Select[Take[p,inverse[[i]]], (# > i)&] ],
                        {i,Length[p]-1}
                ]
        ]
```

Obtaining the Inversion Vector

The inversion vector contains only $n-1$ elements. The ith element can range from 0 to $n-i$, so there are indeed $n!$ distinct inversion vectors, one for each permutation.

```
In[53]:= ToInversionVector[{5,9,1,8,2,6,4,7,3}]
Out[53]= {2, 3, 6, 4, 0, 2, 2, 1}
```

To complete the bijection, observe that if $v_{n-i} = 0$, then the $n-i$ is to the left of all elements greater than it. Each element of the inversion vector identifies how many elements bigger than $n-i$ are to the left of it. Thus we can start with the nth element and repeatedly insert the $(n-i)$th element v_{n-i} positions from the left, until the entire permutation is determined.

```
FromInversionVector[vec_List] :=
        Block[{n=Length[vec]+1,i,p={n}},
                Do [
                        p = Insert[p, i, vec[[i]]+1],
                        {i,n-1,1,-1}
                ];
                p
        ]
```

Inverting the Inversion Vector

Composing these two functions gives an identity operation.

```
In[54]:= FromInversionVector[
ToInversionVector[{5,9,1,8,2,6,4,7,3}] ]

Out[54]= {5, 9, 1, 8, 2, 6, 4, 7, 3}
```

A *permutation graph* G_p is a graph whose edges $\{v_i, v_j\}$ correspond exactly to (i, j) being an inversion in some permutation p; ie., $i < j$ but j occurs before i in p. The structure of permutation graphs permits fast algorithms for certain problems which are intractable for general graphs [AMU88, BK87].

A permutation and its inversion vector provide different representations of the same structure.

```
In[55]:= {p=RandomPermutation[10], ToInversionVector[p]}

Out[55]= {{10, 7, 6, 1, 3, 2, 9, 4, 5, 8},
  {3, 4, 3, 4, 4, 2, 1, 2, 1}}
```

The permutation graph provides a pictorial view of the inversion structure of a permutation. Elements responsible for many inversions have high degree, while vertices of zero degree are in the correct position at either end of the permutation.

```
In[56]:= (h = InversePermutation[p];
ShowLabeledGraph[ g = MakeGraph[Range[10],
((#1<#2 && h[[#1]]>h[[#2]]) || (#1>#2 &&
h[[#1]]<h[[#2]]))&] ]);
```

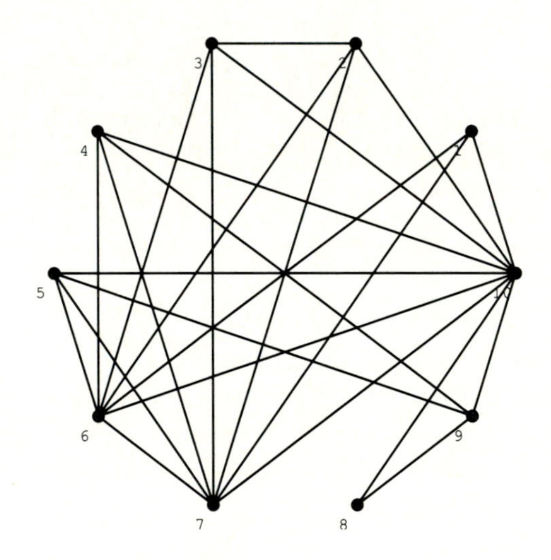

The total number of inversions in a permutation is equal to the number of edges in its permutation graph.

```
In[57]:= {Inversions[p], M[g]}

Out[57]= {24, 24}
```

■ 1.3.2 Counting Inversions

The total number of inversions in a permutation is a classic measure of order [Knu73b, Man85a, Ski88], and can be obtained by summing up the inversion vector.

```
Inversions[p_?PermutationQ] := Apply[Plus,ToInversionVector[p]]
```

Counting Inversions

The number of inversions in a permutation is equal to that of its inverse [Knu73b].

```
In[58]:= {Inversions[p],
Inversions[InversePermutation[p]]}
Out[58]= {24, 24}
```

The number of inversions in a permutation ranges from 0 to $\binom{n}{2}$, where the largest number of inversions comes from the reverse of the total order.

```
In[59]:= Inversions[Reverse[Range[10]]]
Out[59]= 45
```

Every one of these numbers is realizable as an inversion total for some permutation.

```
In[60]:= Union[ Map[Inversions, Permutations[{1,2,3,4,5}]]
]
Out[60]= {0, 1, 2, 3, 4, 5, 6, 7, 8, 9, 10}
```

A neat proof that there are an average of $n(n-1)/4$ inversions per permutation is that the number of inversions in a permutation and its reverse permutation always total to $\binom{n}{2}$.

```
In[61]:= ( p = RandomPermutation[10];
Inversions[p] + Inversions[Reverse[p]] )
Out[61]= 45
```

■ 1.3.3 The Index of a Permutation

The *index* of a permutation p is the sum of all subscripts j such that $p_j > p_{j+1}$, $1 \leq j < n$. MacMahon [Mac60] proved that the number of permutations of size n having index k is the same as the number having exactly k inversions.

```
Index[p_?PermutationQ]:=
        Block[{i},
                Sum[ If [p[[i]] > p[[i+1]], i, 0], {i,Length[p]-1} ]
        ]
```

Computing the Index of a Permutation

These are the six permutations of length 4 with index 3.

```
In[62]:= Select[Permutations[{1,2,3,4}], (Index[#]==3)&]
Out[62]= {{1, 2, 4, 3}, {1, 3, 4, 2}, {2, 3, 4, 1},
  {3, 2, 1, 4}, {4, 2, 1, 3}, {4, 3, 1, 2}}
```

As MacMahon proved, there are an equal number of permutations of length 4 with 3 inversions. A bijection between the two sets of permutations is not obvious [Knu73b].

```
In[63]:= Select[Permutations[{1,2,3,4}],
(Inversions[#]==3)&]

Out[63]= {{1, 4, 3, 2}, {2, 3, 4, 1}, {2, 4, 1, 3},
  {3, 1, 4, 2}, {3, 2, 1, 4}, {4, 1, 2, 3}}
```

■ 1.3.4 Runs and Eulerian Numbers

Reading from left to right, any permutation can be partitioned into a set of ascending sequences or *runs*. A sorted permutation consists of one run, while a reverse permutation consists of n runs, each of one element. For this reason, the number of runs is also used as a measure of the presortedness of a permutation [Man85a].

It is interesting to note that the expected length of the first run in a random permutation is shorter than the second run [Gas67, Knu73b]. The first element of the first run is whatever happens to be the first element of the permutation, while the first element of the second run must be less than the element before it in the permutation. Thus there is more "room" for successive elements to be greater than the head, and so the run should be longer.

```
Runs[p_?PermutationQ] :=
        Map[
                (Apply[Take,{p,{#[[1]]+1,#[[2]]}}])&,
                Partition[
                        Join[
                                {0},
                                Select[Range[Length[p]-1], (p[[#]]>p[[#+1]])&],
                                {Length[p]}
                        ],
                        2,
                        1
                ]
        ]

Eulerian[0,k_Integer] := If [k==1, 1, 0]
Eulerian[n_Integer,k_Integer] := Eulerian[n,k] =
        k Eulerian[n-1,k] + (n-k+1) Eulerian[n-1,k-1]
```

Finding and Counting Runs in Permutations

The *Eulerian* numbers $\left\langle {n \atop k} \right\rangle$ count the number of permutations of length n with exactly k runs. A recurrence for the Eulerian numbers can be formulated by considering each permutation p of $1, ..., n-1$. There are n places to insert n, and each either splits an existing run of p or occurs immediately after the last element of an existing run, meaning the run count is preserved. Thus $\left\langle {n \atop k} \right\rangle = k \left\langle {n-1 \atop k} \right\rangle + (n-k+1) \left\langle {n-1 \atop k-1} \right\rangle$. Dynamic programming has been used to make `Eulerian` reasonably efficient.

There are eleven permutations of length four with exactly two runs.

```
In[64]:=
Select[Permutations[Range[4]],(Length[Runs[#]]==2)&]

Out[64]= {{1, 2, 4, 3}, {1, 3, 2, 4}, {1, 3, 4, 2},
  {1, 4, 2, 3}, {2, 1, 3, 4}, {2, 3, 1, 4}, {2, 3, 4, 1},
  {2, 4, 1, 3}, {3, 1, 2, 4}, {3, 4, 1, 2}, {4, 1, 2, 3}}
```

The correct definition of Eulerian numbers depends upon which of Knuth's books you choose [GKP89, Knu73b]. We have used the older one, which counts runs instead of *ascents*, adjacent positions which are out of order. The difference is that k ascents imply $k+1$ runs.

```
In[65]:= Eulerian[4,2]

Out[65]= 11
```

1.4 Special Classes of Permutations

In this section, we discuss several interesting classes of permutations and some of the properties which make them interesting.

1.4.1 Involutions

Involutions are permutations which are their own multiplicative inverses. A permutation is an involution if and only if its cycle structure consists exclusively of fixed points and transpositions. The number of involutions on n elements is exactly the number of distinct Young tableaux on n elements, as will be discussed in Section 2.3.2.

```
InvolutionQ[p_?PermutationQ] := p[[p]] == Range[Length[p]]
```

Testing for Involutions

There are ten involutions on four elements.

```
In[66]:= Select[ Permutations[{1,2,3,4}], InvolutionQ ]
Out[66]= {{1, 2, 3, 4}, {1, 2, 4, 3}, {1, 3, 2, 4},
  {1, 4, 3, 2}, {2, 1, 3, 4}, {2, 1, 4, 3}, {3, 2, 1, 4},
  {3, 4, 1, 2}, {4, 2, 3, 1}, {4, 3, 2, 1}}
```

Since these are involutions, squaring them gives the identity permutation.

```
In[67]:= Map[(Permute[#,#])&, %]
Out[67]= {{1, 2, 3, 4}, {1, 2, 3, 4}, {1, 2, 3, 4},
  {1, 2, 3, 4}, {1, 2, 3, 4}, {1, 2, 3, 4}, {1, 2, 3, 4},
  {1, 2, 3, 4}, {1, 2, 3, 4}, {1, 2, 3, 4}}
```

The number of involutions is given by the recurrence $t_n = t_{n-1} + (n-1)t_{n-2}$, since the cycle structure consists exclusively of fixed points and transpositions. No simple closed form for this is known, but [Knu73b] gives the following equivalent formula.

```
NumberOfInvolutions[n_Integer] :=
        n! Sum[1/((n - 2k)! 2^k k!), {k, 0, Quotient[n, 2]}]
```

The Number of Involutions on n Elements

Involutions are exactly the permutations with a maximum cycle length of two.

```
In[68]:= Map[ ToCycles, Select[ Permutations[{1,2,3,4}],
InvolutionQ ] ]

Out[68]= {{{1}, {2}, {3}, {4}}, {{1}, {2}, {4, 3}},
  {{1}, {3, 2}, {4}}, {{1}, {4, 2}, {3}},
  {{2, 1}, {3}, {4}}, {{2, 1}, {4, 3}},
  {{3, 1}, {2}, {4}}, {{3, 1}, {4, 2}},
  {{4, 1}, {2}, {3}}, {{4, 1}, {3, 2}}}
```

As shown above, there are ten involutions on four elements.

```
In[69]:= NumberOfInvolutions[4]

Out[69]= 10
```

■ 1.4.2 Derangements

Derangements are permutations p with no element in its proper position, meaning there exists no $\{p_i = i,\ 1 \leq i \leq n\}$. Thus the derangements are permutations without a fixed point, having no cycle of length one.

The recurrence $d_n = (n-1)(d_{n-1} + d_{n-2})$ gives the number of derangements of size n, as shown by the following argument. Construct all permutations where the first $n-1$ elements are a derangement of $1, ..., n-1$, and the nth element is n. Exchanging the nth element with any other gives a derangement of size n. Further, any of the $(n-1)d_{n-2}$ permutations of $n-1$ elements with exactly one fixed point can be turned into a derangement by similarly exchanging the fixed point with n.

```
DerangementQ[p_?PermutationQ] :=
        !(Apply[ Or, Map[( # == p[[#]] )&, Range[Length[p]]] ])

NumberOfDerangements[0] = 1;
NumberOfDerangements[n_] := n * NumberOfDerangements[n-1] + (-1)^n

Derangements[n_Integer] := Derangements[Range[n]]
Derangements[p_?PermutationQ] := Select[ Permutations[p], DerangementQ ]
```

Obtaining and Counting Derangements

An alternate recurrence for the number of derangements is $d_n = nd_{n-1} + (-1)^n$, which is an easier form to compute.

```
In[70]:= Table[NumberOfDerangements[i], {i,1,10}]

Out[70]= {0, 1, 2, 9, 44, 265, 1854, 14833, 133496,
  1334961}
```

If a confused secretary randomly stuffs n different letters into n pre-addressed envelopes, what is the probability that none of them end up where they are supposed to? The ratio of the number of derangements to the number of permutations converges rapidly to $1/e$. Thus the answer is essentially independent of n.

```
In[71]:= Table[ N[ NumberOfDerangements[i]/(i!) ],
{i,1,10} ]

Out[71]= {0., 0.5, 0.333333, 0.375, 0.366667, 0.368056,
  0.367857, 0.367882, 0.367879, 0.367879}
```

In fact, rounding $n!/e$ gives a nicer way to compute the number of derangements.

```
In[72]:= Table[Round[n!/N[E]], {n,1,10}]

Out[72]= {0, 1, 2, 9, 44, 265, 1854, 14833, 133496,
  1334961}
```

Since the number of derangements is always at least a third of the total number of permutations, we are morally justified in selecting them from a list of possibilities instead of generating them from scratch.

```
In[73]:= Derangements[4]

Out[73]= {{2, 1, 4, 3}, {2, 3, 4, 1}, {2, 4, 1, 3},
  {3, 1, 4, 2}, {3, 4, 1, 2}, {3, 4, 2, 1}, {4, 1, 2, 3},
  {4, 3, 1, 2}, {4, 3, 2, 1}}
```

■ 1.4.3 Josephus' Problem

When times got bad in the Roman Empire, n men of honor would arrange themselves in a circle and repeatedly execute every mth man until one was left, who had a moral obligation to kill himself and finish the job. Josephus was no man of honor, and when times got bad he and a like-minded friend figured out where to stand in order to be the last two left, thus surviving to become a traitor to the other side. The Josephus problem is to return the permutation describing the order in which the men get slain. In keeping with Knuth [GKP89, Knu73a, Knu73b], the returned permutation is the inverse of the order in which they were slaughtered.

```
Josephus[n_Integer,m_Integer] :=
        Block[{live=Range[n],next},
                InversePermutation[
                        Table[
                                next = RotateLeft[live,m-1];
                                live = Rest[next];
                                First[next],
                                {n}
                        ]
                ]
        ]
```

Josephus Permutations

More efficient algorithms for computing the position of an element in this permutation exploit modular arithmetic [GKP89]. The second man is the lucky survivor in this instance.

```
In[74]:= Josephus[17,7]
Out[74]= {16, 17, 5, 3, 14, 7, 1, 11, 10, 12, 9, 4, 6, 2,
  15, 13, 8}
```

Plotting the inverse of a Josephus permutation neatly illustrates the order in which the men get executed as we count around the circle.

```
In[75]:= ListPlot[ InversePermutation[ Josephus[49,4] ] ];
```

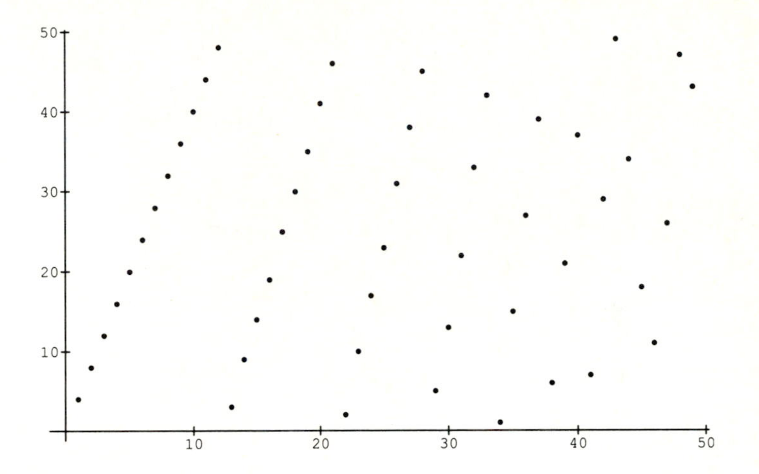

■ 1.4.4 Heaps

A *binary heap* is a permutation p of length n such that $p_i < p_{2i}$ and $p_i < p_{2i+1}$ for all $1 \leq i \leq n/2$. As a consequence of this definition, p_1 must be the smallest element of p. Heaps can be viewed as labeled binary trees such that the label of the ith node is smaller than the labels of any of its descendents.

Floyd's `Heapify` algorithm [Flo64] constructs a heap in linear time through an operation which merges two heaps of height l and one isolated element into a single heap of height $l + 1$. The isolated element starts as the root and is trickled down into the correct position, thus forcing the smallest element into the root. Since a single element always forms a heap, the heap definition is immediately satisfied for the elements in the second half of the permutation, and so applying `Heapify` $\lfloor n/2 \rfloor$ times starting from the $\lfloor n/2 \rfloor$nd element merges these heaps into one.

```
Heapify[p_List] :=
        Block[{j,heap=p},
                Do [
                        heap = Heapify[heap,j],
                        {j,Quotient[Length[p],2],1,-1}
```

```
                ];
                heap
        ]

Heapify[p_List, k_Integer] :=
        Block[{hp=p, i=k, l, n=Length[p]},
                While[ (l = 2 i) <= n,
                        If[ (l < n) && (hp[[l]] > hp[[l+1]]), l++ ];
                        If[ hp[[i]] > hp[[l]],
                                {hp[[i]],hp[[l]]}={hp[[l]],hp[[i]]};
                                i = l,
                                i = n+1
                        ];
                ];
                hp
        ]

RandomHeap[n_Integer] := Heapify[RandomPermutation[n]]
```

Constructing Random Heaps

The number of complete heaps of l levels (or equivalently heaps of $2^l - 1$ elements) is given by the recurrence

$$S_n = \binom{2^l - 2}{2^{l-1} - 1} S_{l-1}^2, \quad S_1 = 1$$

since after the minimum element takes the root position, all heaps are constructed by partitioning the elements and building them into two heaps on $l-1$ levels.

Plotting the value of the ith element in a permutation against i provides a graphical way to reveal its structure. In a random permutation, there isn't much structure to reveal.

In[76]:= **ListPlot[RandomPermutation[127]];**

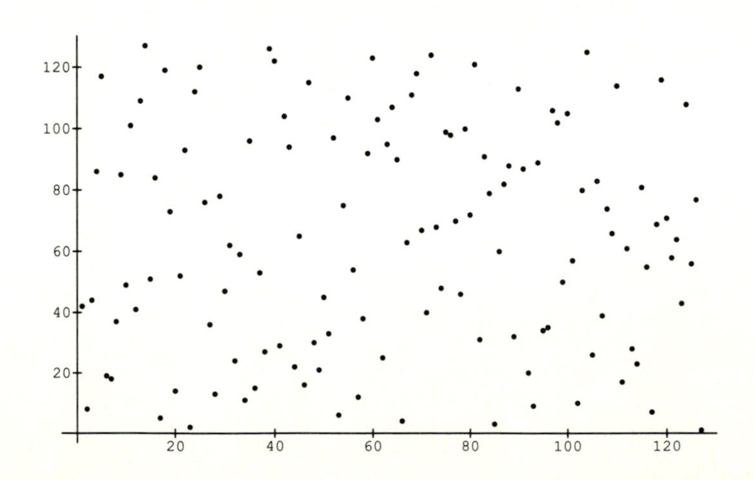

In a random heap, the first element is in the appropriate position, but the permutation gets progressively less ordered as we move to the right. Diagrams such as these illustrating a variety of different sorting algorithms appear in [Sed88].

```
In[77]:= ListPlot[ RandomHeap[127] ];
```

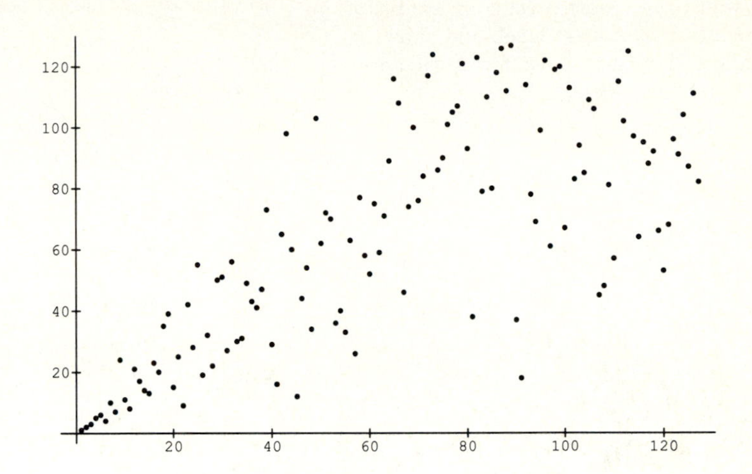

Performing Floyd's heapify algorithm on a random permutation does indeed give a random heap, since depending upon its value, the root node can occupy any position in the merged heap [Knu73b].

```
In[78]:= Distribution[ Table[RandomHeap[5], {200}] ]
Out[78]= {24, 22, 37, 17, 27, 24, 27, 22}
```

The inversion structure of a heap is significantly different from your typical random permutation

```
In[79]:= {(p=RandomHeap[15]), ToInversionVector[p]}
Out[79]= {{1, 2, 3, 7, 8, 4, 6, 10, 14, 11, 13, 12, 5, 9,
   15}, {0, 0, 0, 2, 8, 2, 0, 0, 5, 0, 1, 2, 1, 0}}
```

A different way of making the same point by constructing the permutation graph of a random heap. Most inversions are between elements in the same level, and thus the smallest elements create few inversions.

```
In[80]:= (h = InversePermutation[p];
ShowLabeledGraph[ MakeGraph[Range[15], ((#1<#2 &&
p[[#1]]>p[[#2]]) || (#1>#2 && p[[#1]]<p[[#2]]))&] ]);
```

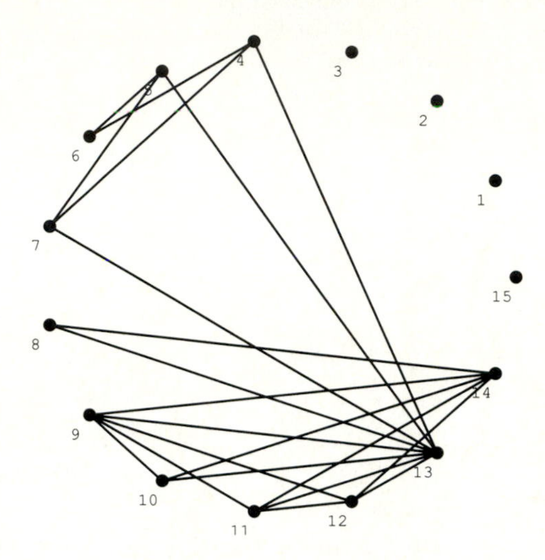

The most important application of heaps is as a *priority queue* data structure for repeatedly extracting the smallest remaining key. For example, in the simulation of a queuing system, new requests may get added to the queue at arbitrary times, but must get processed at the intended times. Heaps support arbitrary insertion and finding/deleting the minimum in $O(\log n)$ time per update.

Since the operations of finding and deleting the smallest element are exactly what is required for selection sort (Section 1.1.6), using a priority queue yields *heapsort*, an efficient $O(n \log n)$ time sorting algorithm.

```
HeapSort[p_List] :=
        Block[{heap=Heapify[p],min},
                Append[
                        Table[
                                min = First[heap];
                                heap[[1]] = heap[[n]];
                                heap = Heapify[Drop[heap,-1],1];
                                min,
                                {n,Length[p],2,-1}
                        ],
                        Max[heap]
                ]
        ]
```

HeapSort

Because of the *Mathematica* model of computation, `HeapSort` does not significantly outperform `SelectionSort`.

```
In[81]:= (p=RandomPermutation[50]; Timing[HeapSort[p];] /
Timing[SelectionSort[p,Less];] )

Out[81]= {1.05952, 1}
```

1.5 Combinations

Perhaps the only combinatorial result more elementary than proving that there are $n!$ distinct permutations is proving that there are 2^n distinct subsets of n elements. As with permutations, subsets can be generated in a variety of interesting orders.

An important subset of the subsets are those containing the same number of elements. In elementary combinatorics, permutations are always mentioned in the same breath as *combinations*, which are collections of elements of a given size where order doesn't matter. Thus a combination is exactly a k-subset for some k, and we will use the terms interchangeably.

1.5.1 Strings

There is a another simple combinatorial object which will prove useful in several different contexts, and which it is appropriate to consider at this time. We define a *string* of length k on an alphabet M to be an arrangement of k, not necessarily distinct, symbols from M. Such strings should not be confused with the character string, which is a data type for text in *Mathematica*. If the alphabet size is m, there are m^k distinct strings, which can all be constructed with a simple recursive algorithm.

```
Strings[l_List,0] := { {} }

Strings[l_List,k_Integer?Positive] :=
        Block[{oneless = Strings[l,k-1],i,n=Length[l]},
                Apply[Join, Table[ Map[(Prepend[#,l[[i]]])&, oneless], {i,n}] ]
        ]
```

Constructing all Strings on an Alphabet

This recursive construction prepends each symbol successively to all strings of length $k-1$ thus yielding the strings in lexicographic order.

```
In[82]:= Strings[{1,2,3},3]
Out[82]= {{1, 1, 1}, {1, 1, 2}, {1, 1, 3}, {1, 2, 1},
  {1, 2, 2}, {1, 2, 3}, {1, 3, 1}, {1, 3, 2}, {1, 3, 3},
  {2, 1, 1}, {2, 1, 2}, {2, 1, 3}, {2, 2, 1}, {2, 2, 2},
  {2, 2, 3}, {2, 3, 1}, {2, 3, 2}, {2, 3, 3}, {3, 1, 1},
  {3, 1, 2}, {3, 1, 3}, {3, 2, 1}, {3, 2, 2}, {3, 2, 3},
  {3, 3, 1}, {3, 3, 2}, {3, 3, 3}}
```

■ 1.5.2 Binary Representation and Random Sets

An element of a set is either in a particular subset or not in it. This binary representation provides a method for generating all subsets, by iterating through the 2^n distinct binary strings of length n and using them as a descriptor of a subset. These strings are indistinguishable from the integers written in binary notation, and so the function `Strings`, defined above, can be used to construct them.

Since we have established a bijection between the integers mod $2^n - 1$ and subsets of an n-set, this gives a ranking which defines the ith subset of a set. The built-in function `Digits` can be used to generate the binary string for any integer, which must be reversed to prevent the first element from appearing in each subset.

```
NthSubset[n_Integer,m_Integer] := NthSubset[n,Range[m]]
NthSubset[n_Integer,l_List] :=
        l[[ Flatten[ Position[Reverse[Digits[ Mod[n,2^Length[l]],2]],1] ] ]]

BinarySubsets[l_List] :=
        Block[{pos=Reverse[Range[Length[l]]], n=Length[l]},
                Map[(l[[ Reverse[Select[pos*#, Positive]] ]])&, Strings[{0,1},n] ]
        ]

NextSubset[set_List,subset_List] := NthSubset[ RankSubset[set,subset], set  ]

RankSubset[set_List,subset_List] :=
        Block[{i,n=Length[set]},
                Sum[ 2^(i-1) * If[ MemberQ[subset,set[[i]]], 1, 0], {i,n}]
        ]
```

Subsets from Bit Strings

Although this method is relatively slow, we can use it to generate all subsets, and even to generate the next or previous subset by performing an arithmetic operation.

```
In[83]:= BinarySubsets[{a,b,c,d}]
Out[83]= {{}, {a}, {b}, {a, b}, {c}, {a, c}, {b, c},
  {a, b, c}, {d}, {a, d}, {b, d}, {a, b, d}, {c, d},
  {a, c, d}, {b, c, d}, {a, b, c, d}}
```

Generating subsets incrementally is efficient when the goal is to find the first subset with a given property, since every subset need not be constructed.

```
In[84]:= Table[NthSubset[n,{a,b,c,d}], {n,0,15}]
Out[84]= {{}, {a}, {b}, {a, b}, {c}, {a, c}, {b, c},
  {a, b, c}, {d}, {a, d}, {b, d}, {a, b, d}, {c, d},
  {a, c, d}, {b, c, d}, {a, b, c, d}}
```

Because n is taken modulo the number of subsets, any positive or negative integer can specify a subset.

```
In[85]:= NthSubset[-10,{a,b,c,d}]
Out[85]= {b, c}
```

The rankings of the binary ordered subsets are equivalent to the integers which generated them.

```
In[86]:= Map[(RankSubset[{1,2,3,4},#])&,
BinarySubsets[{1,2,3,4}]]

Out[86]= {0, 1, 2, 3, 4, 5, 6, 7, 8, 9, 10, 11, 12, 13,
  14, 15}
```

Ranking combinatorial structures makes it easy to generate random instances of them, by picking a random integer between 1 and the number of structures and performing an unranking operation. Picking a random integer between 0 and $2^n - 1$ is equivalent to flipping n coins, each determining whether an element of the set appears in the subset.

```
RandomSubset[set_List] := NthSubset[Random[Integer,2^(Length[set])-1],set]
```

Generating Random Subsets

Since the sizes of random subsets obey a binomial distribution, very large or very small subsets are rare compared to subsets with approximately half the elements.

```
In[87]:= Distribution[ Table[
Length[RandomSubset[Range[10]]], {1024}], Range[0,10] ]

Out[87]= {3, 11, 56, 109, 200, 248, 204, 125, 61, 7, 0}
```

Each subset occurs with roughly equal frequency since `RandomSubset` is truly random.

```
In[88]:= Distribution[ Table[ RandomSubset[{1,2,3,4}],
{1024}], Subsets[{1,2,3,4}] ]

Out[88]= {72, 68, 64, 59, 65, 68, 51, 68, 60, 59, 65, 62,
  66, 70, 60, 67}
```

■ 1.5.3 Gray Codes

We have seen that permutations can be generated to limit the difference between adjacent permutations to one transposition. A binary *Gray code* [Gra53] is a similar ordering of subsets, so that adjacent subsets differ by the insertion or deletion of exactly one element. The original application was to find the best assignment of codes for numerical data such that a transmission error causes only a small difference in magnitude between the transmitted and received data.

There is a beautiful recursive algorithm [NW78, Wil89] for constructing Gray codes which is ideally suited for implementation in *Mathematica*. Consider the elements one at a time. At each step, we take all the subsets to date and concatenate them with a reversed list of the same subsets. Clearly, all subsets differ by one from their neighbors except in the center, where they are identical. Adding the next element to the subsets in the top half maintains the Gray code property while distinguishing the center two subsets by exactly one element.

As we shall see in Section 4.2.5, this construction for Gray codes is equivalent to identifying a Hamiltonian cycle on an n-dimensional hypercube. Other properties of Gray codes are discussed in [Gil58].

```
GrayCode[l_List] := GrayCode[l,{{}}]

GrayCode[{},prev_List] := prev

GrayCode[l_List,prev_List] :=
        GrayCode[
                Rest[l],
                Join[ prev, Map[(Append[#,First[l]])&,Reverse[prev]] ]
        ]

Subsets[l_List] := GrayCode[l]
Subsets[n_Integer] := GrayCode[Range[n]]
```

Constructing Binary Reflected Gray Codes

Each subset differs in exactly one element from its neighbors. Observe that the last eight subsets all contain 4, while none of the first eight do.

```
In[89]:= GrayCode[{1,2,3,4}]

Out[89]= {{}, {1}, {1, 2}, {2}, {2, 3}, {1, 2, 3},
  {1, 3}, {3}, {3, 4}, {1, 3, 4}, {1, 2, 3, 4},
  {2, 3, 4}, {2, 4}, {1, 2, 4}, {1, 4}, {4}}
```

It takes less time to generate the *1024* subsets of ten elements in Gray code order than the 256 subsets of eight elements by generating all bit strings! Thus the generic `Subsets` returns a Gray code.

```
In[90]:= Timing[Length[BinarySubsets[Range[8]]]] /
Timing[Length[GrayCode[Range[10]]]]

                     1
Out[90]= {3.21622, -}
                     4
```

■ 1.5.4 Lexicographically Ordered Subsets

Lexicographically ordered subsets can be constructed using a recurrence with a similar flavor to Gray codes. In lexicographic order, all subsets containing the first element appear before any of the subsets which do not. Thus we can recursively construct all subsets of the rest of the elements lexicographically and make two copies. The leading copy gets the first element inserted in each subset, while the second one does not.

```
LexicographicSubsets[l_List] := LexicographicSubsets[l,{{}}]

LexicographicSubsets[{},s_List] := s

LexicographicSubsets[l_List,subsets_List] :=
```

```
        LexicographicSubsets[
                Rest[l],
                Join[
                        subsets,
                        Map[(Prepend[#,First[l]])&,LexicographicSubsets[Rest[l],{{}}] ]
                ]
        ]
```

Generating Subsets in Lexicographic Order

In lexicographic order, two subsets are ordered by their smallest elements. Thus the 2^{n-1} subsets containing 1 appear together.

```
In[91]:= LexicographicSubsets[{1,2,3,4}]
Out[91]= {{}, {1}, {1, 2}, {1, 2, 3}, {1, 2, 3, 4},
  {1, 2, 4}, {1, 3}, {1, 3, 4}, {1, 4}, {2}, {2, 3},
  {2, 3, 4}, {2, 4}, {3}, {3, 4}, {4}}
```

Lexicographic ordering is not as fast as Gray code ordering in this implementation.

```
In[92]:= Timing[LexicographicSubsets[Range[10]];] /
Timing[GrayCode[Range[10]];]
Out[92]= {8.02703, 1}
```

1.5.5 Generating k-Subsets

A *k-subset* is a subset with exactly k elements in it. The simplest way to construct all k-subsets of a set uses one of the previously discussed algorithms to construct all the subsets and then filters out those of the wrong length. However, since there are 2^n subsets and only $\binom{n}{k}$ k-subsets, this is very wasteful when k is small.

```
KSubsets[l_List,0] := { {} }
KSubsets[l_List,1] := Partition[l,1]
KSubsets[l_List,k_Integer?Positive] := {l} /; (k == Length[l])
KSubsets[l_List,k_Integer?Positive] := {}  /; (k > Length[l])

KSubsets[l_List,k_Integer?Positive] :=
        Join[
                Map[(Prepend[#,First[l]])&, KSubsets[Rest[l],k-1]],
                KSubsets[Rest[l],k]
        ]
```

Generating Lexicographically Ordered k-Subsets

Our recursive construction starts from the observation that each k-subset on n elements either contains the first element of the set or it does not. Prepending the first element to each $(k-1)$-subset of the other $n-1$ elements gives the former, and building all the k-subsets of the other $n-1$ elements gives the latter. The terminating conditions are when $k = 1$, so each element forms its own subset, and when $k = n$, since the set itself is the only n-subset.

Since the lead element is placed in first, the k-subsets are given in lexicographic order.

```
In[93]:= KSubsets[{1,2,3,4,5},3]
Out[93]= {{1, 2, 3}, {1, 2, 4}, {1, 2, 5}, {1, 3, 4},
  {1, 3, 5}, {1, 4, 5}, {2, 3, 4}, {2, 3, 5}, {2, 4, 5},
  {3, 4, 5}}
```

The *next* k-subset in lexicographical order can be constructed by eliminating the elements in the subset which represent the tail of the set, and replacing them with the appropriate number of elements.

```
NextKSubset[set_List,subset_List] :=
        Take[set,Length[subset]] /; (Take[set,-Length[subset]] === subset)

NextKSubset[set_List,subset_List] :=
        Block[{h=1, x=1},
                While [set[[-h]] == subset[[-h]], h++];
                While [set[[x]] =!= subset[[-h]], x++];
                Join[ Drop[subset,-h], Take[set, {x+1,x+h}] ]
        ]
```

Generating the Next Lexicographically Ordered k-Subset

These subsets are generated, as above, in lexicographic order.

```
In[94]:= ( p = Take[Range[5],-3];
Table[ p = NextKSubset[Range[5],p], {Binomial[5,3]} ] )
Out[94]= {{1, 2, 3}, {1, 2, 4}, {1, 2, 5}, {1, 3, 4},
  {1, 3, 5}, {1, 4, 5}, {2, 3, 4}, {2, 3, 5}, {2, 4, 5},
  {3, 4, 5}}
```

To construct a random k-subset, it is sufficient to select the first k elements in a random permutation of the set, then sort them to restore the canonical order. A more subtle and efficient algorithm appears in [NW78].

```
RandomKSubset[n_Integer,k_Integer] := RandomKSubset[Range[n],k]

RandomKSubset[set_List,k_Integer] :=
        Block[{s=Range[Length[set]],i,n=Length[set],x},
                set [[
                        Sort[
                                Table[
                                        x=Random[Integer,{1,i}];
                                        {s[[i]],s[[x]]} = {s[[x]],s[[i]]};
                                        s[[i]],
```

```
                                        {i,n,n-k+1,-1}
                                 ]
                          ]
                   ]]
            ]
```

Generating Random k-subsets

Here we generate random subsets of increasing size.

```
In[95]:= TableForm[ Table[ RandomKSubset[10,i], {i,10}] ]
                    {4}
                    {9, 10}
                    {2, 4, 6}
                    {6, 7, 8, 10}
                    {1, 4, 5, 6, 7}
                    {2, 3, 4, 6, 8, 9}
                    {1, 3, 4, 5, 7, 8, 10}
                    {1, 3, 4, 5, 6, 7, 8, 10}
                    {1, 2, 3, 4, 5, 6, 7, 8, 10}
Out[95]//TableForm= {1, 2, 3, 4, 5, 6, 7, 8, 9, 10}
```

The distribution of 3-subsets appears to be uniform.

```
In[96]:= Distribution[ Table[RandomKSubset[5,3], {200}] ]
Out[96]= {18, 22, 14, 24, 25, 19, 20, 17, 18, 23}
```

1.6 Exercises and Research Problems

Exercises

1. Plot the time complexity of the following different subset construction functions: `BinarySubsets`, `GrayCode`, and `LexicographicSubsets`. Do they appear to have the same asymptotic complexity? Can you write a more efficient subset generator?

2. Perform experiments on the distribution of the length of the ith run in random permutations. How much longer, on average, is the first run than the second, and what value does the expected length converge to for large i?

3. Based on the recurrence relation enumerating the number of complete heaps on l levels, develop and implement an algorithm for constructing all heaps on the elements $\{1, ..., n\}$.

4. How many permutations on n elements have inversion vectors which consist of $n-1$ distinct integers? Can you characterize the permutations with this property?

5. Implement a more general form of Polya's theorem that generates the structures instead of just counting them.

6. Characterize the permutations which are identical to their hidden cycle representation.

7. Write versions of `ToCycles` and `FromCycles` which use no local variables. How do their length and efficiency compare with the iterative versions?

8. Study the distribution of how many permutations on n elements contain i inversions, for each integer $0 \leq i \leq \binom{n}{2}$.

9. Write a function to expand a generator into a full permutation group.

10. The insertions and deletions between successive elements of a Gray code for $n = 4$ defines the sequence $\{+,+,-,+,+,-,-,+,+,+,-,-,+,-,-,-\}$, where "+" is an insertion and "−" a deletion. Study the properties of such sequences, and decide whether they can be generated without constructing the Gray code.

11. If you look back on the n people of the opposite sex you have dated over the course of a lifetime, these people can be ranked from 1 to n in terms of desirability. Thus if these people are ordered by the time you started dating

them, they define a permutation. We can use this model to decide how to marry the most wonderful person possible.

After you start dating person p_i, you must make a decision on whether to marry that person, or else reject him or her to start dating person p_{i+1}. We assume that anyone you ask will marry you unless you have previously rejected him or her, so there is no going back for seconds. The problem is how to decide whether the person you are dating is the best, or whether it is likely that someone better lies down the road. A good strategy is to date several people and reject them, just to get a feel for the distribution, and then accept the first person who is better than anyone you have seen. Experiment with strategies to maximize the chances of ending up with the best possible spouse, as well as strategies to maximize the expected value of your spouse. This problem has been extensively studied in the context of hiring a secretary [Fre83, SL87].

Research Problems

1. Characterize which permutations are Josephus permutations.

2. Find a linear time algorithm for constructing random permutations under the applicative model of computation [PMN88].

3. Assume you are given a stack of n numbered pancakes, and a spatula with which you can reverse the order of the top k pancakes, for $2 \leq k \leq n$. The *pancake sorting* problem asks how many such prefix reversals are sufficient to sort an arbitrary stack. Non-trivial upper and lower bounds were obtained by Gates and Papadimitriou [GP79]. Eli Goodman asks how many arc reversals are necessary and sufficient to sort a circular permutation of n elements.

Chapter 2.

Partitions, Compositions, and Young Tableaux

Permutations and subsets are one-dimensional structures, comprising the simple selection and arrangement of objects. This chapter considers combinatorial objects in two dimensions: partitions, compositions, and Young tableaux.

A *partition* of an integer n is a set of k positive integers which sum up to n. A *composition* of n into k parts represents n as the sum of k non-negative integers where order matters. Thus there are only 3 partitions of six into three parts but 28 compositions of six into three parts. *Young tableaux* are arrangements of n integers whose shape can be described by a partition of n, such that each row and column is sorted.

About the illustration overleaf:

In a random partition, the expected size of the largest part is the same as the expected number of parts. Ferrers diagrams provide a convenient way for us to visualize integer partitions. This Ferrers diagram of a random partition was constructed with the following command: `FerrersDiagram[RandomPartition[1000]];`.

2.1 Partitions

Not content with having founded graph theory, Euler is also the father of the study of partitions. A *partition* of a positive integer n is a set of k strictly positive integers which sum up to n. By convention, partitions are listed in non-increasing order.

```
PartitionQ[p_List] := (Min[p]>0) && Apply[And, Map[IntegerQ,p]]
```

Identifying an Integer Partition

Partitions are important in number theory, and we shall use them to represent shapes of other structures, particularly Young tableaux.

2.1.1 Generating Partitions

Like permutations and subsets, all the partitions of n can be generated efficiently using a recursive definition, this time in reverse lexicographic order. The computation is best done by solving a more general problem, constructing all partitions of n with a largest part of at most k. Since the elements of a partition are strictly positive, the largest element must be at most n, so `Partitions[n,n]` gives all the partitions of n.

Any partition of n with a largest part of at most k either contains a part of size k or it does not. The partitions that do can be constructed by prepending k to all partitions of $n-k$ with largest part at most k. The partitions that don't are all the partitions of n with largest part at most $k-1$.

```
Partitions[n_Integer] := Partitions[n,n]

Partitions[n_Integer,_] := {} /; (n<0)
Partitions[0,_] := { {} }
Partitions[n_Integer,1] := { Table[1,{n}] }
Partitions[_,0] := {}

Partitions[n_Integer,maxpart_Integer] :=
        Join[
                Map[(Prepend[#,maxpart])&, Partitions[n-maxpart,maxpart]],
                Partitions[n,maxpart-1]
        ]
```

Generating all Partitions of n

Here are the eleven partitions of six. Observe that they are given in reverse lexicographic order.

```
In[1]:= Partitions[6]
Out[1]= {{6}, {5, 1}, {4, 2}, {4, 1, 1}, {3, 3},
  {3, 2, 1}, {3, 1, 1, 1}, {2, 2, 2}, {2, 2, 1, 1},
  {2, 1, 1, 1, 1}, {1, 1, 1, 1, 1, 1}}
```

Most of these partitions do not contain a part bigger than three.

```
In[2]:= Partitions[6,3]
Out[2]= {{3, 3}, {3, 2, 1}, {3, 1, 1, 1}, {2, 2, 2},
  {2, 2, 1, 1}, {2, 1, 1, 1, 1}, {1, 1, 1, 1, 1, 1}}
```

Although the number of partitions grows exponentially, it does so more slowly than permutations or subsets, so complete tables can be generated for larger values of n.

```
In[3]:= Length[Partitions[20]]
Out[3]= 627
```

The set of integer partitions of n define a lattice under the relation of *dominance*, where partition p_a dominates p_b if for all k, the sum of the k largest parts of p_a is greater than or equal to the sum of the k largest parts of p_b. For $n = 7$, $\{7\}$ dominates all other partitions, with $\{1,1,1,1,1,1,1\}$ dominated by all others. $\{3,1,1,1,1\}$ and $\{2,2,2,1\}$ are examples of permutations which do not dominate each other [SW86].
In this figure, each partition is represented by a vertex of the graph, and p_a dominates p_b if there is a path from a to b where no edge goes up the page.

```
In[4]:= ShowGraph[ HasseDiagram[ MakeGraph[ Partitions[7],
(Block[{l}, l=Min[Length[#1],Length[#2]]; Min[
Accumulate[Plus,Take[#1,l]] - Accumulate[Plus,Take[#2,l]]
] >= 0])&] ] ];
```

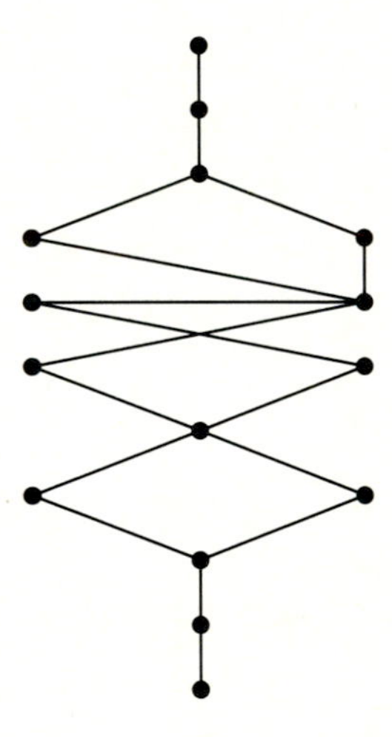

The rules for generating the successor to a partition in reverse lexicographic order are fairly straightforward. If the smallest part p_1 is greater than one, peel off one from it, thus increasing the number of parts by one. If not, find the smallest k such that part $p_k > 1$, and replace parts $p_1, ..., p_k$ by $\lfloor (p_k + k - 1)/(p_k - 1) \rfloor$ copies of $p_k - 1$, with a last element containing any remainder. The wrap-around condition occurs when the partition is all ones.

```
NextPartition[p_List] := Join[Drop[p,-1],{Last[p]-1,1}]  /; (Last[p] > 1)

NextPartition[p_List] := {Apply[Plus,p]}  /; (Max[p] == 1)

NextPartition[p_List] :=
        Block[{index,k,m},
                {index} = First[ Position[p,1] ];
                k = p[[index-1]] - 1;
                m = Apply[Plus,Drop[p,index-1]] + k + 1;
                Join[
                        Take[p,index-2],
                        Table[k,{Quotient[m,k]}],
                        If [Mod[m,k] == 0, {}, {Mod[m,k]}]
                ]
        ]
```

Constructing the Next Partition

Calling `NextPartition` repeatedly generates the complete set of partitions in reverse lexicographic order, although it will be slower than `Partitions`.

```
In[5]:= ( p=Table[1,{6}];
  Table[p=NextPartition[p],{NumberOfPartitions[6]}] )
Out[5]= {{6}, {5, 1}, {4, 2}, {4, 1, 1}, {3, 3},
  {3, 2, 1}, {3, 1, 1, 1}, {2, 2, 2}, {2, 2, 1, 1},
  {2, 1, 1, 1, 1}, {1, 1, 1, 1, 1, 1}}
```

Partitions can be sequenced in a Gray code or minimum change order just as permutations and subsets [Sav89] can. Here the notion of minimum change is decrementing one part of the partition and incrementing another. For $n = 5$, such a sequence is $\{1,1,1,1,1\}$, $\{2,1,1,1\}$, $\{2,2,1\}$, $\{3,1,1\}$, $\{3,2\}$, $\{4,1\}$, and $\{5\}$.

■ 2.1.2 Ferrers Diagrams

Ferrers diagrams represent partitions as patterns of dots. They provide a useful tool for visualizing partitions, because moving the dots around provides a mechanism for proving bijections between classes of partitions. For example, the increment/decrement operation of the minimum change ordering on partitions is equivalent to moving a dot from one row to another.

There are at least two good ways to spell "Ferrers", and probably more. Andrews [And76], a primary reference on the theory of partitions, uses "Ferrars". However, [Liu68, Sta86, SW86] all use "Ferrers", and so we will surrender to the majority opinion.

```
FerrersDiagram[p1_List] :=
        Block[{i,j,n=Length[p1],p=Sort[p1]},
                Show[
                        Graphics[
                                Join[
                                        {PointSize[ Min[0.05,1/(2 Max[p])] ]},
                                        Table[Point[{i,j}], {j,n}, {i,p[[j]]}]
                                ],
                                {AspectRatio -> 1, PlotRange -> All}
                        ]
                ]
        ]
```

Drawing the Ferrers Diagram of a Partition

Mathematica graphics can scale any drawing to fill the window associated with it.

```
In[6]:= FerrersDiagram[Table[i,{i,20,1,-1}]];
```

A smaller partition causes larger dots. Uglier diagrams can result from very skewed partitions.

In[7]:= `FerrersDiagram[{8,6,4,4,3,1}]`

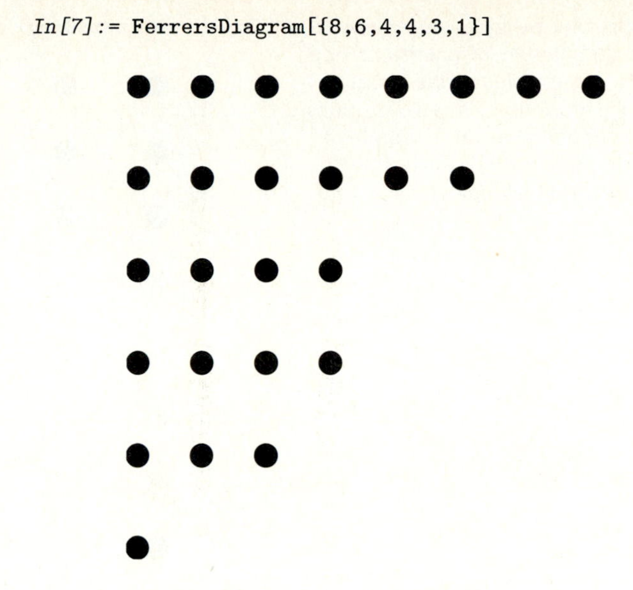

The Ferrers diagram can be generalized by replacing the underlying square grid by other regular patterns in the plane, with interesting results [Pro89].

■ 2.1.3 Bijections between Partitions

Ferrers diagrams are useful in bijective proofs, because the dots can be moved around to convert one partition into another. One of the most common such operations *transposes* the partition, or simply exchanges rows for columns.

```
TransposePartition[p_List] :=
        Block[{s=Select[p,(#>0)&], i, row, r},
                row = Length[s];
                Table [
                        r = row;
                        While [s[[row]]<=i, row--];
                        r,
                        {i,First[s]}
                ]
        ]
```

Transposing a Partition

The Ferrers diagram provides an easy way to prove that the number of partitions of n where m is the largest element is equal to the number of partitions of n with m parts, since transposing one gives an example of the other. This Ferrers diagram is a transpose of the previous one.

```
In[8]:= FerrersDiagram[TransposePartition[{8,6,4,4,3,1}]];
```

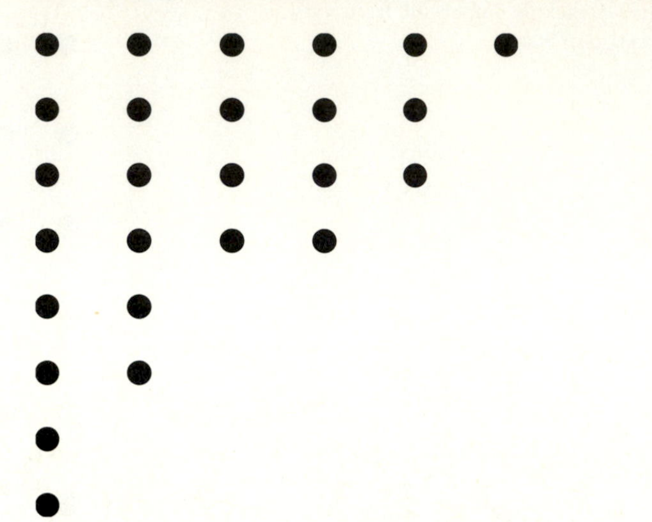

A Ferrers diagram proof shows that the number of partitions of n with all odd parts ...

```
In[9]:= Select[Partitions[7], (Apply[And,Map[OddQ,#]])&]

Out[9]= {{7}, {5, 1, 1}, {3, 3, 1}, {3, 1, 1, 1, 1},
  {1, 1, 1, 1, 1, 1, 1}}
```

... equals the number of partitions of n into distinct parts [SW86].

```
In[10]:= Select[Partitions[7], (Length[#] ==
Length[Union[#]])&]

Out[10]= {{7}, {6, 1}, {5, 2}, {4, 3}, {4, 2, 1}}
```

Another basis of many bijections is the *Durfee square*, the largest enclosed square of dots in the Ferrers diagram.

```
DurfeeSquare[s_List] :=
        Block[{i,max=1},
                Do [
                        If [s[[i]] >= i, max=i],
                        {i,2,Min[Length[s],First[s]]}
                ];
                max
        ]

DurfeeSquare[{}] := 0
```

The Durfee Square of a Partition

The size of the Durfee square remains unchanged in transposing the partition.

```
In[11]:= {DurfeeSquare[p=RandomPartition[20]],
DurfeeSquare[TransposePartition[p]]}
Out[11]= {2, 2}
```

2.1.4 Counting Partitions

The number p_n of distinct partitions of n is known as the *partition function* and is of importance in number theory. There are built-in *Mathematica* functions to count the total number of partitions of n as well as partitions into distinct parts, and published tables of these functions appear in [GGM58].

Euler showed that the number of partitions of n, p_n, is given by the recurrence

$$p_n = \sum_{m=1}^{\infty} (-1)^{m+1} (p_{n-m(3m-1)/2} + p_{n-m(3m+1)/2})$$

where $p_0 = 1$ and $p_k = 0$ for $k < 0$. Although this is an infinite summation, we can use it to compute p_n since only the first $O(\sqrt{n})$ terms are non-zero.

This implementation uses dynamic programming to compute the recurrence, and fully exploits it by summing the p_i's in order of increasing i.

```
NumberOfPartitions[n_Integer] := 0  /; (n < 0)
NumberOfPartitions[n_Integer] := 1  /; (n == 0)

NumberOfPartitions[n_Integer] := NumberOfPartitions[n] =
        Block[{m},
                Sum[ (-1)^(m+1) NumberOfPartitions[n - m (3m-1)/2] +
                        (-1)^(m+1) NumberOfPartitions[n - m (3m+1)/2],
                        {m, Ceiling[ (1+Sqrt[1.0 + 24n])/6 ], 1, -1}
                ]
        ]
```

Computing the Number of Integer Partitions

The total number of partitions of n can be computed using the built-in *Mathematica* function `PartitionsP` instead of `NumberOfPartitions`.

```
In[12]:= {PartitionsP[10], NumberOfPartitions[10]}
Out[12]= {42, 42}
```

Both functions can be tested by generating and counting all the partitions of n.

```
In[13]:= NumberOfPartitions[10] == Length[Partitions[10]]
Out[13]= True
```

The number of partitions of n into distinct parts is given by the built-in *Mathematica* function `PartitionsQ`. Because the partitions into distinct parts are selected from the set of all partitions, they appear in reverse lexicographic order.

```
In[14]:= PartitionsQ[10] && Select[Partitions[10],
(Length[#] == Length[Union[#]])&]

Out[14]= 10 && {{10}, {9, 1}, {8, 2}, {7, 3}, {7, 2, 1},
   {6, 4}, {6, 3, 1}, {5, 4, 1}, {5, 3, 2}, {4, 3, 2, 1}}
```

■ 2.1.5 Random Partitions

An enumeration formula for counting the number of different structures of a given type usually leads to an algorithm for constructing random instances of the structure. For each of the possible ways to create the first element of a structure, the enumeration formula can be used to give the number of ways to complete it. Using these counts, we can randomly select the first part in such a way that the complete structures are uniformly distributed.

As we have seen, the number of partitions $p_{n,k}$ of n with a largest part of at most k is given by the recurrence

$$p_{n,k} = p_{n-k,k} + p_{n,k-1} \,.$$

The largest part l in a random partition can be found by selecting a random number x such that $1 \leq x \leq p_{n,n}$, with l determined by $p_{n,l-1} < x \leq p_{n,l}$.

Unfortunately, this function requires tabulating values of a two-dimensional function, which can be expensive for large values of n. Nijenhuis and Wilf [NW78] provide the following more efficient algorithm, which randomly selects the magnitude of a part d and its multiplicity j from n, recurring to find a random partition of $n - dj$.

```
RandomPartition[n_Integer?Positive] :=
        Block[{mult = Table[0,{n}],j,d,m = n},
                While[ m != 0,
                        {j,d} = NextPartitionElement[m];
                        m -= j d;
                        mult[[d]] += j;
                ];
                Flatten[Map[(Table[#,{mult[[#]]}])&,Reverse[Range[n]]]]
        ]

NextPartitionElement[n_Integer] :=
        Block[{d=0,j,m,z=Random[] n PartitionsP[n],done=False,flag},
                While[!done,
                        d++; m = n; j = 0; flag = False;
                        While[ !flag,
                                j++; m -=d;
```

```
                        If[ m > 0,
                                z -= d PartitionsP[m];
                                If[ z <= 0, flag=done=True],
                                flag = True;
                                If[m==0, z -=d; If[z <= 0, done = True]]
                        ];
                ];
        ];
        {j,d}
]
```

Constructing Random Partitions

The rectangle defined by the Ferrers diagram of a random partition for any value of $n > 3$ can be expected to contain more empty space than dots, even though exchanging the roles of dots and the empty positions defines another Ferrers diagram.

In[15]:= **FerrersDiagram[RandomPartition[100]];**

```
• • • • • • • • • • • • • • • • • • • • • • • •
• • • • • • • • • • • • • • • • • • • •
• • • • • • • • • • • • • • • •
• • • • • • • • •
• • • • •
• • • •
• • • •
• • • •
• • •
• • •
• •
• •
• •
•
•
```

Repeating an experiment from [NW78] illustrates that each partition is equally likely to occur with this implementation.

In[16]:= **Distribution[Table[RandomPartition[6], {880}]]**

Out[16]= {82, 76, 68, 74, 74, 85, 90, 88, 77, 85, 81}

2.2 Compositions

A *composition* of n is a particular arrangement of non-negative integers which sum up to n. Compositions are perhaps a little less pleasing than partitions, since order matters and zero elements can be included. There are an infinite number of compositions of an integer unless the number of parts is bounded. Compositions are easier to generate and count than partitions, however.

2.2.1 Random Compositions

The number of compositions of n into k parts follows from the observation that it is equivalent to inserting $k-1$ dividers in a row of n dots and summing the dots between them.

```
NumberOfCompositions[n_,k_] := Binomial[ n+k-1, n ]
```

Counting Compositions

This observation can be used to generate a random composition by using a random $(k-1)$-subset to select the dividers, since we already have a function to generate them.

```
RandomComposition[n_Integer,k_Integer] :=
        Map[
                (#[[2]] - #[[1]] - 1)&,
                Partition[Join[{0},RandomKSubset[Range[n+k-1],k-1],{n+k}], 2, 1]
        ]
```

Generating Random Compositions

There are 28 compositions of six into three parts.

```
In[17]:= NumberOfCompositions[6,3]
Out[17]= 28
```

Calling random compositions repeatedly will eventually generate all compositions, but it is likely there will be duplicates.

```
In[18]:= Length[ Union[ Table[RandomComposition[6,3],
{28}] ] ]
Out[18]= 16
```

The quality of the random compositions reflects the quality of the random k-subsets generator.

```
In[19]:= Distribution[ Table[RandomComposition[6,3],
{200}] ]
Out[19]= {4, 5, 8, 5, 2, 6, 9, 12, 9, 5, 4, 11, 5, 6, 8,
  8, 14, 9, 5, 10, 5, 8, 12, 5, 6, 5, 5, 9}
```

2.2.2 Generating Compositions

The bijection between compositions and k-subsets exploited above also provides a method for listing all the compositions of an integer. By using k-subsets generated in lexicographic order, the resulting compositions will also be lexicographically ordered.

```
Compositions[n_Integer,k_Integer] :=
        Map[
                (Map[(#[[2]]-#[[1]]-1)&, Partition[Join[{0},#,{n+k}],2,1] ])&,
                KSubsets[Range[n+k-1],k-1]
        ]
```

Constructing all Compositions

Here every composition is generated exactly once.

```
In[20]:= Compositions[6,3]
Out[20]= {{0, 0, 6}, {0, 1, 5}, {0, 2, 4}, {0, 3, 3},
  {0, 4, 2}, {0, 5, 1}, {0, 6, 0}, {1, 0, 5}, {1, 1, 4},
  {1, 2, 3}, {1, 3, 2}, {1, 4, 1}, {1, 5, 0}, {2, 0, 4},
  {2, 1, 3}, {2, 2, 2}, {2, 3, 1}, {2, 4, 0}, {3, 0, 3},
  {3, 1, 2}, {3, 2, 1}, {3, 3, 0}, {4, 0, 2}, {4, 1, 1},
  {4, 2, 0}, {5, 0, 1}, {5, 1, 0}, {6, 0, 0}}
```

Rather than converting a composition to a k-subset, using `NextKSubset` and then converting back to a composition, the construction of the next composition can be done directly.

```
NextComposition[l_List] :=
        Block[{c=l, h=1, t},
                While[c[[h]] == 0, h++];
                {t,c[[h]]} = {c[[h]],0};
                c[[1]] = t - 1;
                c[[h+1]]++;
                c
        ]

NextComposition[l_List] :=
        Join[{Apply[Plus,l]},Table[0,{Length[l]-1}]] /; Last[l]==Apply[Plus,l]
```

Constructing the Next Composition

Here are the same compositions constructed earlier, although in a different order. Specifically, they are the reversals of the lexicographically sequenced compositions.

```
In[21]:= (c = {0,0,6};
Table[ c = NextComposition[c], {28}])

Out[21]= {{6, 0, 0}, {5, 1, 0}, {4, 2, 0}, {3, 3, 0},
  {2, 4, 0}, {1, 5, 0}, {0, 6, 0}, {5, 0, 1}, {4, 1, 1},
  {3, 2, 1}, {2, 3, 1}, {1, 4, 1}, {0, 5, 1}, {4, 0, 2},
  {3, 1, 2}, {2, 2, 2}, {1, 3, 2}, {0, 4, 2}, {3, 0, 3},
  {2, 1, 3}, {1, 2, 3}, {0, 3, 3}, {2, 0, 4}, {1, 1, 4},
  {0, 2, 4}, {1, 0, 5}, {0, 1, 5}, {0, 0, 6}}
```

As with permutations, subsets and partitions, compositions can be sequenced in a minimum change order [Kli82].

2.3 Young Tableaux

As we have seen, permutations are linear configurations of objects. Many combinatorial problems are better interpreted as two-dimensional structures. For example, suppose we are interested in counting the number of ways to arrange n left parentheses and n right parentheses so that they make balanced formulae. For $n = 3$, there are five distinct formulae: ((())), ()(()), (())(), (()()), and ()()().

Now consider a $2 \times n$ matrix containing the numbers $1, ..., 2n$, such that each row and column is in increasing order. There is a bijection between the balanced formulae and these matrices, by interpreting the $(1, i)$th entry in the matrix to be the position of the ith left parenthesis, and the $(2, j)$th entry to be the position of the jth right parenthesis. Since the columns are ordered, the left parenthesis always comes before the corresponding right parenthesis.

This example illustrates the importance of ordered structures of ordered lists. A *Young tableau* is a structure of integers $1, ..., n$ where the number of elements in each row is defined by an integer partition of n. Further, the elements of each row and column are in increasing order, and the rows are left justified. Young tableaux represent a very general class of objects, for the *shape* defined by a partition can run from a rectangular array to an arbitrarily jagged profile.

```
TableauQ[{}] = True
TableauQ[t_List] :=
        And [
                Apply[ And, Map[(Apply[LessEqual,#])&,t] ],
                Apply[ And, Map[(Apply[LessEqual,#])&,TransposeTableau[t]] ],
                Apply[ GreaterEqual, Map[Length,t] ],
                Apply[ GreaterEqual, Map[Length,TransposeTableau[t]] ]
        ]

TransposeTableau[tb_List] :=
        Block[{t=Select[tb,(Length[#]>=1)&],row},
                Table[
                        row = Map[First,t];
                        t = Map[ Rest, Select[t,(Length[#]>1)&] ];
                        row,
                        {Length[First[tb]]}
                ]
        ]

ShapeOfTableau[t_List] := Map[Length,t]
```

Identifying Young Tableaux

This is a tableau.

```
In[22]:= TableauQ[{{1,2,5},{3,4,5},{6}}]
Out[22]= True
```

But this isn't. Note the first two elements of the second row.

```
In[23]:= TableauQ[{{1,2,5,9,10},{5,4,7,13},{4,8,12},{11}}]
Out[23]= False
```

Young tableaux can be properly displayed as two-dimensional structures using `TableForm`. The singular is tableau, the plural tableaux.

```
In[24]:= TableForm[ {{1,2,5},{3,4,5},{6}} ]
                    {1, 2, 5}
                    {3, 4, 5}
Out[24]//TableForm= {6}
```

Our interest in Young tableaux is combinatorial and follows Knuth [Knu73b], although their original application was algebraic. Young tableaux are also known as *plane partitions*. Another detailed treatment is [Sta71], which includes open problems.

■ 2.3.1 Insertion and Deletion

There is an interesting bijection between permutations and pairs of Young tableaux of the same shape. This bijection uses an insertion algorithm, which proves useful in several different contexts.

Young tableaux can be constructed from permutations using a row-to-row insertion or "bumping" procedure. If the element x is greater than the last element of the row, it can be placed at the end, resulting in a new tableau. If not, x cannot just be inserted in the middle of a row, because it shifts over the subsequent entries on the row, and thus the columns may not remain sorted. Instead, x "bumps" the least element greater than x into the next row, where the insertion procedure starts anew with the bumped element. Eventually, an element either ends up on the end of a row, or we run out of rows and the item must reside on a new row.

```
InsertIntoTableau[e_Integer,{}] := { {e} }

InsertIntoTableau[e_Integer, t1_?TableauQ] :=
        Block[{item=e,row=0,col,t=t1},
                While [row < Length[t],
                        row++;
                        If [Last[t[[row]]] <= item,
                                AppendTo[t[[row]],item];
                                Return[t]
                        ];
                        col = Ceiling[ BinarySearch[t[[row]],item] ];
                        {item, t[[row,col]]} = {t[[row,col]], item};
```

```
            ];
            Append[t, {item}]
    ]
```

Inserting a New Element into a Young Tableau

We can use the insertion algorithm to construct a Young tableau associated with a permutation, by starting with an empty tableau and inserting each element into the tableau based on its position in the permutation.

```
ConstructTableau[p_List] := ConstructTableau[p,{}]

ConstructTableau[{},t_List] := t

ConstructTableau[p_List,t_List] :=
        ConstructTableau[Rest[p], InsertIntoTableau[First[p],t]]
```

Constructing Young Tableaux

Each permutation generates a tableau, which is not necessarily unique.

```
In[25]:= ConstructTableau[{6,4,9,5,7,1,2,8}]
Out[25]= {{1, 2, 7, 8}, {4, 5}, {6, 9}}
```

In particular, this permutation generates the same tableau as the previous one.

```
In[26]:= ConstructTableau[{6,4,9,5,7,8,1,2}]
Out[26]= {{1, 2, 7, 8}, {4, 5}, {6, 9}}
```

The inverse operation for insertion is *deletion*, but just deleting an element from the middle of a row could shift columns so that the structure is not a tableau. Thus to delete an element, we specify the last item of a particular row and reverse the bumping procedure, with each iteration kicking a single item up exactly one row until an element of the first row gets deleted. To delete a particular element from the first row, we must know the correct row to start on. Starting the deletion from a row which is the same length as the one below it yields a non-tableau, so identifying the correct row is the key for all deletion operations.

```
DeleteFromTableau[t1_?TableauQ,r_Integer]:=
        Block [{t=t1, col, row, item=Last[t1[[r]]]},
                col = Length[t[[r]]];
                If[col == 1, t = Drop[t,-1], t[[r]] = Drop[t[[r]],-1]];
                Do [
                        While [t[[row,col]]<=item && Length[t[[row]]]>col, col++];
                        If [item < t[[row,col]], col--];
                        {item,t[[row,col]]} = {t[[row,col]],item},
                        {row,r-1,1,-1}
                ];
```

```
                t
        ]
```

Deleting an Element From a Young Tableau

Observe that this tableau is missing 3.

```
In[27]:= TableForm[ ConstructTableau[{6,4,9,5,7,1,2,8}] ]
                    {1, 2, 7, 8}
                    {4, 5}
Out[27]//TableForm= {6, 9}
```

Inserting a 3 into the tableau bumps 7 from the first row, and thus adds an element to the second row.

```
In[28]:= TableForm[ InsertIntoTableau[3,%] ]
                    {1, 2, 3, 8}
                    {4, 5, 7}
Out[28]//TableForm= {6, 9}
```

Deleting the last element from the second row restores the tableau to its initial state.

```
In[29]:= TableForm[ DeleteFromTableau[%,2] ]
                    {1, 2, 7, 8}
                    {4, 5}
Out[29]//TableForm= {6, 9}
```

2.3.2 Permutations and Pairs of Tableaux

The insertion and deletion operations are inverses if the deletion row is correctly specified. This information can be conveniently represented in a tableau Q, of the same shape as the tableau P built up by insertions. When the ith element p_i of the permutation is inserted into the first row of P, there will be exactly one new position created at the end of some row of P. The corresponding position of Q is filled with i. Since i is greater than any previous element of Q and at the end of a row and column, Q remains a tableau after this operation. Thus the largest number in Q gives the starting row for the deletion operation which eliminates the last element inserted.

This interpretation gives a bijection between ordered pairs of Young tableaux (P,Q) of the same shape and permutations [Sch61]. Further, when p is the permutation associated with (P,Q), the inverse of p is associated with (Q,P).

```
TableauxToPermutation[p1_?TableauQ,q1_?TableauQ] :=
        Block[{p=p1, q=q1, row, firstrow},
                Reverse[
                        Table[
                                firstrow = First[p];
```

```
                            row = Position[q, Max[q]] [[1,1]];
                            p = DeleteFromTableau[p,row];
                            q[[row]] = Drop[ q[[row]], -1];
                            If[ p == {},
                                    First[firstrow],
                                    First[Complement[firstrow,First[p]]]
                            ],
                            {Apply[Plus,ShapeOfTableau[p1]]}
                    ]
            ]
    ] /; ShapeOfTableau[p1] === ShapeOfTableau[q1]
```

Permutations from Pairs of Tableaux

To illustrate this bijection, we start with a random permutation and its inverse.

```
In[30]:= {(p=RandomPermutation[10]),
InversePermutation[p]}
Out[30]= {{2, 4, 1, 3, 8, 5, 10, 6, 9, 7},
  {3, 1, 4, 2, 6, 8, 10, 5, 9, 7}}
```

Observe that the tableaux for p and its inverse are both the same shape.

```
In[31]:= {P,Q} = {ConstructTableau[p],
ConstructTableau[InversePermutation[p]]}
Out[31]= {{{1, 3, 5, 6, 7}, {2, 4, 8, 9}, {10}},
  {{1, 2, 5, 7, 9}, {3, 4, 6, 8}, {10}}}
```

As promised, this pair of tableaux correspond to the original permutation and its inverse, depending upon their order.

```
In[32]:= {TableauxToPermutation[P,Q],
TableauxToPermutation[Q,P]}
Out[32]= {{2, 4, 1, 3, 8, 5, 10, 6, 9, 7},
  {3, 1, 4, 2, 6, 8, 10, 5, 9, 7}}
```

Which permutations correspond to the pair of tableaux (P, P)? Since reversing the order of tableaux yields the inverse permutation, these tableaux generate involutions, or permutations which are self-inverses.

```
In[33]:= Select[Permutations[{1,2,3,4}],InvolutionQ]
Out[33]= {{1, 2, 3, 4}, {1, 2, 4, 3}, {1, 3, 2, 4},
  {1, 4, 3, 2}, {2, 1, 3, 4}, {2, 1, 4, 3}, {3, 2, 1, 4},
  {3, 4, 1, 2}, {4, 2, 3, 1}, {4, 3, 2, 1}}
```

This property defines a bijection between Young tableaux and involutions.

```
In[34]:= Map[(TableauxToPermutation[ConstructTableau[#],
ConstructTableau[#]])&, %]
Out[34]= {{1, 2, 3, 4}, {1, 2, 4, 3}, {1, 3, 2, 4},
  {1, 4, 3, 2}, {2, 1, 3, 4}, {2, 1, 4, 3}, {3, 2, 1, 4},
  {3, 4, 1, 2}, {4, 2, 3, 1}, {4, 3, 2, 1}}
```

2.3.3 Generating Young Tableaux

Systematically constructing all the Young tableaux of a given shape is a more complex problem than building the other combinatorial objects we have seen. Part of the problem is defining a logical sequence for the two-dimensional structures. We follow the construction of [NW78].

```
LastLexicographicTableau[s_List] :=
        Block[{c=0},
                Map[(c+=#; Range[c-#+1,c])&, s]
        ]

FirstLexicographicTableau[s_List] :=
        TransposeTableau[ LastLexicographicTableau[ TransposePartition[s] ] ]
```

The First and Last Lexicographic Tableaux

We define the first lexicographic tableau to consist of *columns* of contiguous integers.

```
In[35]:= TableForm[FirstLexicographicTableau[{4,3,3,2}]]
                      {1, 5, 9, 12}
                      {2, 6, 10}
                      {3, 7, 11}
Out[35]//TableForm= {4, 8}
```

The last lexicographic tableau consists of *rows* of contiguous integers.

```
In[36]:= TableForm[ LastLexicographicTableau[{4,3,3,2}] ]
                      {1, 2, 3, 4}
                      {5, 6, 7}
                      {8, 9, 10}
Out[36]//TableForm= {11, 12}
```

Observe that in the lexicographically last tableau all rows consist of a contiguous range of integers. Thus for each integer k, the subtableau defined by the elements $1,...,k$ has k on the last row of the subtableau t. In general, to construct the lexicographically next tableau, we identify the the smallest k such that k is *not* on the last row of its subtableau. Say this subtableau of k elements has shape $\{s_1,...,s_m\}$. The next tableaux will be the lexicographically first tableau with k in the next right-most corner of the subtableaux, with the elements of t which are greater than k appended to the corresponding rows of the new subtableau.

```
NextTableau[t_?TableauQ] :=
        Block[{s,y,row,j,count=0,tj,i,n=Max[t]},
                y = TableauToYVector[t];
                For [j=2, (j<n)  && (y[[j]]>=y[[j-1]]), j++, ];
                If [y[[j]] >= y[[j-1]],
```

```
                Return[ FirstLexicographicTableau[ ShapeOfTableau[t] ] ]
            ];
            s = ShapeOfTableau[ Table[Select[t[[i]],(#<=j)&], {i,Length[t]}] ];
            {row} = Last[ Position[ s, s[[ Position[t,j] [[1,1]] + 1 ]] ] ];
            s[[row]] --;
            tj = FirstLexicographicTableau[s];
            If[ Length[tj] < row,
                tj = Append[tj,{j}],
                tj[[row]] = Append[tj[[row]],j]
            ];
            Join[
                Table[
                    Join[tj[[i]],Select[t[[i]],(#>j)&]],
                    {i,Length[tj]}
                ],
                Table[t[[i]],{i,Length[tj]+1,Length[t]}]
            ]
        ]

Tableaux[s_List] :=
        Block[{t = LastLexicographicTableau[s]},
                Table[ t = NextTableau[t], {NumberOfTableaux[s]} ]
        ]

Tableaux[n_Integer?Positive] := Apply[ Join, Map[ Tableaux, Partitions[n] ] ]
```

Constructing all Tableaux of a Given Shape

To identify the pivot element k, it is convenient to define an alternate representation for tableaux, the *Y-vector*, such that $Y[i]$ is the row containing the ith smallest element of the tableau. Thus the position of an element j on row r is the number of elements of $Y[1,j] = r$. It is easily verified that k is the smallest integer such that $Y[k] < Y[k-1]$.

```
YVectorToTableau[y_List] :=
        Block[{k},
                Table[ Flatten[Position[y,k]], {k,Length[Union[y]]}]
        ]

TableauToYVector[t_?TableauQ] :=
        Block[{i,y=Table[1,{Length[Flatten[t]]}]},
                Do [ Scan[ (y[[#]]=i)&, t[[i]] ], {i,2,Length[t]} ];
                y
        ]
```

The Y-vector Representation of Tableaux

The list of tableaux of shape $\{3, 2, 1\}$ illustrates the amount of freedom available to tableau structures. The smallest element is always in the upper left-hand corner, but the largest element is free to be the rightmost position of the last row defined by the *distinct* parts of the partition.

```
In[37]:= Tableaux[{3,2,1}]
Out[37]= {{{1, 4, 6}, {2, 5}, {3}},
  {{1, 3, 6}, {2, 5}, {4}}, {{1, 2, 6}, {3, 5}, {4}},
  {{1, 3, 6}, {2, 4}, {5}}, {{1, 2, 6}, {3, 4}, {5}},
  {{1, 4, 5}, {2, 6}, {3}}, {{1, 3, 5}, {2, 6}, {4}},
  {{1, 2, 5}, {3, 6}, {4}}, {{1, 3, 4}, {2, 6}, {5}},
  {{1, 2, 4}, {3, 6}, {5}}, {{1, 2, 3}, {4, 6}, {5}},
  {{1, 3, 5}, {2, 4}, {6}}, {{1, 2, 5}, {3, 4}, {6}},
  {{1, 3, 4}, {2, 5}, {6}}, {{1, 2, 4}, {3, 5}, {6}},
  {{1, 2, 3}, {4, 5}, {6}}}
```

By iterating through the different integer partitions as shapes, all tableaux of a particular size can be constructed.

```
In[38]:= Tableaux[3]
Out[38]= {{{1, 2, 3}}, {{1, 3}, {2}}, {{1, 2}, {3}},
  {{1}, {2}, {3}}}
```

■ 2.3.4 Counting Tableaux by Shape

Each position p within a Young tableau defines an L-shaped *hook*, consisting of p, all the elements below p, and all the elements to the right of p. The *hook length formula* gives the number of tableaux of a given shape as $n!$ divided by the product of the hook length of each position, where n is the number of positions in the tableau. A convincing argument that the formula works is, of the $n!$ ways to label a tableau of given shape, only those where the minimum element in each hook is in the corner can be tableaux, so for each hook the probability that the tableau condition is satisfied is one over the hook length. Unfortunately, this argument is bogus because these probabilities are not independent, but correct proofs that the formula works appear in [FZ82, Knu73b, NW78].

```
NumberOfTableaux[{}] := 1
NumberOfTableaux[s_List] :=
        Block[{row,col,transpose=TransposePartition[s]},
                (Apply[Plus,s])! /
                Product [
                        (transpose[[col]]-row+s[[row]]-col+1),
                        {row,Length[s]}, {col,s[[row]]}
                ]
        ]

NumberOfTableaux[n_Integer] := Apply[Plus, Map[NumberOfTableaux, Partitions[n]]]
```

The Hook Length Formula

The hook length formula can be used to count the number of tableaux for any shape, which for small sizes can be confirmed by constructing and counting them.

```
In[39]:= {NumberOfTableaux[{3,2,1}], Length[Tableaux[{3,2,1}]]}
Out[39]= {16, 16}
```

Using the hook length formula over all partitions of n computes the number of tableaux on n elements.

```
In[40]:= NumberOfTableaux[10]
Out[40]= 9496
```

A *biparential heap* [MS80] is a triangular data structure where each element is less than both its parents. Pascal's triangle is an example of a biparential heap. The structure of a biparential heap is determined by a Young tableau of the same shape, so here we compute the number of biparential heaps on 55 distinct elements.

```
In[41]:= NumberOfTableaux[ Reverse[Range[10]] ]
Out[41]= 44261486084874072183645699204710400
```

Our study of Young tableaux was motivated by the number of well-formed formulae which can be made from n sets of parentheses, which we showed is the number of distinct tableaux of shape $\{n, n\}$. Since any balanced set of parentheses has a leftmost point $k+1$ at which the number of left and right parentheses are equal, peeling off the first left parenthesis and the $k+1$th right parenthesis leaves two balanced sets k and $n-1-k$ parentheses, which leads to the following recurrence:

$$C_n = \sum_{k=0}^{n-1} C_k C_{n-1-k}$$

This recurrence defines the *Catalan numbers*, which occur in a surprising number of problems in combinatorics, from the number of triangulations of a convex polygon to the number of paths across a lattice which do not rise above the main diagonal. This recurrence can be solved using generating functions to reveal a nice closed form.

```
CatalanNumber[n_] := Binomial[2n,n]/(n+1)        /; (n>=0)
```

Computing the Catalan Numbers

This function is named CatalanNumber to avoid conflict with the built-in *Mathematica* function Catalan, which returns Catalan's constant.

```
In[42]:= Table[ CatalanNumber[i], {i,2,20} ]
Out[42]= {2, 5, 14, 42, 132, 429, 1430, 4862, 16796,
  58786, 208012, 742900, 2674440, 9694845, 35357670,
  129644790, 477638700, 1767263190, 6564120420}
```

As we have seen, the Catalan numbers are a special case of the hook length formula.

```
In[43]:= Table[ NumberOfTableaux[{i,i}], {i,2,20}]
Out[43]= {2, 5, 14, 42, 132, 429, 1430, 4862, 16796,
  58786, 208012, 742900, 2674440, 9694845, 35357670,
  129644790, 477638700, 1767263190, 6564120420}
```

2.3.5 Random Tableaux

Constructing tableaux of a given shape randomly with a uniform distribution is another stickly problem solved in [NW78]. The key observation is that the largest element in the tableau must go in a right-most corner of the shape, and that once the appropriate corner is selected the construction can proceed to position all smaller elements.

Selecting a corner at random from the list of candidates will not give a uniform construction, since there are different numbers of ways to complete the tableau depending upon the shape left after the insertion. Using `NumberOfTableaux`, this can be computed for each of the possible shapes, and so a truly random corner selection can be made.

`RandomTableau` implements this procedure, but in a non-obvious way. Starting from a random square in the tableau, it makes a sequence of random jumps down or to the right until a corner point is reached. See [NW78] for a proof of the correctness of this construction.

```
RandomTableau[shape_List] :=
	Block[{n=Apply[Plus,shape],done,l,m,i=j=n,h=1,k,y,p=shape},
		y= Join[TransposePartition[shape],Table[0,{n - Max[shape]}]];
		Do[
			{i,j} = RandomSquare[y,p]; done = False;
			While [!done,
				h = y[[j]] + p[[i]] - i - j;
				If[ h != 0,
					If[ Random[] < 0.5,
						j = Random[Integer,{j,p[[i]]}],
						i = Random[Integer,{i,y[[j]]}]
					],
					done = True
				];
			];
			p[[i]]--; y[[j]]--;
			y[[m]] = i,
			{m,n,1,-1}
		];
		YVectorToTableau[y]
```

```
        ]

RandomSquare[y_List,p_List] :=
        Block[{i=Random[Integer,{1,First[y]}], j=Random[Integer,{1,First[p]}]},
                While[(i > y[[j]]) || (j > p[[i]]),
                        i = Random[Integer,{1,First[y]}];
                        j = Random[Integer,{1,First[p]}]
                ];
                {i,j}
        ]
```

Constructing Random Tableaux of a Given Shape

Each of the 117,123,756,750 tableaux of this shape will be selected with equal likelihood.

```
In[44]:= TableForm[ RandomTableau[{6,5,5,4,3,2}] ]
                     {1, 3, 4, 10, 13, 16}
                     {2, 6, 8, 11, 18}
                     {5, 7, 12, 14, 20}
                     {9, 15, 17, 25}
                     {19, 21, 22}
Out[44]//TableForm= {23, 24}
```

Repeating the experiment in [NW78], each of the 16 tableaux occurs with roughly equal frequency.

```
In[45]:= Distribution[ Table[RandomTableau[{3,2,1}],
{640}] ]
Out[45]= {49, 43, 34, 39, 40, 35, 39, 43, 48, 31, 38, 45,
  39, 39, 38, 40}
```

■ 2.3.6 Longest Increasing Subsequences

When constructing a tableau with the insertion algorithm, every element begins its life in some position in the first row, from which it may later be bumped. The elements which originally entered in the ith column are said to belong to the ith *class* of the tableau, and this class distinction leads to an interesting algorithm to find the longest increasing scattered subsequence of a permutation [Sch61].

```
TableauClasses[p_?PermutationQ] :=
        Block[{classes=Table[{},{Length[p]}],t={}},
                Scan [
                        (t = InsertIntoTableau[#,t];
                         PrependTo[classes[[Position[First[t],#] [[1,1]] ]], #])&,
                        p
                ];
                Select[classes, (# != {})&]
        ]
```

Partitioning a Permutation into Classes

Note that $4, 5, 7, 8$ form a longest increasing (scattered) subsequence of the following tableaux, which contains the same number of columns.

```
In[46]:= TableauClasses[{6,4,9,5,7,1,2,8,3}]
Out[46]= {{1, 4, 6}, {2, 5, 9}, {3, 7}, {8}}
```

The reverse permutation has $3, 5, 9$ as a longest increasing subsequence, making it the longest decreasing subsequence of the original permutation.

```
In[47]:= TableauClasses[{3,8,2,1,7,5,9,4,6}]
Out[47]= {{1, 2, 3}, {4, 5, 7, 8}, {6, 9}}
```

This connection between the number of classes and the length of the longest scattered subsequence is not a coincidence. For an element of the permutation to enter the tableau in the kth column, there had to be elements creating $k-1$ columns before it. A column is created whenever the inserted element is greater than the last element of the first row, so clearly the elements which start new columns form an increasing subsequence. That is it is the *longest* increasing subsequence is a consequence of the first class corresponding to the left-to-right minima of the permutation, and the kth class being the left-to-right minima of the permutation minus the elements of the first $k-1$ classes.

```
LongestIncreasingSubsequence[p_?PermutationQ] :=
        Block[{c,x,xlast},
                c = TableauClasses[p];
                xlast = x = First[ Last[c] ];
                Append[
                        Reverse[
                                Map[
                                        (x = First[ Intersection[#,
                                              Take[p, Position[p,x][[1,1]] ] ] ])&,
                                        Reverse[ Drop[c,-1] ]
                                ]
                        ],
                        xlast
```

```
                    ]
        ]

LongestIncreasingSubsequence[{}] := {}
```

Identifying the Longest Increasing Subsequence in a Permutation

The longest increasing *contiguous* subsequence can be found by selecting the largest of the runs in the permutation. Finding the largest scattered subsequence is a much harder problem.

```
In[48]:= First[ Sort[ Runs[p=RandomPermutation[50]],
(Length[#1] > Length[#2])&] ]

Out[48]= {3, 33, 37}
```

A pigeonhole result [ES35] states that any sequence of n^2+1 distinct integers must contain either an increasing or decreasing scattered subsequence of length $n+1$. Thus at least one of these sequences must be at least eight integers long.

```
In[49]:= {LongestIncreasingSubsequence[p],
LongestIncreasingSubsequence[Reverse[p]]}

Out[49]= {{6, 14, 17, 23, 24, 28, 31, 35, 38, 50},
  {1, 2, 7, 9, 12, 14, 20, 22, 36, 41, 46}}
```

2.3.7 Encroaching List Sets

An *encroaching list set* [Ski88] is a structure, similar in certain ways to a Young tableau, consisting of an ordered set of sorted lists such that the head and tail entries of later lists nest within earlier ones.

For any encroaching list set $\{s_1, ..., s_m\}$ and new key k, there is at most one place to insert k at either end of a sublist of s while remaining an encroaching list set. If k is less than the head of s_m it will be prepended to some list s_i, while if k is greater than the tail of s_m it will be appended to a particular list. If neither of these conditions holds, the only way to maintain the encroaching list structure is to create a new list s_{m+1} consisting only of k.

```
AddToEncroachingLists[k_Integer,{}] := {{k}}

AddToEncroachingLists[k_Integer,l_List] :=
        Append[l,{k}]  /; (k > First[Last[l]]) && (k < Last[Last[l]])

AddToEncroachingLists[k_Integer,l1_List] :=
        Block[{i,l=l1},
                If [k <= First[Last[l]],
                        i = Ceiling[ BinarySearch[l,k,First] ];
                        PrependTo[l[[i]],k],
                        i = Ceiling[ BinarySearch[l,-k,(-Last[#])&] ];
                        AppendTo[l[[i]],k]
                ];
```

```
            l
      ]
```

Insertion into Encroaching Lists

Combinatorially, the interesting structure is the encroaching list set resulting from applying the insertion procedure repeatedly to elements of a permutation. The size of the encroaching list set generated by the insertion procedure appears to be an excellent measure of presortedness.

```
EncroachingListSet[l_List] := EncroachingListSet[l,{}]
EncroachingListSet[{},e_List] := e

EncroachingListSet[l_List,e_List] :=
      EncroachingListSet[Rest[l], AddToEncroachingLists[First[l],e] ]
```

Constructing Encroaching Lists

Observe that each subset is in sorted order, and further the heads and tails of the successive lists are also ordered around the structure.

```
In[50]:= TableForm[
EncroachingListSet[{6,7,1,8,2,5,9,3,4}] ]

                       {1, 6, 7, 8, 9}

                       {2, 5}

Out[50]//TableForm= {3, 4}
```

It can be shown that the expected length of the first list created by the insertion procedure is $O(\log n)$, and that the expected number of lists for a random permutation is $O(\sqrt{n})$.

```
In[51]:= Table[
Length[EncroachingListSet[RandomPermutation[i]]],
{i,10,100,10}]

Out[51]= {3, 5, 5, 6, 7, 7, 9, 10, 12, 11}
```

2.4 Exercises and Research Problems

Exercises

1. An alternate recurrence [NW78] for the number of integer partitions of n is:

$$p_n = \frac{1}{n} \sum_{m=0}^{n-1} \sigma(n-m) p_m$$

   where $p_0 = 1$ and $\sigma(k)$ is the sum of the divisors of k. Does this recurrence give a better way to calculate p_n than `NumberOfPartitions`? You may use the built-in *Mathematica* function `DivisorSigma[1,k]` to compute $\sigma(k)$.

2. Design and implement an efficient algorithm for constructing all partitions of n with distinct parts. Compare it to using `Select` and `Partitions`.

3. Repeat the previous exercise, constructing all partitions of n into odd parts, even parts, distinct odd parts, and finally all *self-conjugate* partitions, meaning they are equal to their transpose.

4. Develop and implement a recursive algorithm to construct all Young tableaux of shape $\{s_1, ..., s_m\}$, by constructing all subsets of size s_1 and then prepending it to all tableaux of shape $\{s_2, ..., s_m\}$. How does this algorithm perform compared to `Tableaux`?

5. Use the recurrence for the number of partitions of n with largest part at most k to design an algorithm for constructing random partitions. How does this algorithm perform compared to `RandomPartition`?

6. For a given Young tableau, only certain rows can be the start of a deletion operation. Which rows are these? For small n, determine the expected number of possible deletion rows for a tableau on n elements.

7. Experiment with `NumberOfTableaux` to determine the shape which, for a given number of elements, maximizes the number of Young tableaux over all partitions of n.

8. A partition p *contains* partition q if the Ferrers diagram of p contains the Ferrers diagram of q. For example, $\{3,3,2\}$ contains both $\{3,3,1\}$ and $\{3,2,2\}$ as well as many smaller partitions. *Young's lattice* Y_p [SW86] is the partial order of the partitions contained within p ordered by containment. Use `MakeGraph` to construct Young's lattice for an arbitrary partition p.

9. `ConstructTableau` takes a permutation to a Young tableau which is not necessarily unique. For the permutations of length n, which tableaux are the most frequently generated?

10. Experiment to determine the expected amount of vacant dots in the rectangle defined by the Ferrers diagram of a random partition of size n.

11. Implement a function [Sav89] to generate the partitions of n in Gray code order, as defined in the chapter.

Research Problems

1. What is the expected number of encroaching lists associated with a random permutation of size n? It is conjectured to be approximately $\sqrt{2n}$ for sufficiently large n [Ski88].

2. What is the expected size of the Durfee square in a random partition of n?

3. What is the correct way to spell Ferrers diagram? How many distinct spellings have appeared in print?

Chapter 3.

Representing Graphs

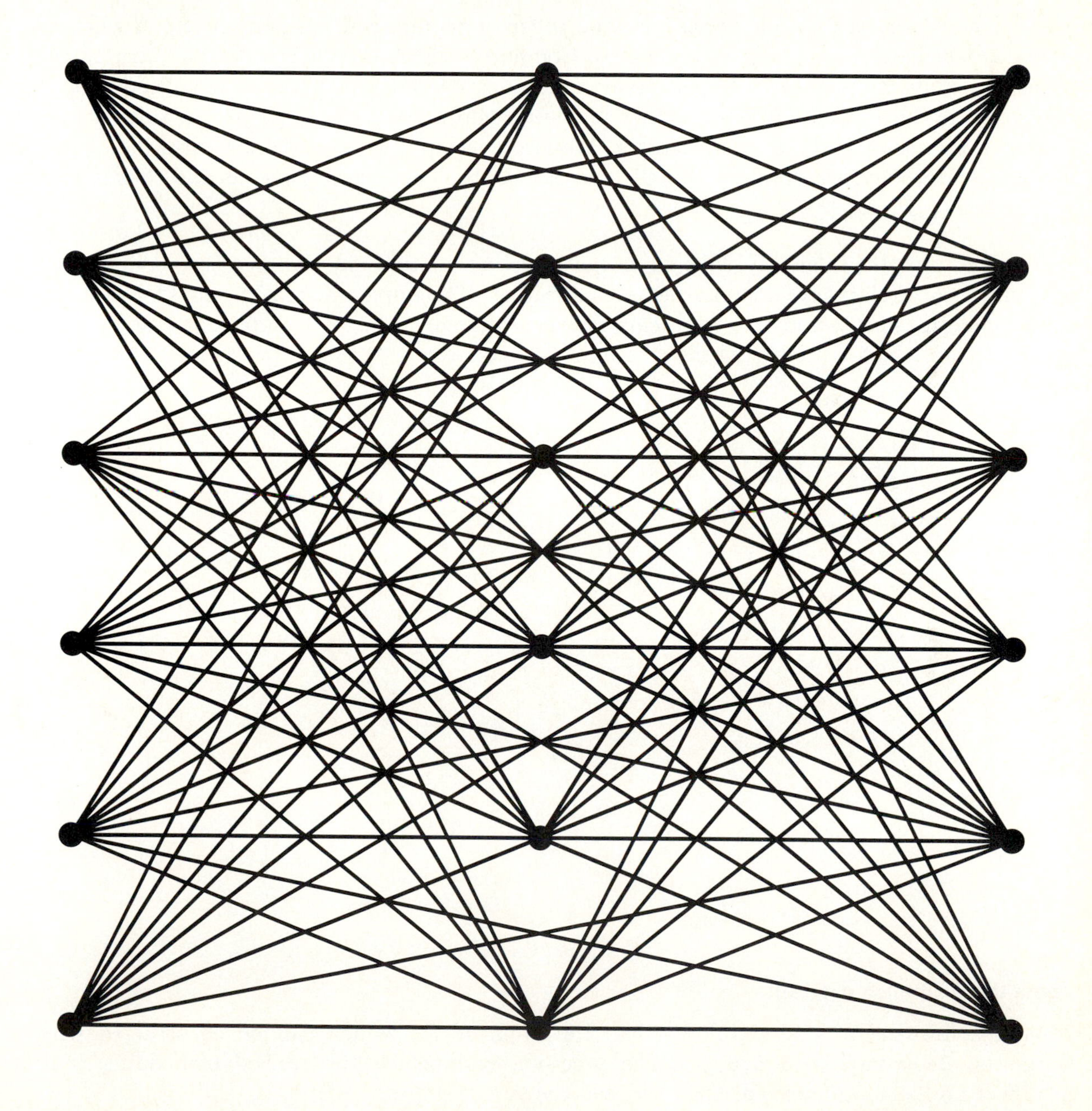

We define a *graph* to be a set of *vertices* with a set of pairs of these vertices called *edges*. This terminology is readily understood if not quite universal, since *point*, *node*, *junction*, and *0-simplex* have been used as synonyms for vertex and *line*, *arc*, *branch*, and *1-simplex* as synonyms for edge [Har69].

The representation of graphs takes on different requirements depending upon whether the intended consumer is a person or a machine. Computers digest graphs best as data structures such as adjacency matrices or lists. People prefer a visualization of the structure as a collection of points connected by lines, which implies adding geometric information to the graph. In this chapter, we discuss a variety of graph data structures. We also develop a variety of graph drawing algorithms to exploit *Mathematica* graphics. Finally, we provide tools for storing graphs and thus interfacing with other systems.

This chapter is mostly devoted to nut and bolt routines we will use throughout the text, but which have relatively little mathematical content. The impatient reader should give it a quick skim, while those interested in *Mathematica* programming will see examples of graphics and file manipulation which do not exist elsewhere in the book.

About the illustration overleaf:

This is a drawing of the complete tripartite graph $K_{6,6,6}$, which can be constructed with the command `ShowGraph[K[6,6,6]]`. The precision with which the ranked embedding is drawn produces an illusion of vertical lines between the vertex stages.

3.1 Data Structures for Graphs

Two traditional data structures used to represent graphs are *adjacency matrices* and *adjacency lists*. An adjacency matrix is an $n \times n$ matrix, where n is the number of vertices and edge (x, y) is a member of graph G if and only if the array entry $[x, y]$ is one. Adjacency matrices easily generalize to weighted and directed graphs, and permit constant time answers to the query "is (x, y) an edge?" Adjacency lists define a graph G as n lists of vertices, so edge (x, y) is a member of G if and only if y is a member of the xth list. We shall find that a third representation, ordered pairs, is often especially convenient in *Mathematica*.

3.1.1 Adjacency Matrices

In this book, the adjacency matrix will be the default representation. Weighted graphs and multigraphs are represented by having the table entry contain the weight or multiplicity of the edge. This is a source of potential ambiguity, since fractional weights have meaning for weighted graphs but not for multigraphs.

Although undirected graphs can be represented as triangular matrices, saving almost half the space, for convenience we will represent all graphs as directed graphs. Undirected edges appear twice in the matrix.

We assign the header `Graph` to each graph structure to avoid confusion with other objects in the system. Each graph consists of an adjacency matrix, a list of n lists of length n each where n is the number of vertices, and an *embedding*, a list of n pairs of $\{x, y\}$ positions of vertices in the plane.

```
Edges[Graph[e_,_]] := e

Vertices[Graph[_,v_]] := v

V[Graph[e_,_]] := Length[e]

M[Graph[g_,_],___] := Apply[Plus, Map[(Apply[Plus,#])&,g] ] / 2
M[Graph[g_,_],Directed] := Apply[Plus, Map[(Apply[Plus,#])&,g] ]

ChangeVertices[g_Graph,v_List] := Graph[ Edges[g], v ]

ChangeEdges[g_Graph,e_List] := Graph[ e, Vertices[g] ]
```

Definition of Graph Extraction Primitives

Simple primitives can be used to extract the adjacency and embedding information from a graph. Since the functions themselves are very simple, and they are primarily intended for internal use, the names were deliberately selected to be short.

In the *complete graph* on five vertices each vertex is adjacent to all other vertices. K[n] is a function which constructs the complete graph on n vertices.

```
In[1]:= ShowGraph[ K[5] ];
```

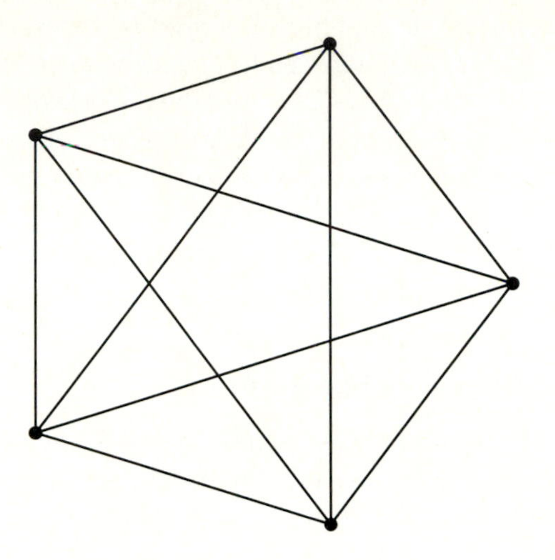

The adjacency matrix of K_5 shows that each vertex is incident to all other vertices. The main diagonal consists of zeros, since there are no *self-loops* in the complete graph, meaning edges from a vertex to itself.

```
In[2]:= TableForm[ Edges[K[5]] ]
                   0  1  1  1  1
                   1  0  1  1  1
                   1  1  0  1  1
                   1  1  1  0  1
Out[2]//TableForm= 1  1  1  1  0
```

The standard embedding of K_5 consists of five vertices equally spaced on a circle.

```
In[3]:= Vertices[ K[5] ]
Out[3]= {{0.309017, 0.951057}, {-0.809017, 0.587785},
  {-0.809017, -0.587785}, {0.309017, -0.951057}, {1., 0}}
```

The number of vertices in a graph is termed the *order* of the graph.

```
In[4]:= V[ K[5] ]
Out[4]= 5
```

With an optional argument to specify whether we count directed or undirected edges, M returns the number of edges in a graph.

```
In[5]:= {M[K[5]], M[K[5],Directed]}
Out[5]= {10, 20}
```

One advantage to making the drawing information a part of the representation of the graph is that it is usually difficult to determine what the "right" drawing of a graph should be. This varies according to the structure and application of the graph in question. Thus once we have a desired embedding, it will be viewed as a part of the graph. Throughout this book, we will see many operations which construct graphs from other graphs, and for constructions such as line graphs and graph products a reasonable embedding can be found using the embeddings of the component graphs.

Since graphs are built from vertices and edges, we need primitives to manipulate them at this level. Adding and deleting edges from adjacency matrices is a simple matter of addition or subtraction, if the edge is actually there.

```
AddEdge[Graph[g_,v_],{x_,y_},Directed] :=
        Block[ {gnew=g},
                gnew[[x,y]] ++;
                Graph[gnew,v]
        ]

AddEdge[g_Graph,{x_,y_},flag_:Undirected] :=
        AddEdge[ AddEdge[g, {x,y}, Directed], {y,x}, Directed]

DeleteEdge[Graph[g_,v_],{x_,y_},Directed] :=
        Block[ {gnew=g},
                If [ g[[x,y]] > 1, gnew[[x,y]]--, gnew[[x,y]] = 0];
                Graph[gnew,v]
        ]

DeleteEdge[g_Graph,{x_,y_},flag_:Undirected] :=
        DeleteEdge[ DeleteEdge[g, {x,y}, Directed], {y,x}, Directed]
```

Edge Insertion and Deletion Primitives

A *star* is a tree with one vertex of degree $n-1$. Adding any new edge to a star produces a cycle of length three.

In[6]:= **ShowGraph[AddEdge[Star[10], {1,2}]];**

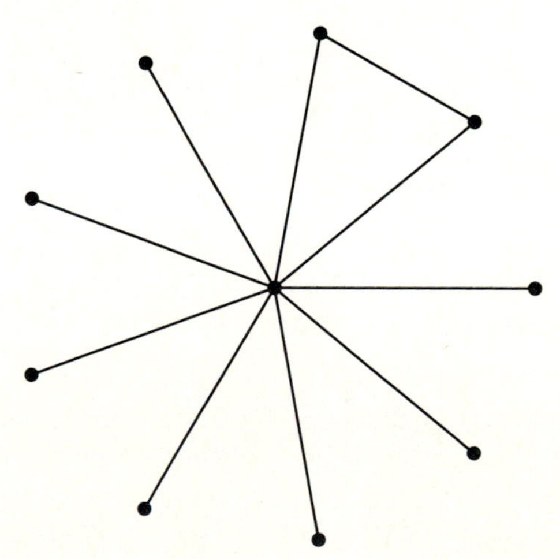

The deleted directed edge from the undirected graph shows up as an edge with only one arrow. These adjacency matrix manipulation primitives will not prove as useful as they might appear, because it is often easier and more efficient to make large changes to the graph at once instead of an edge at a time. As previously discussed, *Mathematica* often rewrites the entire structure when a part of it is changed.

```
In[7]:= ShowGraph[ DeleteEdge[Cycle[5],{1,2},Directed],
Directed]
```

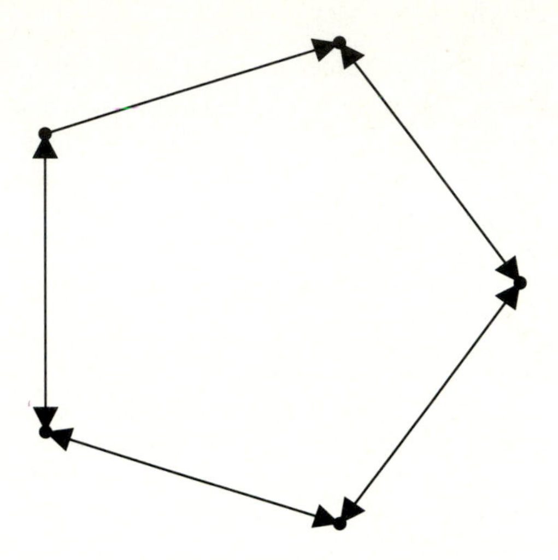

Adding and deleting vertices instead of edges is a useful operation, and can be achieved as special cases of two other functions we shall define later. `GraphUnion` concatenates two disjoint graphs, so `GraphUnion[g,K[1]]` adds a disconnected vertex. `InduceSubgraph` returns the graph formed by any subset of vertices, and so can be used to delete any vertex.

```
AddVertex[g_Graph] := GraphUnion[g, K[1]]

DeleteVertex[g_Graph,v_Integer] := InduceSubgraph[g,Complement[Range[V[g]],{v}]]
```

Vertex Insertion and Deletion Primitives

Adding a vertex to a star creates a graph with two connected components.

```
In[8]:= ConnectedComponents[ AddVertex[Star[10]] ]

Out[8]= {{1, 10, 2, 3, 4, 5, 6, 7, 8, 9}, {11}}
```

Deleting a vertex deletes all edges incident upon it, and so deleting the center of a star leaves an empty graph.

```
In[9]:= ShowGraph[DeleteVertex[Star[10],10]];
```

Eigenvalues

The matrix representation has inspired the study of graphs using linear algebra [Big74, CDS80], considering such properties as the eigenvalues of a graph. The set of eigenvalues for a particular graph is called its *spectrum*.

```
Spectrum[Graph[g_,_]] := Eigenvalues[g]
```

Eigenvalues of a Graph

As these two examples show, non-isomorphic graphs can share the same spectrum. They prove that connectivity is not a property which can be tested from the spectrum [Wil85a].

```
In[10]:= {Spectrum[ GraphUnion[Cycle[4],K[1]] ],
Spectrum[Star[5]]}
Out[10]= {{2, -2, 0, 0, 0}, {2, -2, 0, 0, 0}}
```

The maximum degree Δ of a connected graph G is an eigenvalue of G if and only if G is regular.

```
In[11]:= Spectrum[ RealizeDegreeSequence[{4,4,4,4,4,4}] ]
Out[11]= {4, -2, -2, 0, 0, 0}
```

For any bipartite graph G, the spectrum $\chi(G) = -\chi(G)$.

```
In[12]:= Spectrum[K[3,4]]
Out[12]= {2 Sqrt[3], -2 Sqrt[3], 0, 0, 0, 0, 0}
```

■ 3.1.2 Adjacency Lists

The adjacency list representation of a graph consists of n lists, one list for each vertex v_i, $1 \leq i \leq n$, which records the vertices v_i is adjacent to. Although slower than adjacency matrices for edge existence queries, adjacency lists use space proportional to the size of the graph and so are more efficient for algorithms on sparse graphs.

```
ToAdjacencyLists[Graph[g_,_]] :=
        Map[ (Flatten[ Position[ #, _?(Function[n, n!=0])] ])&, g ]

FromAdjacencyLists[e_List] :=
        Block[{blanks = Table[0,{Length[e]}] },
                Graph[
                        Map [ (MapAt[ 1&,blanks,Partition[#,1]])&, e ],
                        CircularVertices[Length[e]]
                ]
        ]

FromAdjacencyLists[e_List,v_List] := ChangeVertices[FromAdjacencyLists[e], v]
```

Adjacency List Conversion Routines

Each vertex in the complete graph is adjacent to all other vertices.

```
In[13]:= TableForm[ ToAdjacencyLists[K[5]] ]
                   2  3  4  5
                   1  3  4  5
                   1  2  4  5
                   1  2  3  5
Out[13]//TableForm= 1  2  3  4
```

Reconstructing the adjacency matrix from edge lists is an identity operation on a (0,1) matrix, but observe that the original embedding information of the star is lost.

```
In[14]:= ShowGraph[ FromAdjacencyLists[ ToAdjacencyLists[
Star[5] ]] ];
```

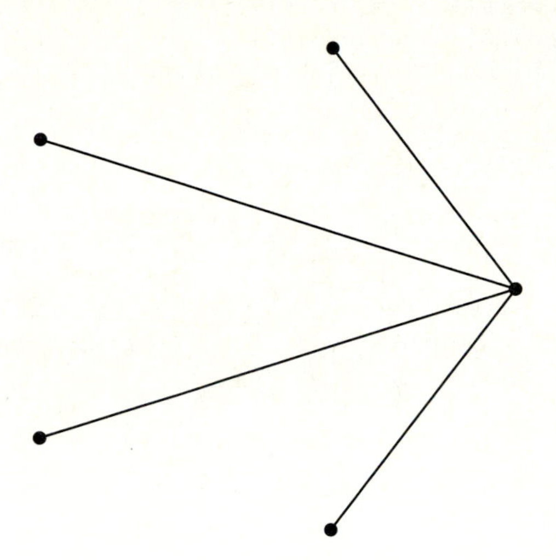

Although the adjacency matrix is necessary to maintain the edge weights in weighted graphs, the adjacency list representation will prove natural for a variety of algorithms.

■ 3.1.3 Ordered Pairs

Our third graph representation treats the edges as a list of ordered or unordered pairs of vertices. They can be constructed by iterating through each edge list but will prove more natural for certain applications.

There is an inherent ambiguity in the number of vertices in a graph which is represented in terms of ordered pairs. How can we know if there are isolated vertices? `FromOrderedPairs` accepts an embedding as an optional argument, which it uses to specify the number of vertices. If no embedding is given, the largest vertex number within an ordered pair determines the order of the graph.

```
ToOrderedPairs[g_Graph] := Position[ Edges[g], _?(Function[n,n != 0]) ]

ToUnorderedPairs[g_Graph] := Select[ ToOrderedPairs[g], (#[[1]] < #[[2]])& ]

FromOrderedPairs[l_List] :=
        Block[{n=Max[l]},
                Graph[
                        MapAt[1&, Table[0,{n},{n}],l],
                        CircularVertices[n]
```

```
                ]
        ]
FromOrderedPairs[{}] := Graph[{},{}]
FromOrderedPairs[l_List,v_List] :=
        Graph[ MapAt[1&, Table[0,{Length[v]},{Length[v]}], l], v]

FromUnorderedPairs[l_List] := MakeUndirected[ FromOrderedPairs[l] ]
FromUnorderedPairs[l_List,v_List] := MakeUndirected[ FromOrderedPairs[l,v] ]
```

Ordered Pair Conversion Routines

There are $n(n-1)$ ordered pairs defined by a complete graph of order n.

```
In[15]:= ToOrderedPairs[ K[5] ]
Out[15]= {{1, 2}, {1, 3}, {1, 4}, {1, 5}, {2, 1}, {2, 3},
  {2, 4}, {2, 5}, {3, 1}, {3, 2}, {3, 4}, {3, 5}, {4, 1},
  {4, 2}, {4, 3}, {4, 5}, {5, 1}, {5, 2}, {5, 3}, {5, 4}}
```

Once again, the embedding and weight information is lost in the conversion between representations, unless explicitly maintained and restored.

```
In[16]:= ShowGraph[ FromOrderedPairs[ ToOrderedPairs[
Star[5] ] ] ];
```

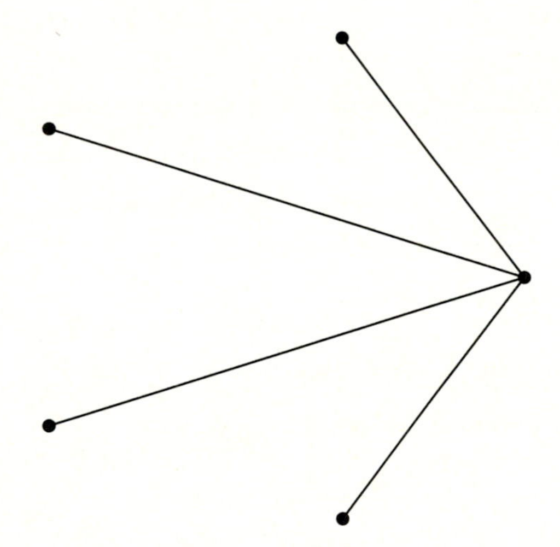

Undirected graphs are unambiguously represented by *unordered pairs*. It is often more efficient to work with half as many unordered pairs than with ordered pairs.

```
In[17]:= ToUnorderedPairs[ K[5] ]
Out[17]= {{1, 2}, {1, 3}, {1, 4}, {1, 5}, {2, 3}, {2, 4},
  {2, 5}, {3, 4}, {3, 5}, {4, 5}}
```

3.2 Elementary Graph Operations

The rest of this book concerns itself with operations on graphs. The operations of mathematical interest are built up using more primitive functions. In this section, we present some elementary routines which will be frequently used. They provide a gentle introduction to the more sophisticated functions we will see in later chapters.

3.2.1 Classifying Simple Graphs

Many of the algorithms we will use work correctly only on a subset of the possible graph structures, and so it is important to have predicates to test whether a particular graph is of a particular type. A *simple graph* is an unweighted, undirected graph which contains no self-loops. A *multigraph* is a graph which contains repeated edges, while graphs with self-loops are called *pseudographs*. Much of classical graph theory applies only to simple graphs. But when graphs are used to model networks of roads or pipes, each edge represents a distance or capacity and we have a weighted graph. The other aspect of the definition, concerning self-loops, eliminates a graph theoretic nuisance.

```
PseudographQ[Graph[g_,_]] :=
        Block[{i},
                Apply[Or, Table[ g[[i,i]]!=0, {i,Length[g]} ]]
        ]

UnweightedQ[Graph[g_,_]] := Apply[ And, Map[(#==0 || #==1)&, Flatten[g] ] ]

SimpleQ[g_Graph] := (!PseudographQ[g]) && (UnweightedQ[g])

RemoveSelfLoops[g_Graph] :=
        Block[{i,e=Edges[g]},
                Do [ e[[i,i]]=0, {i,V[g]} ];
                Graph[e, Vertices[g]]
        ]

EmptyQ[g_Graph] := Edges[g] == Table[0, {V[g]}, {V[g]}]

CompleteQ[g_Graph] := Edges[RemoveSelfLoops[g]] == Edges[ K[V[g]] ]
```

Elementary Graph Predicates

The complete graph is simple ...

```
In[18]:= SimpleQ[ K[5] ] && CompleteQ[ K[5] ]
Out[18]= True
```

... but adding any edge to K_5 turns it into a multigraph ...

```
In[19]:= UnweightedQ[ AddEdge[K[5],{1,4}] ]
Out[19]= False
```

... or creates a self-loop.

```
In[20]:= PseudographQ[ AddEdge[K[5],{3,3}] ]
Out[20]= True
```

■ 3.2.2 Induced Subgraphs

An *induced subgraph* of a graph G is a subset of the vertices of G together with any edges whose endpoints are both in this subset. Deleting a vertex from a graph is identical to inducing a subgraph with the $n-1$ other vertices.

```
InduceSubgraph[g_Graph,{}] := Graph[{},{}]

InduceSubgraph[Graph[g_,v_],s_List] :=
        Graph[Transpose[Transpose[g[[s]]] [[s]] ],v[[s]]] /; (Length[s]<=Length[g])
```

Inducing Subgraphs

An induced subgraph which is complete is called a *clique*. Any subset of the vertices in a complete graph defines a clique. This drawing presents an interesting illusion, for although the points seem irregularly spaced they all are defined as lying on a circle.

```
In[21]:= ShowGraph[
InduceSubgraph[K[20],RandomSubset[Range[20]]] ];
```

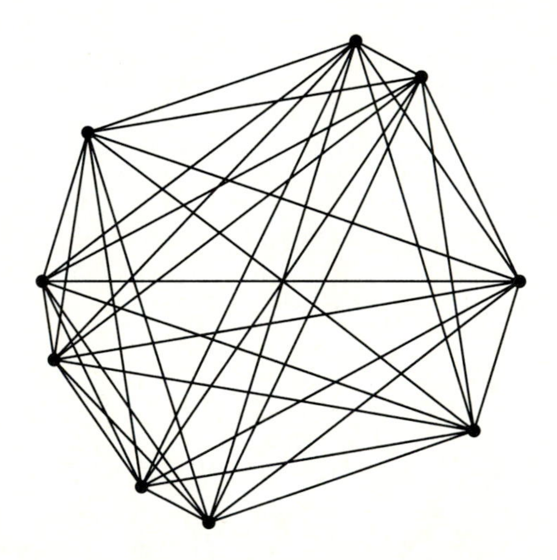

This routine can also be used to permute the vertices of a graph, for example putting the center of this *wheel* first instead of last. Note that the embedding is permuted appropriately.

```
In[22]:= ShowLabeledGraph[ InduceSubgraph[ Wheel[9], Reverse[Range[9]] ] ];
```

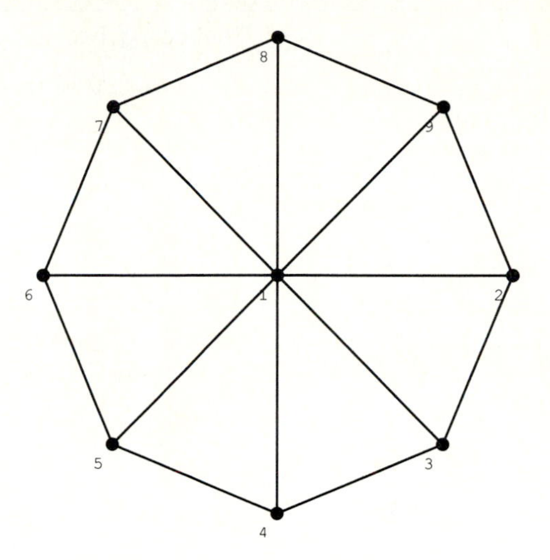

Another operation which makes graphs smaller is *contracting* an edge $\{v_1, v_2\}$, replacing the two vertices by one vertex v such that v is adjacent to anything v_1 or v_2 had been. This implementation works by inducing the subgraph of the $n-2$ vertices not affected by the contraction, and then appending the adjacency information for v to the end of the adjacency matrix.

```
Contract[g_Graph,{u_Integer,v_Integer}] :=
        Block[{o,e,i,n=V[g],newg,range=Complement[Range[V[g]],{u,v}]},
                newg = InduceSubgraph[g,range];
                e = Edges[newg]; o = Edges[g];
                Graph[
                        Append[
                                Table[
                                        Append[e[[i]],
                                                If[o[[range[[i]],u]]>0 ||
                                                        o[[range[[i]],v]]>0,1,0] ],
                                        {i,n-2}
                                ],
                                Append[
                                        Map[(If[o[[u,#]]>0||o[[v,#]]>0,1,0])&,range],
                                        0
                                ]
                        ],
                        Join[Vertices[newg], {(Vertices[g][[u]]+Vertices[g][[v]])/2}]
                ]
        ] /; V[g] > 2
```

```
Contract[g_Graph,_] := K[1]           /; V[g] == 2
```

Contracting Two Vertices in a Graph

The vertices of a *bipartite* graph have the property that they can be partitioned into two sets such that no edge connects two vertices of the same set. Contracting an edge in a bipartite graph can ruin its bipartiteness.

```
In[23]:= ShowGraph[ Contract[ K[6,6],{1,7} ] ];
```

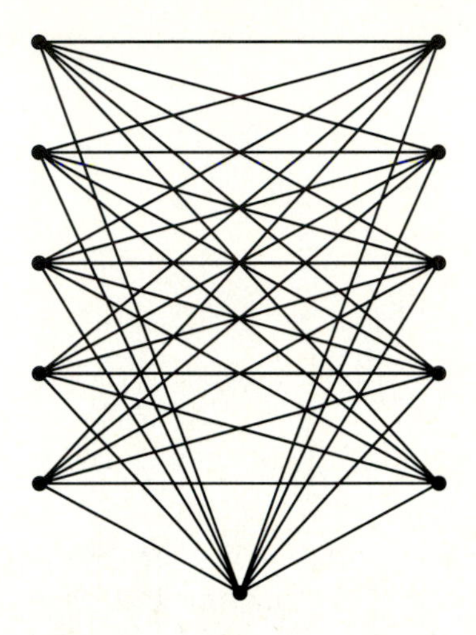

Repeatedly contracting vertex pairs reduces the number of vertices to one and the number of edges to zero. The number of edges removed with each contraction is not uniform, but depends upon the degrees of the vertices involved.

```
In[24]:= ( g = RandomGraph[21,0.5]; ListPlot[ Table[ g =
Contract[g, NthPair[Random[Integer ,{1,Binomial[V[g],2]}]
] ]; M[g], {20}] ];)
```

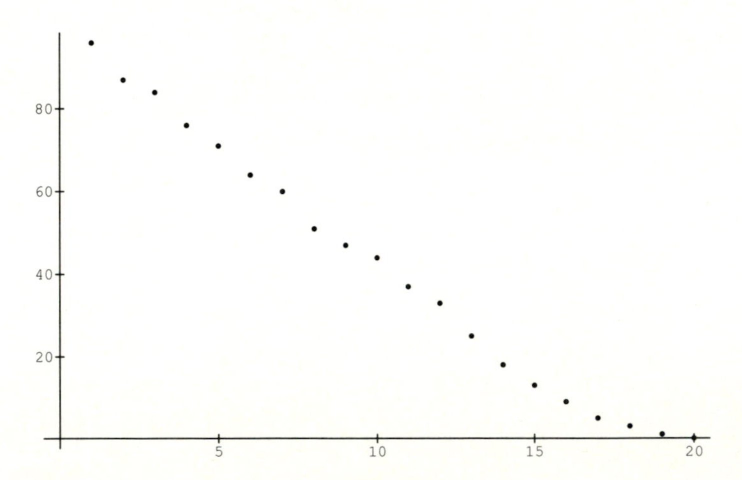

3.2.3 The Complement of a Graph

The *complement* G' of a graph $G = (V, E)$ is the simple graph with the same vertex set V, with two vertices adjacent if and only if they are not adjacent in G. It can be thought of as giving you nothing for something and vice versa.

```
GraphComplement[Graph[g_,v_]] :=
        RemoveSelfLoops[ Graph[ Map[ (Map[ (If [#==0,1,0])&, #])&, g], v ] ]
```

Constructing the Complement of a Graph

The complement of a complete k-partite graph contains edges only between vertices of the same stage.

```
In[25]:= ShowGraph[ GraphComplement[ K[2,2,2,2] ] ];
```

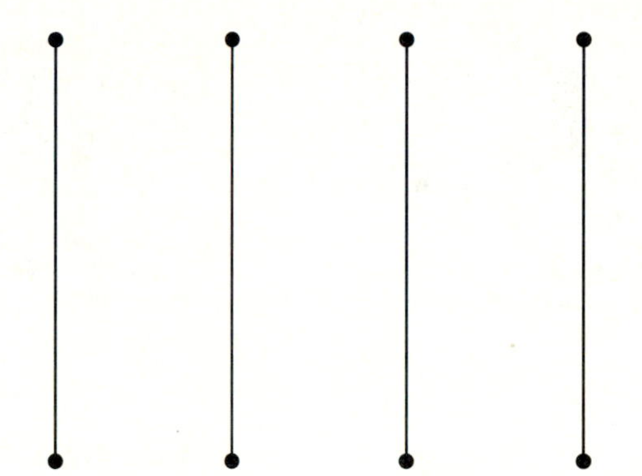

3.2.4 Undirected Graphs

A graph is *undirected* if and only if its adjacency matrix is symmetric, meaning the (i, j)th element equals the (j, i)th element. Many important graph algorithms work correctly only on undirected graphs.

It is often convenient to construct a graph one directed edge at a time, and later convert it into an undirected graph by duplicating directed edges. If there is at most one directed edge between two vertices, a slick way to make it undirected is to add the adjacency matrix to its transpose. The more general approach used here checks each element of the adjacency matrix.

```
MakeUndirected[Graph[g_,v_]] :=
        Block[{i,j,n=Length[g]},
                Graph[ Table[If [g[[i,j]]!=0 || g[[j,i]]!=0,1,0],{i,n},{j,n}], v ]
        ]

UndirectedQ[Graph[g_,_]] := (Apply[Plus,Apply[Plus,Abs[g-Transpose[g]]]] == 0)

MakeSimple[g_Graph] := MakeUndirected[RemoveSelfLoops[g]]
```

Undirecting Directed Graphs

Leaving out one half of an undirected edge in a graph means it isn't undirected any more.

```
In[26]:= UndirectedQ[ DeleteEdge[K[20],{1,2},Directed] ]
Out[26]= False
```

Undirected graphs are a special case of directed graphs, but when taken literally this causes anomalies. For example, any undirected edge forms a directed cycle of length two.

```
In[27]:= UndirectedQ[K[20]]
Out[27]= True
```

A *strongly connected* graph is a directed graph such that there is a directed path between every pair of vertices. Orienting a graph is the process of assigning each edge in an undirected graph a direction so that it is strongly connected.

```
In[28]:= ShowGraph[g=OrientGraph[GridGraph[3,3]],
Directed];
```

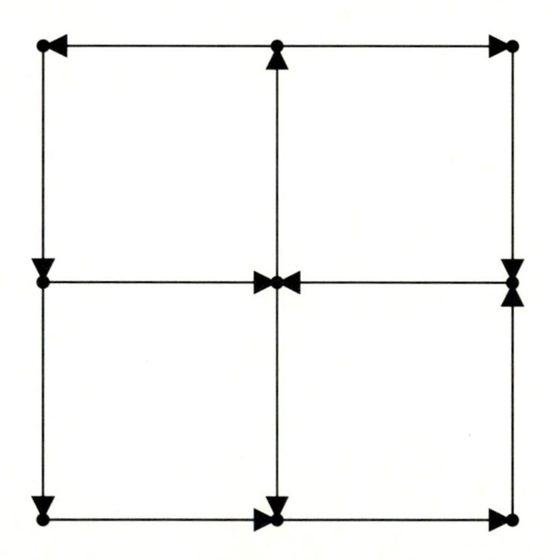

Making an oriented graph simple disorients it.

```
In[29]:= UndirectedQ[ MakeSimple[g] ]
Out[29]= True
```

■ 3.2.5 Breadth-First and Depth-First Search

To determine a property of a graph, it is often necessary to explore each of the edges and vertices of the graph in some reasonable order. Two traversals are particularly important from an algorithmic point of view, *breadth-first* and *depth-first* search. Breadth-first search will be used for such problems as coloring bipartite graphs and finding the shortest cycle of a graph. Depth-first search will be applied to finding connected and biconnected components and testing if a graph is acyclic.

Breadth-First Search

A *breadth-first* search of a graph explores all the vertices adjacent to the current vertex before moving on. In an acyclic graph, we encounter each vertex exactly once. However, for cyclic graphs care must be taken to insure that no vertices are repeated. All vertices in the connected component are eventually traversed.

```
BFS[g_Graph,start_Integer] :=
        Block[{e,bfi=Table[0,{V[g]}],cnt=1,edges={},queue={start}},
                e = ToAdjacencyLists[g];
                bfi[[start]] = cnt++;
                While[ queue != {},
                        {v,queue} = {First[queue],Rest[queue]};
                        Scan[
                                (If[ bfi[[#]] == 0,
                                        bfi[[#]] = cnt++;
                                        AppendTo[edges,{v,#}];
                                        AppendTo[queue,#]
                                ])&,
                                e[[v]]
                        ];
                ];
                {edges,bfi}
        ]

BreadthFirstTraversal[g_Graph,s_Integer,Edge] := First[BFS[g,s]]

BreadthFirstTraversal[g_Graph,s_Integer,___] := InversePermutation[Last[BFS[g,s]]]
```

Breadth-First Traversal of a Graph

A breadth-first traversal of a simple cycle alternates sides as it wraps around the cycle.

```
In[30]:= BreadthFirstTraversal[Cycle[20],1]
Out[30]= {1, 2, 20, 3, 19, 4, 18, 5, 17, 6, 16, 7, 15, 8,
  14, 9, 13, 10, 12, 11}
```

The order in which the edges are visited defines the breadth-first traversal. Observe that the edge traversal is incomplete, since it stops when all vertices have been visited.

```
In[31]:= BreadthFirstTraversal[K[2,2,2],1,Edge]
Out[31]= {{1, 3}, {1, 4}, {1, 5}, {1, 6}, {3, 2}}
```

Starting a breadth-first search from the center of a star finds all vertices in the first step.

```
In[32]:= BreadthFirstTraversal[Star[9],9]
Out[32]= {9, 1, 2, 3, 4, 5, 6, 7, 8}
```

Depth-First Search

In a *depth-first search*, the children of the first son of a vertex are explored before visiting his brothers. Hopcroft and Tarjan [HT73, Tar72] showed that depth-first search leads to linear time algorithms for many fundamental graph problems.

Every edge in an undirected graph on which a depth-first search has been performed is either in the depth-first search tree or is a *back edge* to an ancestor of the vertex. This property makes it easy to detect cycles. The situation is somewhat more complicated for directed graphs. The depth-first search partitions the edges into four classes: tree edges, back edges to ancestors in the tree, *forward edges* to descendents which have already been visited, and finally *cross-edges* to vertices which are not ancestors but have been previously visited.

```
DFS[v_Integer] :=
        ( dfi[[v]] = cnt++;
          AppendTo[visit,v];
          Scan[ (If[dfi[[#]]==0,AppendTo[edges,{v,#}];DFS[#] ])&, e[[v]] ] )

DepthFirstTraversal[g_Graph,start_Integer,flag_:Vertex] :=
        Block[{visit={},e=ToAdjacencyLists[g],edges={},dfi=Table[0,{V[g]}],cnt=1},
                DFS[start];
                If[ flag===Edge, edges, visit]
        ]
```

Depth-First Traversal of a Graph

Both `BreadthFirstTraversal` and `DepthFirstTraversal` visit each vertex in the appropriate connected component. In a disconnected graph, only one component is traversed.

```
In[33]:= DepthFirstTraversal[GraphUnion[K[3],K[4]], 1]
Out[33]= {1, 2, 3}
```

The depth-first traversal differs from the breadth-first traversal above, proceeding directly around the cycle.

```
In[34]:= DepthFirstTraversal[Cycle[20], 1]
Out[34]= {1, 2, 3, 4, 5, 6, 7, 8, 9, 10, 11, 12, 13, 14,
  15, 16, 17, 18, 19, 20}
```

The edges visited in the depth-first traversal also differ from the breadth-first search. Again the edge traversal is incomplete, since the search stops when all vertices have been visited.

```
In[35]:= DepthFirstTraversal[K[2,2,2], 1, Edge]
Out[35]= {{1, 3}, {3, 2}, {2, 4}, {4, 5}, {4, 6}}
```

Observe how the *Mathematica* scope rules operate. `Block` sets up a local enviroment which includes a new copy of `Edges`. Thus when `DFS` is called it uses the new copy and restores the old when `Block` is exited.

```
In[36]:= edges
Out[36]= edges
```

Both traversals also work on directed graphs.

```
In[37]:= DepthFirstTraversal[OrientGraph[GridGraph[3,3]],
1]
Out[37]= {1, 2, 3, 6, 5, 8, 7, 4, 9}
```

We will be using these search strategies often, even though there will be few explicit calls to `DepthFirstTraversal` and `BreadthFirstTraversal`. Instead, these traversals will be an implicit part of the algorithm.

■ 3.3 Graph Embeddings

Different drawings or embeddings of a graph can reveal different aspects of its structure. In this book, we play somewhat loose with the terminology with respect to drawings and embeddings. Strictly speaking, an *embedding* of a graph on a surface is a drawing on a surface so that no edges cross. Most of the algorithms in this section produce drawings which are not planar embeddings. However, "drawing" does not really capture the idea that the geometric information is part of the graph. SPREMB [EFK88] calls these beasts "computational geometry objects", which is even less satisfactory.

To see the effect of different graph embeddings, consider the following two drawings of $K_{6,6}$:

This ranked drawing clearly shows that $K_{6,6}$ is a bipartite graph. This is how it would usually be represented in a textbook.

```
In[38]:= ShowGraph[ K[6,6] ];
```

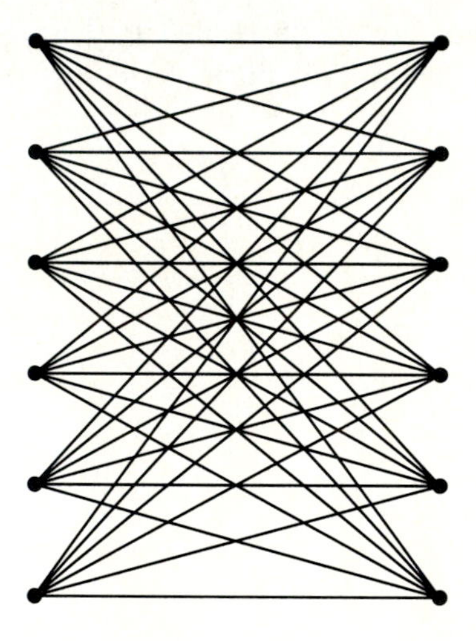

A circular drawing of the same graph shows that every vertex is connected to alternate vertices of the graph. It clearly illustrates that $K_{n,n}$ is a *circulant graph*, meaning that each row of the adjacency matrices are identical except for circular shifts.

```
In[39]:= ShowGraph[ CircularVertices[ InduceSubgraph[
K[6,6], {1,7,2,8,3,9,4,10,5,11,6,12} ] ] ];
```

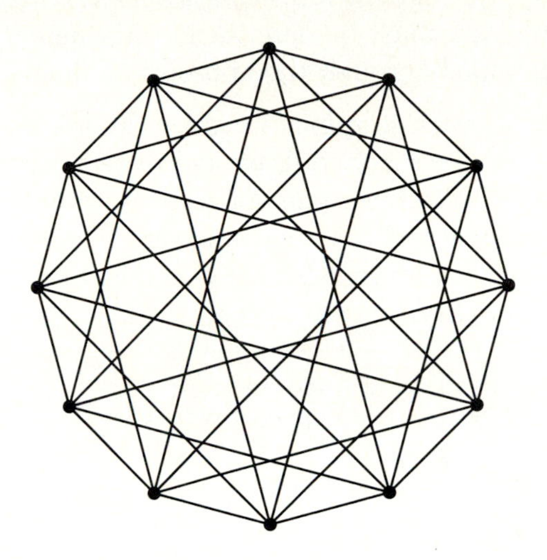

This graph and its circular drawing arose in studying the sizes of *cut sets* of graphs, sets of edges which, if cut, disconnect the graph. For a particular drawing of the graph, we can consider the specific cut sets defined by a straight line through the graph, acting as a knife. The circular embedding of $K_{n,n}$ has the property [Ski89] that all $\binom{n}{2}$ distinct cut set sizes are required for reconstruction, which can be proven by constructing a non-isomorphic graph with identical cut set sizes along any $\binom{n}{2} - 1$ lines through the embedding. The intuition of why so many cut sets are necessary to determine this graph is that every cut set is half the size of that of the complete graph, so a minimum amount of additional information is obtained from each cut set.

The moral of this story is that it is not a well-defined problem to generate the right drawing for a graph, since it becomes an aesthetic question instead of a mathematical one. Different algorithms are appropriate for different types of graphs. Here we consider a variety of different drawing algorithms, providing the means to draw graphs in an attractive fashion. Since many of the well-defined questions, such as minimizing the number of lines which cross or minimizing the variation in line length, are NP-complete, heuristic methods are necessary in practice.

Eades and Tamassia [ET89] provide an extensive annotated bibliography on graph drawing algorithms.

■ 3.3.1 Displaying Graphs

The PostScript graphics of *Mathematica* makes displaying an embedding quite convenient. The graph must be described in *Mathematica*'s graphics language and the result then pumped through the appropriate display routine.

The *Mathematica* graphics language knows about points and lines, which is all that is really necessary to draw undirected graphs. Although using arbitrary or piecewise curves to connect pairs of vertices would be more general, it would require different algorithms and a more complicated description. Straight-line drawings are sufficient for most of our needs. For example, Fáry [Far48] proved that every planar graph has a straight-line embedding.

```
ShowGraph[g1_Graph,type_:Undirected] :=
        Block[{g=NormalizeVertices[g1]},
                Show[
                        Graphics[
                                Join[
                                        PointsAndLines[g],
                                        If[SameQ[type,Directed],Arrows[g],{}]
                                ]
                        ],
                        {AspectRatio->1, PlotRange->FindPlotRange[Vertices[g]]}
                ]
        ]

MinimumEdgeLength[v_List,pairs_List] :=
        Max[ Select[
                Chop[ Map[(Sqrt[ N[(v[[#[[1]]]]-v[[#[[2]]]]) .
                        (v[[#[[1]]]]-v[[#[[2]]]])] ])&,pairs] ],
                (# > 0)&
        ], 0.001 ]

FindPlotRange[v_List] :=
        Block[{xmin=Min[Map[First,v]], xmax=Max[Map[First,v]],
                        ymin=Min[Map[Last,v]], ymax=Max[Map[Last,v]]},
                { {xmin - 0.05 Max[1,xmax-xmin], xmax + 0.05 Max[1,xmax-xmin]},
                  {ymin - 0.05 Max[1,ymax-ymin], ymax + 0.05 Max[1,ymax-ymin]} }
        ]

PointsAndLines[Graph[e_List,v_List]] :=
        Block[{pairs=ToOrderedPairs[Graph[e,v]]},
                Join[
                        {PointSize[ 0.025 ]},
                        Map[Point,Chop[v]],
                        Map[(Line[Chop[ v[[#]] ]])&,pairs]
                ]
```

```
        ]

Arrows[Graph[e_,v_]] :=
        Block[{pairs=ToOrderedPairs[Graph[e,v]], size, triangle},
                size = Min[0.05, MinimumEdgeLength[v,pairs]/3];
                triangle={ {0,0}, {-size,size/2}, {-size,-size/2} };
                Map[
                        (Polygon[
                                TranslateVertices[
                                        RotateVertices[
                                                triangle,
                                                Arctan[Apply[Subtract,v[[#]]]]+Pi
                                        ],
                                        v[[ #[[2]] ]]
                                ]
                        ])&,
                        pairs
                ]
        ]
```

Going from Graphs to Graphics

To display the orientation of edges in a directed graph, traditionally the line segment associated with an edge is turned into an arrow. An arrow is constructed by drawing a filled triangle at the point. Some trigonometry is involved to orient the arrowhead correctly. With the point initially placed at the origin, it is rotated to the proper edge orientation before being translated to the vertex location.

There are at least two different philosophies to scaling the graphs so they fit appropriately in the window. Should the x- and y-axes be scaled the same, so the graph looks exactly as it was embedded, or should each be selected independently to fill the window? Setting `PlotRange->All` converts `ShowGraph` from the first philosophy to the second.

The graphics are displayed with the `Show` command. For convenience, the `Show` and `PointsAndLines` functions have been combined into `ShowGraph`.

```
In[40]:= PointsAndLines[ K[3] ]
Out[40]= {PointSize[0.025], Point[{-0.5, 0.866025}],
  Point[{-0.5, -0.866025}], Point[{1., 0}],
  Line[{{-0.5, 0.866025}, {-0.5, -0.866025}}],
  Line[{{-0.5, 0.866025}, {1., 0}}],
  Line[{{-0.5, -0.866025}, {-0.5, 0.866025}}],
  Line[{{-0.5, -0.866025}, {1., 0}}],
  Line[{{1., 0}, {-0.5, 0.866025}}],
  Line[{{1., 0}, {-0.5, -0.866025}}]}
```

The default presentation for graphs is undirected, meaning that each directed edge is shown as a simple straight line. `ShowGraph` automatically scales the graph so it nicely fills the window.

```
In[41]:= ShowGraph[ K[2,2,4,2] ];
```

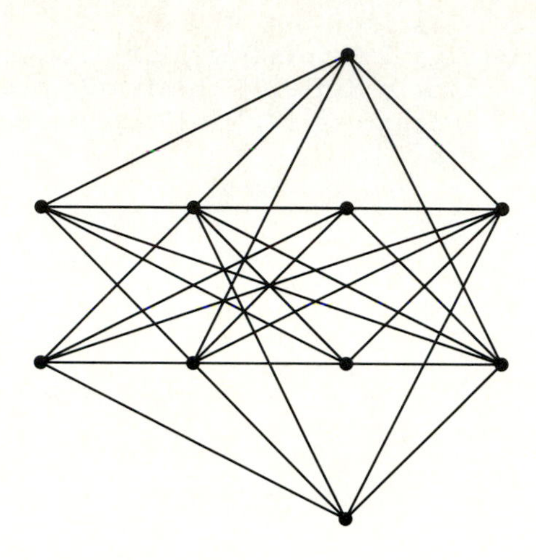

Undirected edges are represented by double-headed arrows, when displayed as a directed graph.

```
In[42]:= ShowGraph[ GridGraph[3,3], Directed ];
```

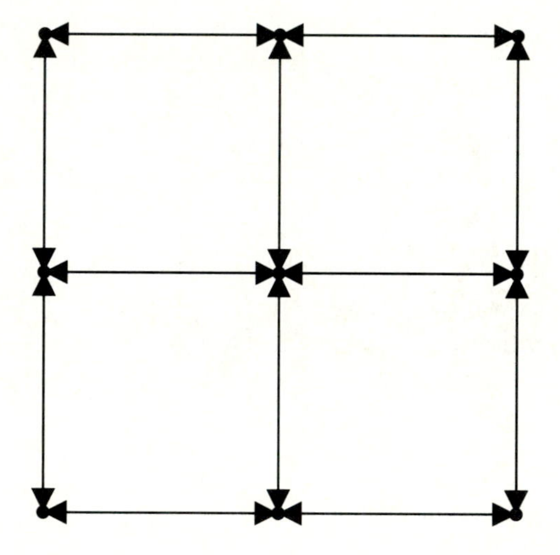

■ 3.3.2 Displaying Labeled Graphs

It is often important to identify each vertex of a graph with a label, particularly when the graph is a structure representing some function or binary relation. The default labels are the vertex numbers, although arbitrary labels can be specified when appropriate.

```
ShowLabeledGraph[g_Graph] := ShowLabeledGraph[g,Range[V[g]]]
ShowLabeledGraph[g1_Graph,labels_List] :=
        Block[{pairs=ToOrderedPairs[g1], g=NormalizeVertices[g1], v},
                v = Vertices[g];
                Show[
                        Graphics[
                                Join[
                                        PointsAndLines[g],
                                        Map[(Line[Chop[ v[[#]] ]])&, pairs],
                                        GraphLabels[v,labels]
                                ]
                        ],
                        {AspectRatio->1, PlotRange->FindPlotRange[v]}
                ]
        ]

GraphLabels[v_List,l_List] :=
        Block[{i},
                Table[ Text[ l[[i]],v[[i]]-{0.03,0.03},{0,1} ],{i,Length[v]}]
        ]
```

Displaying Labeled Graphs

Placing a semicolon after a graphics command stifles the production of the graphics descriptor, giving the picture without the code which generated it. This is somewhat faster, unless you need the descriptor for other reasons. Labeled graphs are by default drawn with each vertex replaced by its number, although the labels can be arbitrarily selected.

In[43]:= **ShowLabeledGraph[K[2,2,4,2], {"A","B","C","D","E","F","G","H","I","J"}];**

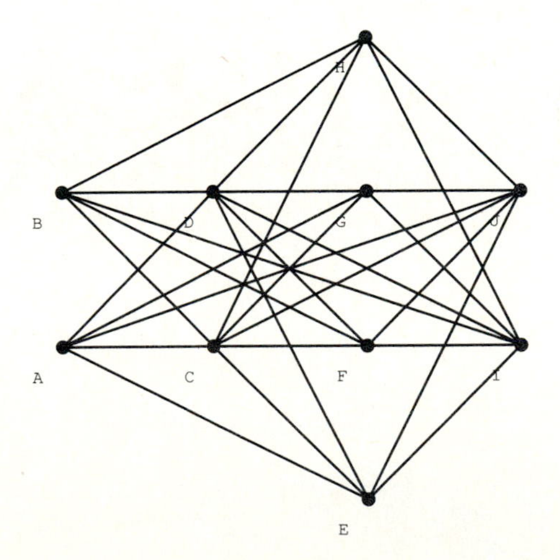

3.3.3 Circular Embeddings

Simple geometric structures lend themselves to attractive drawings. A *circular drawing* positions the vertices as evenly spaced points on the unit circle and is often instructive when the graph has a regular structure. Perhaps the most important property of circular embeddings is that no three vertices are colinear, so each edge is unambiguously represented. Computing evenly spaced points on a circle is simple trigonometry.

```
CircularVertices[n_Integer] :=
        Block[{i,x = N[2 Pi / n]},
                Chop[ Table[ N[{ (Cos[x i]), (Sin[x i]) }], {i,n} ] ]
        ]

CircularVertices[Graph[g_,_]] := Graph[ g, CircularVertices[ Length[g] ] ]
```

Constructing a Circular Embedding

Circular embeddings are natural for complete graphs and cycles. Embeddings for other graphs, such as the wheel shown here, are based on circular embeddings. One simple but important property [Ski89] of circular embeddings is that the vertices form a convex set. This should not be confused with *convex embeddings* of planar graphs, where every face in the embedding is convex [ET89]. This embedding happens to be convex.

```
In[44]:= ShowGraph[ Wheel[20] ];
```

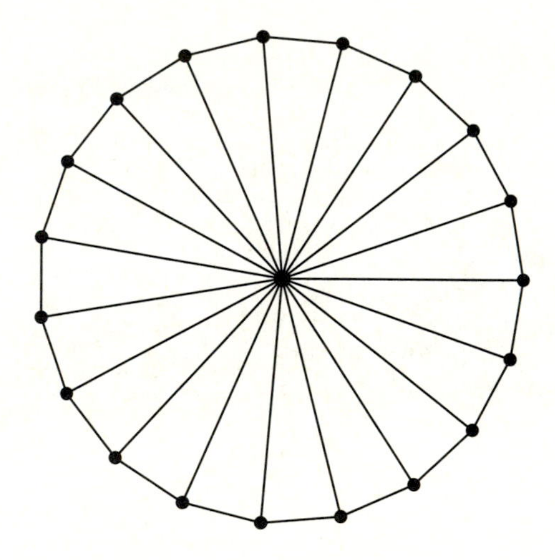

Because they clearly show every edge in the graph, circular embeddings are the default for random graphs.

```
In[45]:= ShowGraph[ RegularGraph[3,12] ];
```

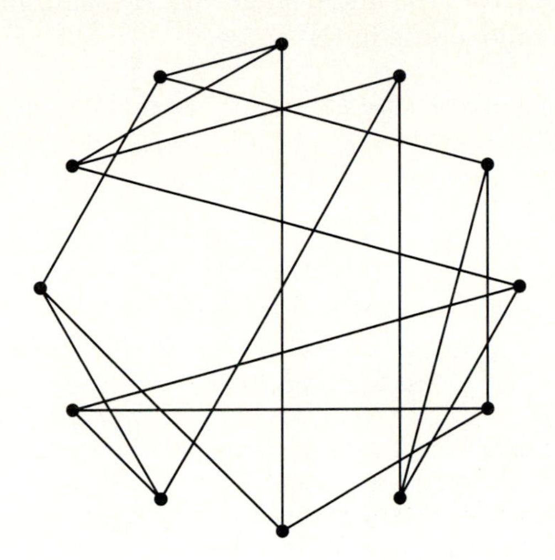

■ 3.3.4 Ranked Embeddings

The vertices in a graph can be partitioned into classes which share a particular property. Given such a partition, it is natural to display the vertices clustered by classes. One important vertex partition categorizes vertices by the shortest geodesic distance to a member of a special subset of the vertices. This we term a *ranked embedding*.

```
RankGraph[g_Graph, start_List] :=
        Block[ {rank = Table[0,{V[g]}],edges = ToAdjacencyLists[g],v,queue,new},
                Scan[ (rank[[#]] = 1)&, start];
                queue = start;
                While [queue != {},
                        v = First[queue];
                        new = Select[ edges[[v]], (rank[[#]] == 0)&];
                        Scan[ (rank[[#]] = rank[[v]]+1)&, new];
                        queue = Join[ Rest[queue], new];
                ];
                rank
        ]
```

Ranking the Vertices

Ranked embeddings are natural for displaying bipartite graphs, since if the initial stage consists of vertices of the same color, then no two vertices in the same class will be connected by an edge. The rankings can be found by a breadth-first search starting from the initial stage vertices, assuming that a representative of each connected component is in the stage.

This ranking information identifies stages for embedding the graph.

```
In[46]:= RankGraph[ GridGraph[5,4], {1,2,3,4} ]
Out[46]= {1, 1, 1, 1, 2, 2, 2, 2, 2, 3, 3, 3, 3, 3, 4, 4,
  4, 4, 4, 5}
```

To make the embedding attractive, each stage consists of a line of equally spaced vertices symmetric about the other axis.

```
RankedEmbedding[g_Graph,start_List] := Graph[ Edges[g],RankedVertices[g,start] ]

RankedVertices[g_Graph,start_List] :=
        Block[{i,m,stages,rank,freq = Table[0,{V[g]}]},
                rank = RankGraph[g,start];
                stages = Distribution[ rank ];
                Table[
                        m = ++ freq[[ rank[[i]] ]];
                        {rank[[i]], (m-1) + (1 - stages[[ rank[[i]] ]])/2 },
                        {i,V[g]}
                ]
        ]

Distribution[l_List] := Distribution[l, Union[l]]
Distribution[l_List, set_List] := Map[(Count[l,#])&, set]
```

Constructing Ranked Embeddings

The default embedding for a grid graph is a ranked embedding from all the vertices on one side. Ranking from the center vertex yields a different but interesting drawing.

```
In[47]:= ShowGraph[RankedEmbedding[GridGraph[5,5],{13}]];
```

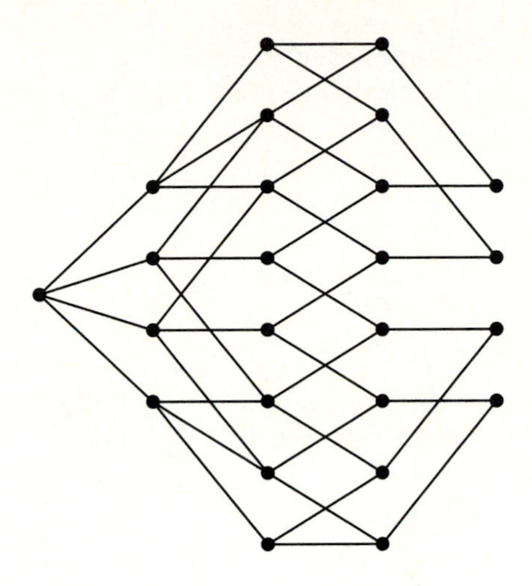

3.3.5 Drawing Trees

Trees are connected graphs without cycles, and are an important class of graphs. Their importance means that it is desirable to find good embedding algorithms for them, while their restricted nature means that we can exploit the structure of trees to do so. Among other properties, trees are bipartite and planar.

Eccentricity and Related Quantities

Free trees are trees without a distinguished root. To properly embed free trees, we must distinguish some vertex, which leads us to consider some properties of paths in graphs.

The *eccentricity* of a vertex v in a graph is the length of the longest shortest-path from v to some other vertex. From the eccentricity come other graph invariants. The *radius* of a graph is the smallest eccentricity of any vertex, while the *center* is the set of vertices whose eccentricity is the radius. The *diameter* of a graph is the maximum eccentricity of any vertex.

```
Eccentricity[g_Graph] := Map[ Max, AllPairsShortestPath[g] ]
Eccentricity[g_Graph,start_Integer] := Map[ Max, Last[Dijkstra[g,start]] ]

Diameter[g_Graph] := Max[ Eccentricity[g] ]
```

```
Radius[g_Graph] := Min[ Eccentricity[g] ]

GraphCenter[g_Graph] :=
        Block[{eccentricity = Eccentricity[g]},
                Flatten[ Position[eccentricity, Min[eccentricity]] ]
        ]
```

Finding the Eccentricity, Diameter, Radius, and Center

As in business, high eccentricity is not characteristic of a wheel.

```
In[48]:= Eccentricity[ Wheel[10] ]

Out[48]= {2, 2, 2, 2, 2, 2, 2, 2, 2, 1}
```

The diameter of a wheel is two, because a path through the center connects any pair of vertices.

```
In[49]:= Diameter[ Wheel[10] ]

Out[49]= 2
```

Because the distance function satisfies the triangle inequality, the diameter of a graph is at most twice its radius.

```
In[50]:= Radius[ Wheel[10] ]

Out[50]= 1
```

The cycle does not have such a well defined center.

```
In[51]:= GraphCenter[ Cycle[10] ]

Out[51]= {1, 2, 3, 4, 5, 6, 7, 8, 9, 10}
```

The center is made the physical center in many graph drawing algorithms.

```
In[52]:= GraphCenter[Wheel[10]]

Out[52]= {10}
```

The center of any tree consists of one vertex or two adjacent vertices. The adjacency of two vertices in a tree cannot be identified simply from the labels assigned to them.

```
In[53]:= GraphCenter[ RandomTree[10] ]

Out[53]= {4, 8}
```

Embedding Trees

Ideally, the center of a graph should be the center of the drawing of that graph. A *radial embedding* is a variation on the idea of a ranked embedding, where we rank all vertices in terms of their distance from the center, and place vertices of equal rank on concentric circles from the origin. By dividing the 2π radians around the center among the vertices of equal rank, we can ensure that the embedding of trees is planar.

```
RadialEmbedding[g_Graph,ct_Integer] :=
        Block[{center=ct,ang,i,da,theta,n,v,positioned,done,next,e=ToAdjacencyLists[g]},
                ang = Table[{0,2 Pi},{n=V[g]}];
                v = Table[{0,0},{n}];
                positioned = next = done = {center};
```

```
                While [next != {},
                        center = First[next];
                        new = Complement[e[[center]], positioned];
                        Do [
                                da = (ang[[center,2]]-ang[[center,1]])/Length[new];
                                ang[[ new[[i]] ]] = {ang[[center,1]] + (i-1)*da,
                                        ang[[center,1]] + i*da};
                                theta = Apply[Plus,ang[[ new[[i]] ]] ]/2;
                                v[[ new[[i]] ]] = v[[center]] +
                                        N[{Cos[theta],Sin[theta]}],
                                {i,Length[new]}
                        ];
                        next = Join[Rest[next],new];
                        positioned = Union[positioned,new];
                        AppendTo[done,center]
                ];
                Graph[Edges[g],v]
        ]

RadialEmbedding[g_Graph] := RadialEmbedding[g,First[GraphCenter[g]]];
```

Drawing Radial Embeddings

The radial embedding of a tree is guaranteed to be planar, but radial embeddings can be used with any graph. However, computing the graph center involves solving the all-pairs shortest path problem, which is an expensive operation. Thus the default on random trees is a radial embedding from an arbitrary vertex.

In[54]:= **ShowGraph[t=RandomTree[10]];**

Wheels are not trees, but are also properly represented by radial embeddings.

```
In[55]:= ShowGraph[ RadialEmbedding[Wheel[10]] ];
```

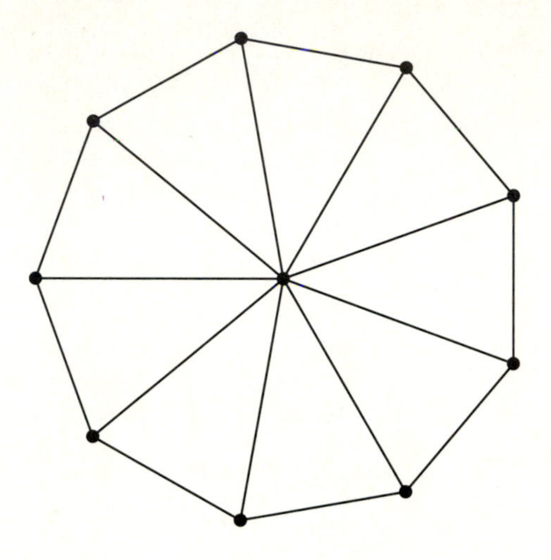

Embedding a tree from its actual center can make a difference in how it looks.

```
In[56]:= ShowGraph[ RadialEmbedding[ RandomTree[20] ] ];
```

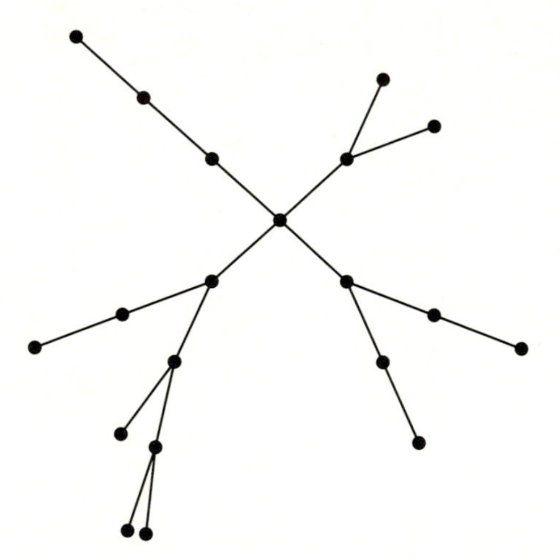

When trees are used to represent a heirarchy, be it a family tree, a data structure, or a corporate ladder, one vertex is selected to be the root of the tree. The most natural way to embed rooted trees is to rank the vertices in terms of distance from the root, and then position the vertices of equal rank on the same line so that no two edges cross.

This implementation assigns vertices of equal rank an equal width for drawing their subtrees. Better heuristics, which attempt to minimize total width while maximizing the minimum spacing between vertices of equal rank have been studied by several researchers [RT81, Vau80, WS79], although under certain aesthetic criteria the problem is NP-complete [SR83].

```
RootedEmbedding[g_Graph,rt_Integer] :=
        Block[{root=rt,pos,i,x,dx,new,n=V[g],v,done,next,e=ToAdjacencyLists[g]},
                pos = Table[{-Sqrt[n],Sqrt[n]},{n}];
                v = Table[{0,0},{n}];
                next = done = {root};
                While [next != {},
                        root = First[next];
                        new = Complement[e[[root]], done];
                        Do [
                                dx = (pos[[root,2]]-pos[[root,1]])/Length[new];
                                pos[[ new[[i]] ]] = {pos[[root,1]] + (i-1)*dx,
                                        pos[[root,1]] + i*dx};
                                x = Apply[Plus,pos[[ new[[i]] ]] ]/2;
                                v[[ new[[i]] ]] = {x,v[[root,2]]-1},
                                {i,Length[new]}
                        ];
                        next = Join[Rest[next],new];
                        done = Join[done,new]
                ];
                Graph[Edges[g],v]
        ]
```

Embedding Rooted Trees

By arbitrarily selecting a root, any tree can be represented as a rooted tree.

```
In[57]:= ShowGraph[ RootedEmbedding[RandomTree[10],1] ];
```

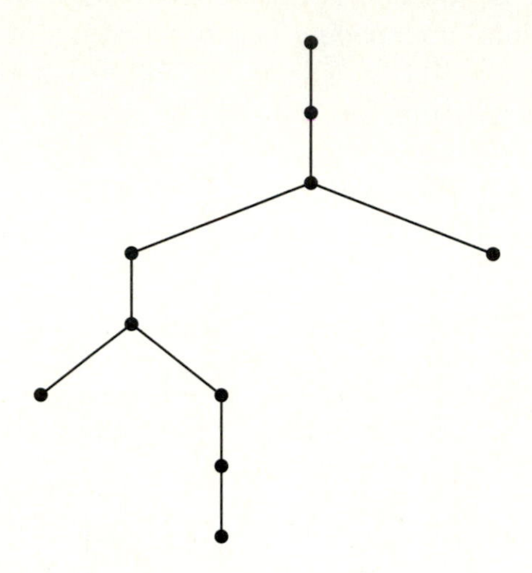

Rooted embeddings can be performed on all graphs. However, the structure of cycles is not always meaningfully represented.

```
In[58]:= ShowGraph[ RootedEmbedding[K[3,3,3],1] ];
```

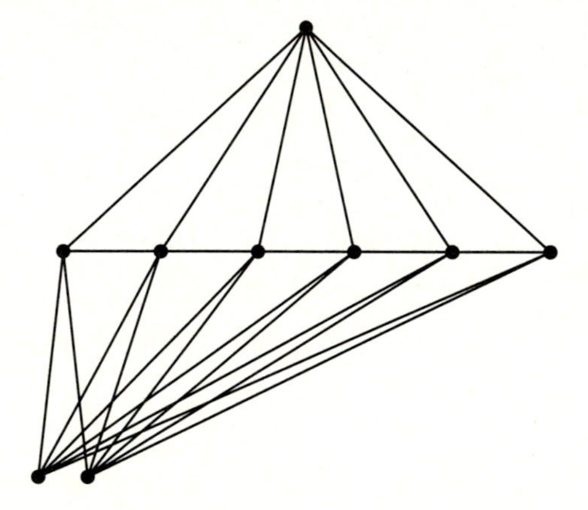

■ 3.3.6 Improving Embeddings

The embedding strategies thus far discussed have been quite rigid and inflexible. To use them to draw arbitrary graphs properly, we need tools to improve the quality of a particular embedding by modifying it according to different criteria.

Translations, Dilations, and Rotations

Performing simple geometric transformations on the set of vertices is useful to change a graph's size, position, or orientation. While these by themselves will not alter the topology of an embedding, they are building blocks for other routines.

```
TranslateVertices[v_List,{x_,y_}] := Map[ (# + {x,y})&, v ]
TranslateVertices[Graph[g_,v_],{x_,y_}] := Graph[g, TranslateVertices[v,{x,y}] ]

DilateVertices[v_List,d_] := (d * v)
DilateVertices[Graph[e_,v_],d_] := Graph[e, DilateVertices[v,d]]

RotateVertices[v_List,t_] :=
        Block[{d,theta},
                Map[
                        (If[# == {0,0}, {0,0},
                                d=Sqrt[#[[1]]^2 + #[[2]]^2];
                                 theta = t + Arctan[#];
                                 N[{d Cos[theta], d Sin[theta]}]
                        ])&,
                        v
                ]
        ]
RotateVertices[Graph[g_,v_],t_] := Graph[g, RotateVertices[v,t]]

Arctan[{x_,y_}] := Arctan1[Chop[{x,y}]]
Arctan1[{0,0}] := 0
Arctan1[{x_,y_}] := ArcTan[x,y]

NormalizeVertices[v_List] :=
        Block[{v1},
                v1 = TranslateVertices[v,{-Min[v],-Min[v]}];
                DilateVertices[v1, 1/Max[v1,0.01]]
        ]

NormalizeVertices[Graph[g_,v_]] := Graph[g, NormalizeVertices[v]]
```

Vertex Transformations

Rotating a k-partite graph by one quarter turn orders stages from top to bottom instead of left to right.

```
In[59]:= ShowGraph[ RotateVertices[K[3,1,2], Pi/2] ];
```

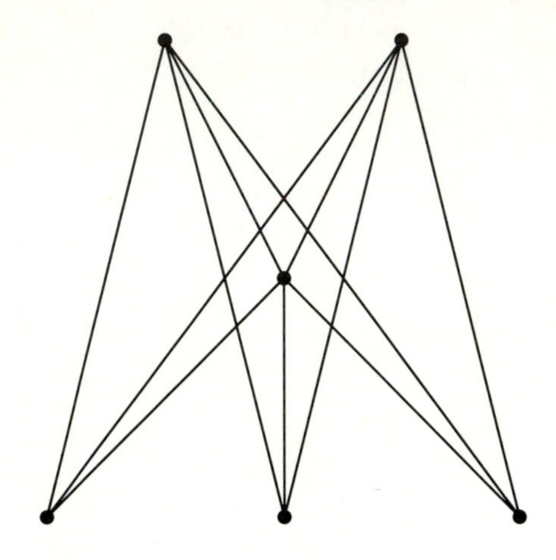

When normalized, $0 \leq x, y \leq 1$ for all vertices (x, y).

```
In[60]:= NormalizeVertices[ Vertices[K[5]] ]
Out[60]= {{0.645842, 0.974914}, {0.0728013, 0.788722},
  {0.0728013, 0.186192}, {0.645842, 0.}, {1., 0.487457}}
```

Random perturbations of an embedding can be useful, for example to place the vertices in general position so that no three are collinear and all edges are unambiguously displayed. A parameter specifies the maximum magnitude of a perturbation, regulating the extent to which the original embedding is preserved.

```
ShakeGraph[Graph[e_List,v_List], fract_:0.1] :=
        Block[{i,d,a},
                Graph[
                        e,
                        Table[
                                d = Random[Real,{0,fract}];
                                a = Random[Real,{0, 2 N[Pi]}];
                                {N[v[[i,1]] + d Cos[a]], N[v[[i,2]] + d Sin[a]]},
                                {i,Length[e]}
                        ]
                ]
        ]
```

Shaking the Vertices of a Graph

In the default embedding of $K_{2,2,2}$, the edges between vertices of the same parity are ambiguous. Is there an edge between v_1 and v_5 or not?

```
In[61]:= ShowLabeledGraph[ K[2,2,2] ];
```

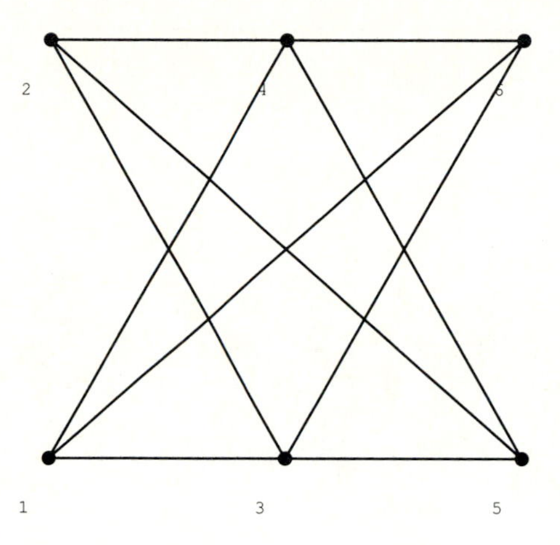

Here, ShakeGraph perturbs $K_{2,2,2}$ enough that all edges are unambiguous. The optional second parameter specifies the maximum distance a vertex can move. Since the magnitude and direction of each vertex perturbation are selected independently, the points on the unit disk are not selected uniformly at random.

```
In[62]:= ShowLabeledGraph[ ShakeGraph[ K[2,2,2], 0.2] ];
```

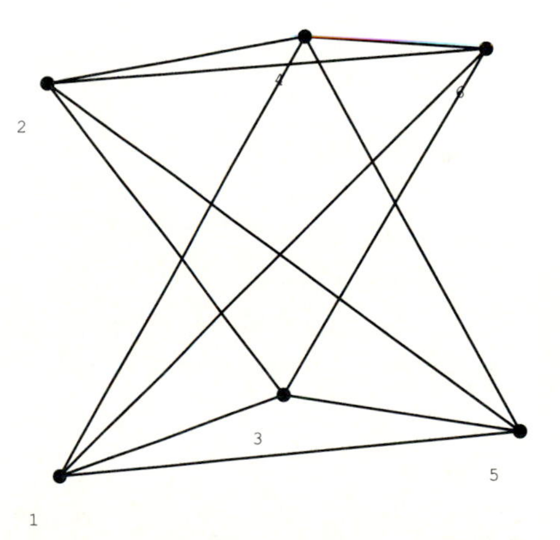

Spring Embeddings

An interesting general heuristic for drawing graphs models the graph as a system of springs and let Hooke's law space the vertices. Eades [Ead84] provides the following implementation of this idea. Adjacent vertices attract each other with a force proportional to the logarithm of their separation, and all non-adjacent vertices repel each other with a force proportional to their separation. These provide incentive for all edges to be as short as possible and yet the vertices be spread apart as far as possible. An elaboration of this technique for drawing graphs appears in [KK89].

```
CalculateForce[u_Integer,g_Graph,em_List] :=
        Block[{n=V[g],stc=0.25,gr=10.0,e=Edges[g],f={0.0,0.0},spl=1.0,v,dsquared},
                Do [
                        dsquared = Max[0.001, Apply[Plus,(em[[u]]-em[[v]])^2] ];
                        f += (1-e[[u,v]]) (gr/dsquared) (em[[u]]-em[[v]])
                                - e[[u,v]] stc Log[dsquared/spl] (em[[u]]-em[[v]]),
                        {v,n}
                ];
                f
        ]

SpringEmbedding[g_Graph,step_:10,inc_:0.15] :=
        Block[{new=old=Vertices[g],n=V[g],i,u,g1=MakeUndirected[g]},
                Do [
                        Do [
                                new[[u]] = old[[u]]+inc*CalculateForce[u,g1,old],
                                {u,n}
                        ];
                        old = new,
                        {i,step}
                ];
                Graph[Edges[g],new]
        ]
```

The Spring Embedding Heuristic

The behavior of such a system can be approximated by determining the force on each vertex at a particular time and then moving the vertex an infinitesimal amount in the appropriate direction, and repeat until the system stablizes.

In Section 1.1.4 we defined the maximum difference permutation graph. With the default circular embedding, many of the edges are long and the structure is obscured.

```
In[63]:= ShowGraph[ g = MakeGraph[Permutations[{1,2,3,4}],
(Count[#1-#2,0]==0)&] ];
```

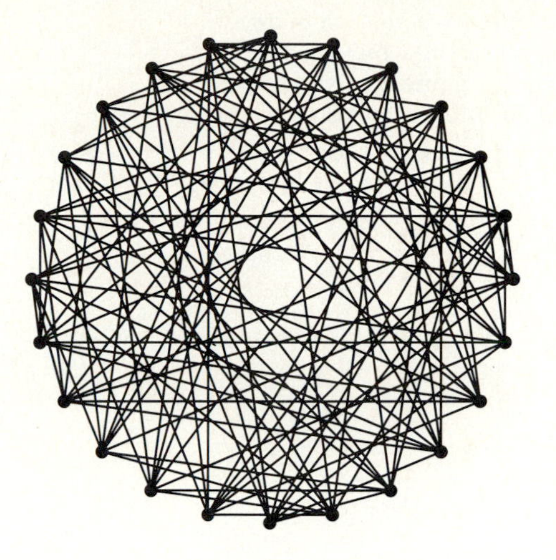

After applying the spring embedding heuristic, it is now apparent that this graph has some kind of structure, which might lead to an efficient algorithm for finding a Hamiltonian cycle. This is equivalent to sequencing permutations in maximum change order.

```
In[64]:= ShowGraph[ SpringEmbedding[g] ];
```

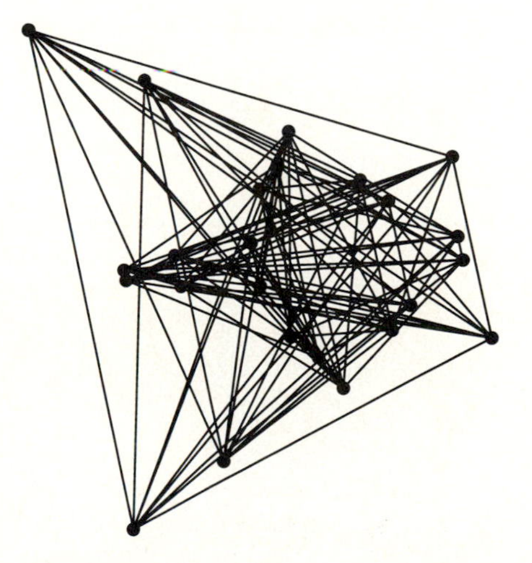

The heuristic tends to perform better on sparse graphs. Here it does a good job illustrating the join operation, where each vertex of K_7 is connected to each of two disconnected vertices. In achieving the minimum energy configuration, these two vertices end up on different sides of K_7.

```
In[65]:= ShowGraph[ SpringEmbedding[ GraphJoin[
EmptyGraph[2], K[7] ] ] ];
```

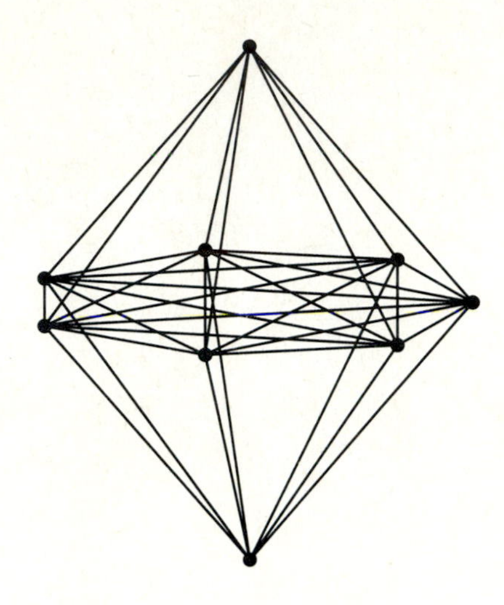

Since non-adjacent vertices repel each other, connected components tend to drift apart from each other.

```
In[66]:= ShowGraph[ SpringEmbedding[ GraphUnion[K[2],K[3]]
] ];
```

3.4 Storage Formats

Over the years, several other systems have been written to work with graphs and embeddings. For example, SPREMB [EFK88] provides a very nice editor GED for entering and modifying graphs. To use these tools, we must be able to read and write graphs in their file format. Given that they got here first, there is no good reason to define a different format, even if the existing one is somewhat inconvenient in *Mathematica*. In SPREMB, a graph is represented by its edge lists. A file consists of n lines, corresponding to the n vertices of the graph. The ith line is of the form $i\ x\ y\ v_1\ v_2\ \ldots\ v_k$, where vertex i is located at position (x, y) in the plane and i is adjacent to $v_1, v_2, \ldots, v_k$. All vertex positions must be within the unit square $0 \leq x, y \leq 1$.

```
1     0.452    0.956    5    4    9    3    2
2     0.046    0.242    9    8    7    3    1
3     0.832    0.246    7    6    5    2    1
4     0.444    0.670    1    10   12   9    5
5     0.566    0.560    3    1    12   4    6
6     0.544    0.458    3    12   11   5    7
7     0.450    0.396    3    2    11   6    8
8     0.350    0.452    2    11   10   9    7
9     0.320    0.564    1    2    10   8    4
10    0.404    0.558    9    4    8    12   11
11    0.442    0.490    7    6    8    12   10
12    0.480    0.558    5    4    6    11   10
```

The Storage Format Illustrated by an Icosahedron

It is important to note that somewhere down the road, Wolfram Research Inc. intends on expanding *Mathematica* in ways to permit editing structures such as graphs without the need for a special purpose graph editor. When this happens, the need for `ReadGraph` and `WriteGraph` will diminish considerably, as there are simpler ways to read and write *Mathematica* objects.

3.4.1 Reading a Graph

Since the storage format is essentially an adjacency list representation, once read, conversion to an adjacency matrix is easy. The subtlety is that the number of edges incident upon a vertex is not given, and so each line must be read until an end-of-line is encountered.

```
ReadGraph[file_] :=
        Block[{edgelist={}, v={},x},
                OpenRead[file];
                While[!SameQ[(x = Read[file,Number]), EndOfFile],
                        AppendTo[v,Read[file,{Number,Number}]];
                        AppendTo[edgelist,
                                Convert[Characters[Read[file,String]]]
                        ];
                ];
                Close[file];
                FromAdjacencyLists[edgelist,v]
        ]

IsDigitQ[ch_String] :=
        ! ((ToASCII[ch] < ToASCII["0"]) || (ToASCII[ch] > ToASCII["9"]))

Convert[l_List] :=
        Block[{ch,num,edge={},i=1},
                While[i <= Length[l],
                        If[ IsDigitQ[ l[[i]] ],
                                num = 0;
                                While[ ((i <= Length[l]) && (IsDigitQ[l[[i]]])),
                                        num = 10 num + ToASCII[l[[i++]]] - ToASCII["0"]
                                ];
                                AppendTo[edge,num],
                                i++
                        ];
                ];
                edge
        ]
```

Reading a Graph from a File

The ReadGraph routine opens the file and takes one line at a time, extracting the embedding and adjacency information before closing the file. To get the adjacency information, each line is exploded into a list of ASCII characters which are manually reassembled into numbers.

Once read, a graph can be displayed or manipulated by any other routine. The *icosahedron* is the Platonic solid on twelve vertices with twenty triangular faces. The graph defined by any convex polyhedron is planar, and these vertex positions represent a planar embedding of the icosahedron.

```
In[67]:= ShowGraph[ ReadGraph["graphs/icosahedron"] ];
```

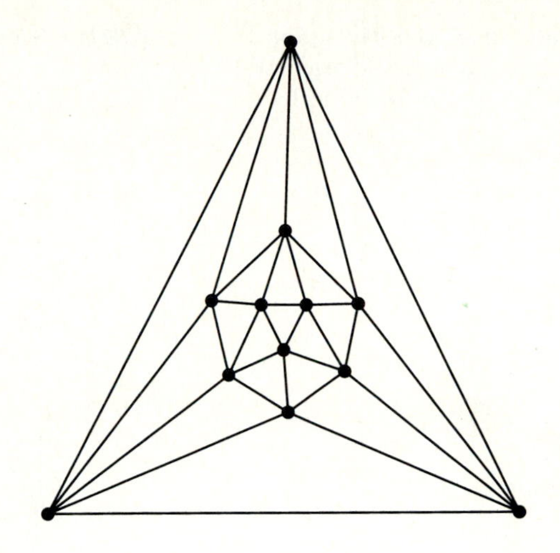

■ 3.4.2 Writing a Graph

Writing a graph is easier than reading, since the degree of each vertex is known. Normalization is done by first translating the vertices to the first quadrant and then scaling them so all coordinates are between 0 and 1.

```
WriteGraph[g_Graph,file_] :=
        Block[{edges=ToAdjacencyLists[g],v=N[NormalizeVertices[Vertices[g]]],i,x,y},
                OpenWrite[file];
                Do[
                        WriteString[file,"        ",ToString[i]];
                        {x,y} = v [[i]];
                        WriteString[file,"        ",ToString[x],"        ",ToString[y]];
                        Scan[
                                (WriteString[file,"        ",ToString[ # ]])&,
                                edges[[i]]
                        ];
                        Write[file],
                        {i,V[g]}
                ];
                Close[file];
        ]
```

Writing a Graph to a File

Write K_5 to a file. The vertices are normalized in the process.

```
In[68]:= WriteGraph[ K[5], "graphs/K5" ]
```

When read in and displayed, the graph appears the same, although the original vertex positions have been replaced by equivalent ones. Displayed graphs are automatically scaled to fit the window.

```
In[69]:= ShowGraph[ ReadGraph[ "graphs/K5" ] ];
```

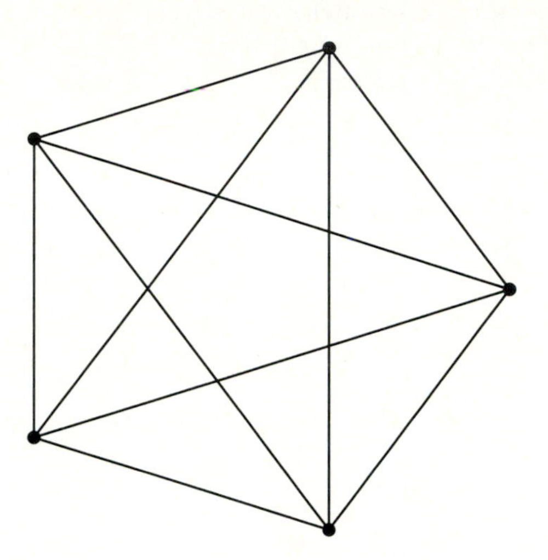

■ 3.4.3 Graph Editors and Related Systems

The idea of a workbench for manipulating graphs is natural enough that many such tools have been built over the years. We have already mentioned SPREMB [EFK88], which is a collection of C language filter programs for operations on embeddings, written by researchers at the University of Queensland in Australia. Particularly nice is their graph editor GED, available for Sun workstations, which uses a mouse to let the user literally draw a graph on the screen and add/delete/move vertices and edges. It is very convenient for entering graphs of interest, and any finite graphs we have included were drawn using GED.

There are several other systems worth mentioning. GRAPH [Cve84, Cve81] is a Yugoslavian system for research in graph theory, with a bibliography and automatic theorem prover as well as FORTRAN programs for determining graph invariants as discussed here. GRAPPLE is a system of Pascal programs for manipulating graphs for IBM PC-compatible computers, built by students at Indiana University, Purdue University at Fort Wayne under the direction of Prof. Marc Lipman. CABRI [DHRT87] is a French program for the Apple Macintosh with a very slick graph editor and a nice but small collection of invariants, while GraphView is an extensible graph editor for the NeXT [BS89]. Nauty [McK90] is a set of C language procedures for determining the automorphism group of a graph. LILA [Ech88] is a programming tool to aid in the creation of combinatorial algorithms. The Stanford Graphbase project [Knu88] has put together a database of interesting non-random graphs, such as the acquaintance graph for characters in long European novels, like *War and Peace*, with the intention of providing standards to empirically compare algorithms against. Older systems include Fortran [Kin72, Rea69] and Algol [CM70] extensions for graph theoretic applications.

An especially interesting program is GRAFFITI [Faj87, Faj88], which generalizes from a collection of graphs to form its own conjectures. Fortunately, mathematicians are still necessary to prove the theorems and write papers, several of which have resulted from GRAFFITI conjectures. The state of all GRAFFITI conjectures (proven, disproven, or open) is recorded in [Faj90]. Another computer project which resulted in theorems is reported in [BD85].

3.5 Exercises and Research Problems

Exercises

1. Using the graph representation functions defined within this chapter, plus regular *Mathematica* constructs, implement five different functions for constructing complete graphs. Experiment to determine which one is most efficient.

2. Write a function to determine whether the given embedding of a graph is planar, by testing whether any two edges intersect.

3. Develop and implement an algorithm for drawing rooted embeddings of trees which does a better job than `RootedEmbedding` of balancing the sizes of the vertical strips associated with each subtree. Thus wide subtrees will not be squashed into a narrow region.

4. Generalize the breadth-first and depth-first traversals to visit and return all edges in the graph, and to partition them into tree and back edges.

5. Experiment with heuristics to permute the vertices in a circular embedding to minimize the total edge length of the embedding.

6. The embeddings we have considered are all in the plane. However, certain embeddings make sense in higher dimensions. VSLI layout problems have led to the study of *book embeddings* [CLR87, Mal88], where all vertices are laid out on a line (the spine of the book) and the edges are printed on *pages* such that no two edges on a page cross. Write functions to find and display book embeddings of graphs.

7. Modify `SpringEmbedding` to stop when the amount of movement per iteration crosses a threshold. Experiment with how many iterations are necessary to obtain a stable drawing.

8. Experiment with repeated contractions of vertex pairs in random graphs, going from n vertices down to one. At what point in the process is the maximum number of edges likely to be deleted in a contraction?

9. Experiment to determine the probability that a random tree has a center of one vertex versus two vertices.

Research Problems

1. Estimate the probability that two randomly chosen graphs on n vertices have the same spectra [Wil85a]. What about for trees? These questions can be decided for small n by using graph generation routines of Section 4.3.

2. How many cut set sizes are neccessary to determine the degree sequence of an embedded graph, when the cut sets are restricted to partitions defined by straight lines in the plane [Ski89]?

3. GRAFFITI conjectures that the number of positive (or negative) eigenvalues of any graph is less than or equal to the sum of the positive eigenvalues. Partial results appear in [Faj88].

4. GRAFFITI conjectures that for any graph, the average distance $\bar{d}$ in the all-pairs shortest path matrix is less than or equal to n/Δ [Faj90]. Also, $\delta \leq n/\bar{d}$.

5. GRAFFITI conjectures that for any graph, the minimum positive eigenvalue is less than or equal to $n/\bar{d}$ [Faj90].

6. GRAFFITI conjectures that for any regular graph, the chromatic number is less than or equal to $n/\bar{d}$ [Faj90].

Chapter 4.

Generating Graphs

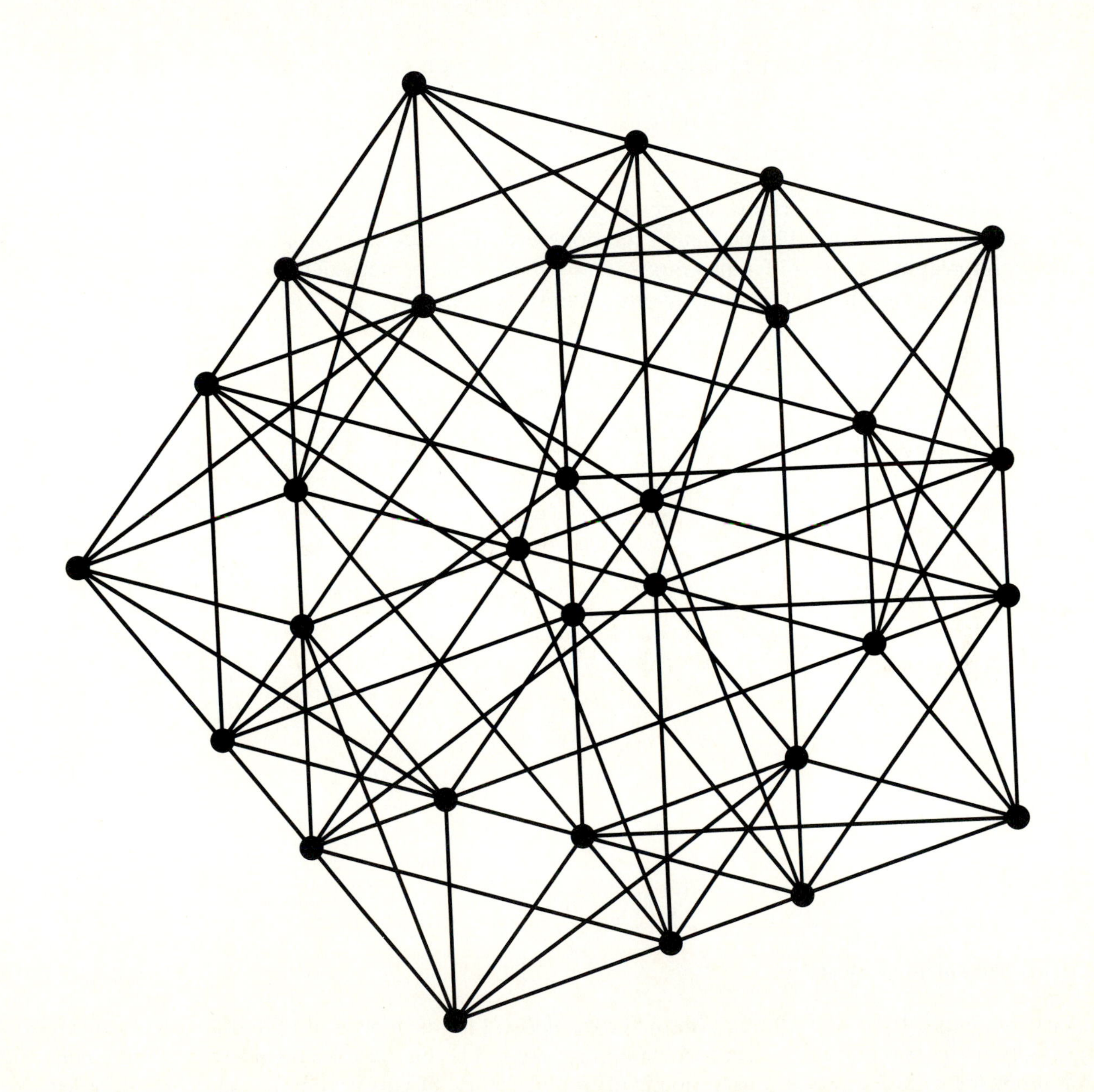

Many graphs consistently prove interesting, in the sense that they are models of important binary relations or have unique graph theoretic properties. Often, these graphs can be parameterized, such as the complete graph on n vertices K_n, giving a concise notation for expressing an infinite class of graphs.

In this chapter, we consider many classes of graphs and how to construct them. The special properties of these graphs will motivate the graph invariants which we compute in later chapters. Many of these graphs are truly beautiful when drawn properly, and they provide a wide range of structures to manipulate and study.

In addition to specific graphs, we provide functions for constructing different types of random graphs, as well as represent arbitrary binary relations as graphs. We start off with several operations which act on graphs to give different graphs, and which together with our parameterized graphs, give us the means to construct essentially any interesting graph.

About the illustration overleaf:

The *line graph* $L(G)$ of a graph G has a vertex of $L(G)$ for each edge of G and an edge of $L(G)$ if and only if the two edges of G share a common vertex. The line graph of K_5 is the complement of the Petersen graph, and the illustration is of the line graph of this graph. It can be constructed by `ShowGraph[ LineGraph[ LineGraph[K[5]] ] ];`.

4.1 Building Graphs from Other Graphs

This section emphasizes operations which take graphs and construct another graphs, such as the induce subgraph and complement graph operations discussed in Section 3.2. Many of the parameterized graphs in the rest of the chapter can be constructed using these operations, and other examples appear in [CM78].

4.1.1 Unions and Intersections

Graphs are sets of vertices and edges, and the most important operations on sets are unions and intersections.

The *union* of two graphs is formed by taking the union of the vertices and edges of the graphs. Thus the union of two graphs is always disconnected.

```
GraphUnion[g_Graph,h_Graph] :=
        Block[{maxg=Max[ Abs[ Map[First,Vertices[g]] ] ]},
                FromOrderedPairs[
                        Join[ ToOrderedPairs[g], (ToOrderedPairs[h] + V[g])],
                        Join[ Vertices[g], Map[({maxg+1,0}+#)&, Vertices[h] ] ]
                ]
        ]

GraphUnion[1,g_Graph] := g
GraphUnion[0,g_Graph] := EmptyGraph[0];
GraphUnion[k_Integer,g_Graph] := GraphUnion[ GraphUnion[k-1,g], g]

ExpandGraph[g_Graph,n_] := GraphUnion[ g, EmptyGraph[n - V[g]] ] /; V[g] <= n
```

The Union of Two Graphs

The union of two connected graphs has two connected components.

```
In[1]:= ShowGraph[ GraphUnion[ K[3], K[5,5] ] ];
```

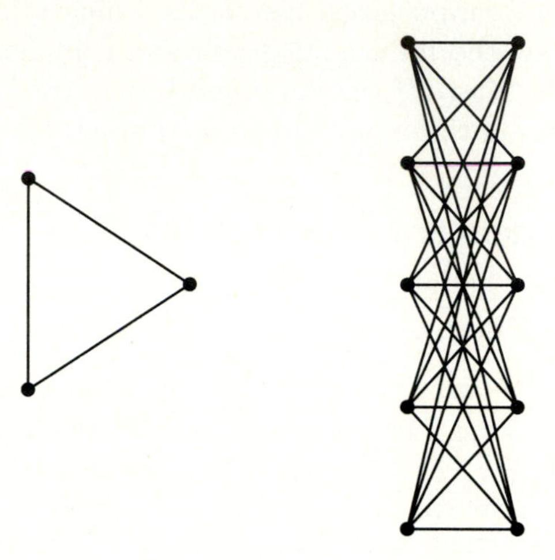

The other form of GraphUnion mimics the standard notation, here specifying five copies of K_3.

```
In[2]:= ShowGraph[ GraphUnion[5, K[3]] ];
```

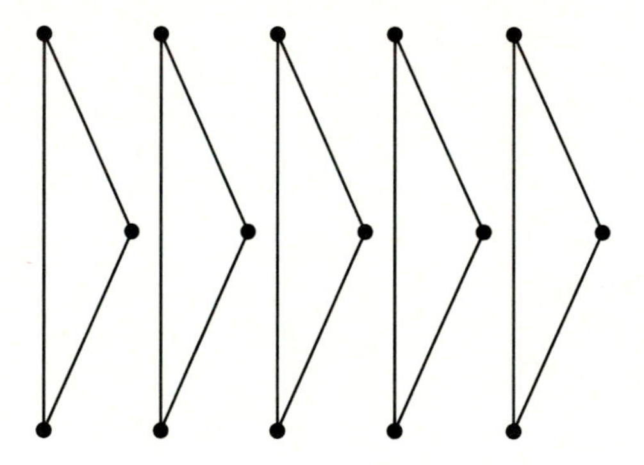

Expanding K_5 to ten vertices means taking the union with the empty graph on five vertices.

```
In[3]:= ConnectedComponents[ ExpandGraph[K[5],10] ]
Out[3]= {{1, 2, 3, 4, 5}, {6}, {7}, {8}, {9}, {10}}
```

The *intersection* of two graphs is formed by taking the intersection of the vertices and edges of the graphs. The labelings of the vertices are implicitly specified by position in the adjacency matrix, and thus the order of the two graphs must be identical.

```
GraphIntersection[g_Graph,h_Graph] :=
        FromOrderedPairs[
                Intersection[ToOrderedPairs[g],ToOrderedPairs[h]],
                Vertices[g]
        ] /; (V[g] == V[h])
```

The Intersection of Two Graphs

Intersection with K_n is an identity operation for any graph of order n.

```
In[4]:= IdenticalQ[ GraphIntersection[Wheel[10],K[10]],
Wheel[10]]
Out[4]= True
```

4.1.2 Sum and Difference

Since graphs are represented by adjacency matrices, they can be added, subtracted, and multiplied in a meaningful way. The multiplication of adjacency matrices, not to be confused with graph products or powers, will be discussed in Section 6.2.3.

```
GraphDifference[g1_Graph,g2_Graph] :=
        Graph[Edges[g1] - Edges[g2], Vertices[g1]] /; V[g1]==V[g2]

GraphSum[g1_Graph,g2_Graph] :=
        Graph[Edges[g1] + Edges[g2], Vertices[g1]] /; V[g1]==V[g2]
```

Finding the Sum and Difference of Two Graphs

The sum of a graph and its complement gives the complete graph.

```
In[5]:= CompleteQ[ GraphSum[ Cycle[10],
GraphComplement[Cycle[10]] ] ]
Out[5]= True
```

The difference of a graph and itself gives the empty graph.

```
In[6]:= EmptyQ[ GraphDifference[Cycle[10],Cycle[10]] ]
Out[6]= True
```

4.1.3 Joins of Graphs

The *join* of two graphs is their union with the addition of edges between all pairs of vertices from different graphs. Many of the graphs we have seen as examples and will implement in this chapter can be specified as the join of two graphs, such as complete bipartite graphs, stars, and wheels.

The *Cartesian product* of two sets is the set of element pairs such that one element is from each set. The edges which get added to the union of the two graphs are exactly the Cartesian product of the two vertex sets.

```
GraphJoin[g_Graph,h_Graph] :=
        Block[{maxg=Max[ Abs[ Map[First,Vertices[g]] ] ]},
                FromUnorderedPairs[
                        Join[
                                ToUnorderedPairs[g],
                                ToUnorderedPairs[h] + V[g],
                                CartesianProduct[Range[V[g]],Range[V[h]]+V[g]]
                        ],
                        Join[ Vertices[g], Map[({maxg+1,0}+#)&, Vertices[h]]]
                ]
        ]

CartesianProduct[a_List,b_List] :=
        Block[{i,j},
                Flatten[ Table[{a[[i]],b[[j]]},{i,Length[a]},{j,Length[b]}], 1]
        ]
```

The Join of Two Graphs

Complete k-partite graphs such as $K_{5,5}$ are naturally described in terms of GraphJoin.

```
In[7]:= ShowGraph[ GraphJoin[ EmptyGraph[5], EmptyGraph[5]
] ];
```

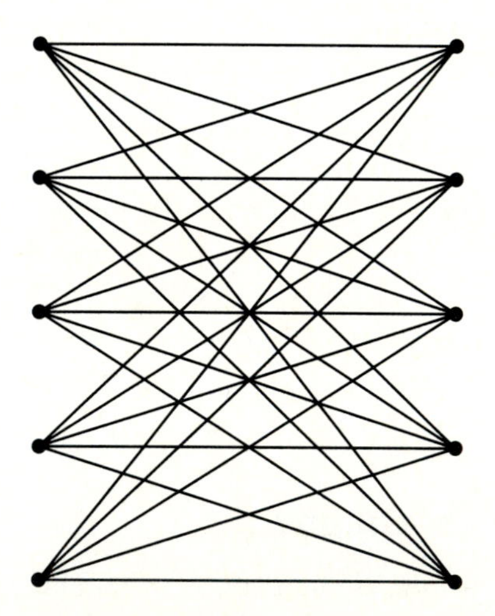

A wheel is the join of a cycle and a single vertex. A star is the join of an empty graph and a single vertex.

```
In[8]:= IsomorphicQ[Wheel[10], GraphJoin[Cycle[9],K[1]]]
Out[8]= True
```

4.1.4 Products of Graphs

The *product* of two graphs $G_1 \times G_2$ has a vertex set defined by the Cartesian product of the vertex sets of G_1 and G_2. There is an edge between (u_1, v_1) and (u_2, v_2) if $u_1 = u_2$ and v_1 is adjacent to v_2 in G_2 or $v_1 = v_2$ and u_1 is adjacent to u_2 in G_1. The intuition in taking the product is that all the vertices of one graph get replaced by instances of the other graph. Products of graphs were first studied in [Sab60].

This implementation uses a neat trick to construct the edges within each copy of the second graph. By numbering the vertices in each copy sequentially, it is sufficient to add the appropriate offset to each ordered pair of the original.

```
GraphProduct[g_Graph,h_Graph] :=
        Block[{k,eg=ToOrderedPairs[g],eh=ToOrderedPairs[h],leng=V[g],lenh=V[h]},
                FromOrderedPairs[
                        Flatten[
                                Join[
                                        Table[eg+(i-1)*leng, {i,lenh}],
                                        Map[ (Table[
                                                {leng*(#[[1]]-1)+k, leng*(#[[2]]-1)+k},
                                                {k,1,leng}
                                              ])&,
                                              eh
                                        ]
                                ],
                                1
                        ],
                        ProductVertices[Vertices[g],Vertices[h]]
                ]
        ]

ProductVertices[vg_,vh_] :=
        Flatten[
                Map[
                        (TranslateVertices[
                                DilateVertices[vg, 1/(Max[Length[vg],Length[vh]])],
                        #])&,
                         RotateVertices[vh,Pi/2]
                ],
                1
        ]
```

Compute the Product of Two Graphs

Graph products can be very interesting. The embedding of a product has been designed to show off its structure, and is formed by shrinking the first graph and translating it to the position of each vertex in the second graph.

```
In[9]:= ShowGraph[ GraphProduct[ K[3], K[5] ] ];
```

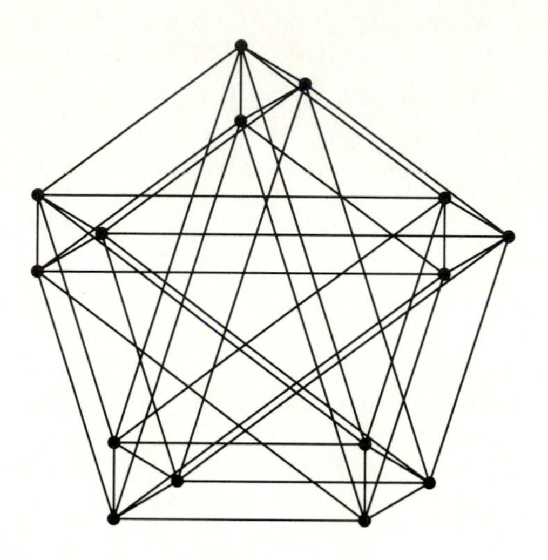

Reversing the order of the two arguments thus dramatically changes the appearance of the product, but the resultant graphs are isomorphic.

```
In[10]:= ShowGraph[ GraphProduct[ K[5], K[3] ] ];
```

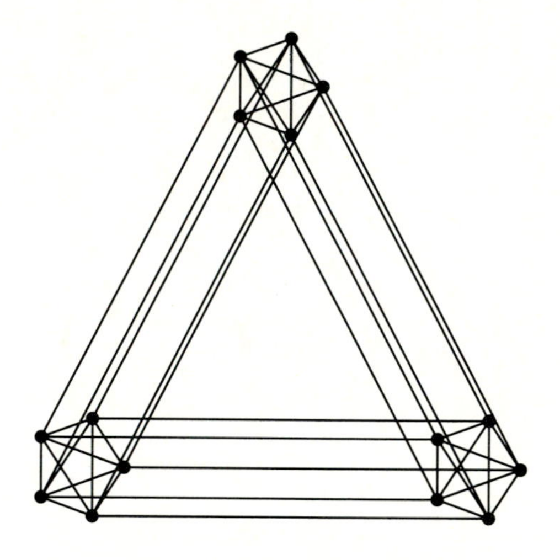

Since both graphs are always isomorphic, the product of two graphs is a commutative operation, even if our embeddings are not!

```
In[11]:= IsomorphicQ[ GraphProduct[K[2],K[3]],
GraphProduct[K[3],K[2]] ]

Out[11]= True
```

Multiplication by K_1 is an identity operation, although there is no corresponding multiplicative inverse.

```
In[12]:= ShowGraph[ GraphProduct[K[1],K[5]] ];
```

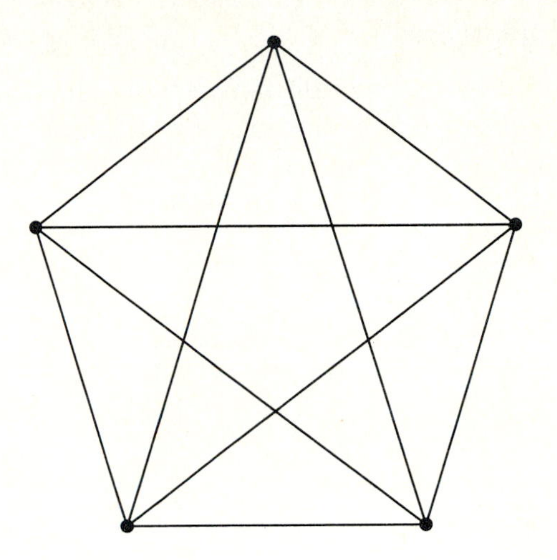

4.1.5 Line Graphs

The *line graph* $L(G)$ of a graph G has a vertex of $L(G)$ associated with each edge of G and an edge of $L(G)$ if and only if the two edges of G share a common vertex. Line graphs are a special type of intersection graph, where each vertex represents a set of size two and each edge connects two sets with a non-empty intersection.

The obvious algorithm for constructing the line graph involves iterating through each pair of edges to decide whether they share a vertex between them, but a better method involves a different graph representation.

```
IncidenceMatrix[g_Graph] :=
        Map[
                ( Join[
                        Table[0,{First[#]-1}], {1},
                        Table[0,{Last[#]-First[#]-1}], {1},
                        Table[0,{V[g]-Last[#]}]
                ] )&,
                ToUnorderedPairs[g]
        ]
```

Constructing the Incidence Matrix of a Graph

The *incidence matrix* of a graph is an $n \times m$ 0-1 matrix with entry $\{v, e\} = 1$ if and only if vertex v is incident on edge e. The incidence matrix was defined by Kirchhoff [Kir47] in the context of a problem on spanning trees. Incidentally, Kirchhoff was a resident of Königsberg [Bol79], the most important city in the history of graph theory.

There is a close connection between the incidence matrix of a graph $I(G)$ and the adjacency matrix of the line graph $L(G)$. Specifically, $L(G) = I(G)^T I(G) - 2I_m$, where I_m is the identity matrix of size m, the number of edges in G.

```
LineGraph[g_Graph] :=
        Block[{b=IncidenceMatrix[g], edges=ToUnorderedPairs[g], v=Vertices[g]},
                Graph[
                        b . Transpose[b] - 2 IdentityMatrix[Length[edges]],
                        Map[ ( (v[[ #[[1]] ]] + v[[ #[[2]] ]]) / 2 )&, edges]
                ]
        ]
```

Constructing Line Graphs

In this incidence matrix, each row represents an edge and each column a vertex. Each row contains two non-zero entries, and each column has four in this case, since K_5 is a regular graph.

```
In[13]:= TableForm[ IncidenceMatrix[K[5]] ]

                     1  1  0  0  0
                     1  0  1  0  0
                     1  0  0  1  0
                     1  0  0  0  1
                     0  1  1  0  0
                     0  1  0  1  0
                     0  1  0  0  1
                     0  0  1  1  0
                     0  0  1  0  1
Out[13]//TableForm=  0  0  0  1  1
```

The line graph of a graph with n vertices and m edges contains m vertices and $\frac{1}{2}\sum_{i=1}^{n} d_i^2 - m$ edges. Thus the number of vertices of the line graph of K_n grows quadratically in n. Proofs of this and most of the results we cite on line graphs appear in [Har69].

```
In[14]:= ShowGraph[ LineGraph[K[5]] ];
```

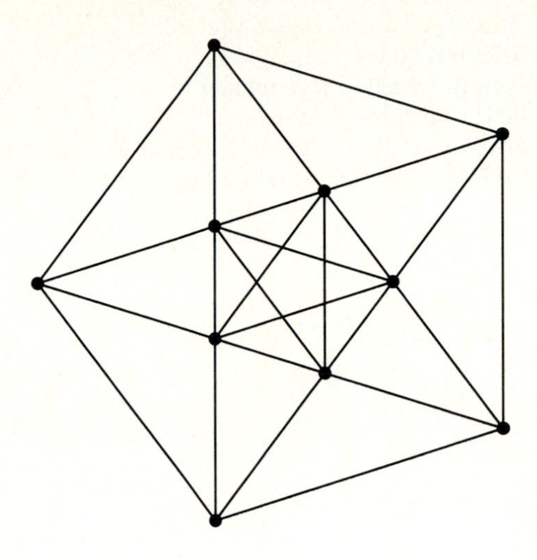

Although some of the edges are ambiguous in the previous embedding, $L(K_5)$ is a 6-regular graph. The coordinates of each vertex in this embedding of $L(G)$ is the average of the coordinates of the vertices associated with the original edge in G.

```
In[15]:= DegreeSequence[ LineGraph[K[5]] ]
Out[15]= {6, 6, 6, 6, 6, 6, 6, 6, 6, 6}
```

The only connected graph which is isomorphic to its line graph is the cycle.

```
In[16]:= ShowGraph[ LineGraph[Cycle[10]] ];
```

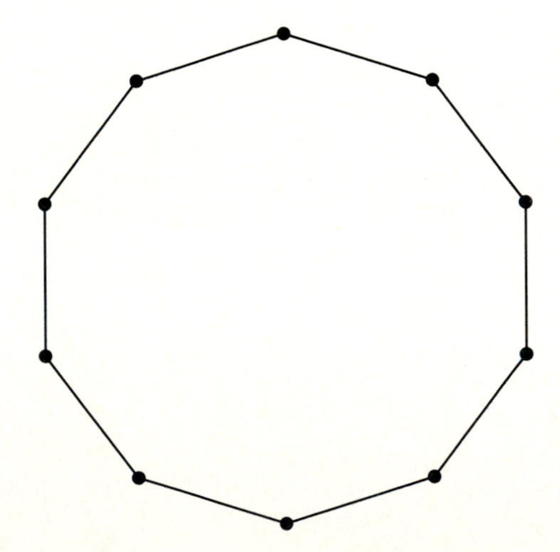

No analogous `FromLineGraph` function is given, because not all graphs represent line graphs of other graphs. An example is the claw $K_{1,3}$. A structural characterization of line graphs is given in [vRW65] and refined by Beineke [Bei68], who showed that a graph is a line graph if and only if it does not contain any of these graphs as induced subgraphs.

```
In[17]:= ShowGraph[ ReadGraph["graphs/nonlinegraphs"] ];
```

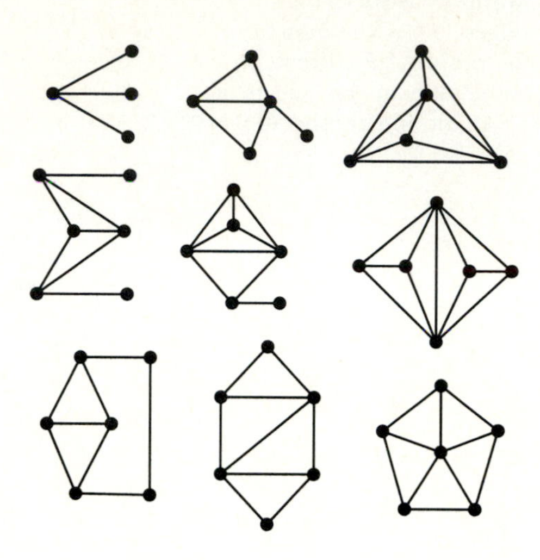

Whitney, in introducing line graphs [Whi32], also showed that with the exception of K_3 and $K_{1,3}$, any two connected graphs with isomorphic line graphs are isomorphic.

```
In[18]:= IsomorphicQ[ LineGraph[K[3]], LineGraph[K[1,3]] ]
Out[18]= True
```

The line graph of an Eulerian graph is both Eulerian and Hamiltonian, while the line graph of a Hamiltonian graph is always Hamiltonian. Further results on cycles in line graphs appear in [HN65, Cha68].

```
In[19]:= EulerianQ[ LineGraph[
RealizeDegreeSequence[{4,4,2,2,2,2}] ] ]
Out[19]= True
```

The complement of the line graph of K_5 is the Petersen graph. This is another example of an attractive but potentially misleading embedding. K_5 has ten edges, so why does this graph have *eleven* vertices? Look carefully, for the centermost "vertex" is just the intersection of five edges. An alternate embedding of the Petersen graph appears in Section 5.3.2.

```
In[20]:= ShowGraph[ GraphComplement[ LineGraph[K[5]] ] ];
```

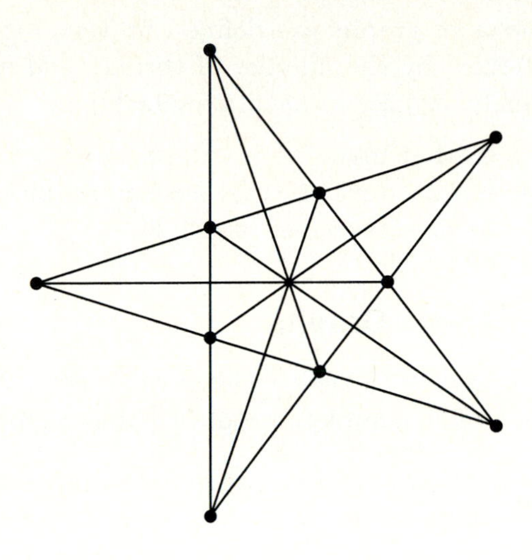

4.2 Regular Structures

Many classes of graphs are defined by very regular structures. They are usually parameterized by the number of vertices and occasionally edges. This regular structure usually suggests a natural embedding.

We have seen that many regular structures can be formulated using such operations as join and product. In this section, we give more efficient, special purpose constructions for several classes of graphs.

4.2.1 Complete Graphs

A binary relation where each element is related to every other element defines a *complete graph*. The complete graph K_n thus contains $\binom{n}{2}$ undirected edges.

Since graph structures are represented by adjacency matrices, it is more efficient to construct matrices directly than to specify ordered pairs and convert. For complete graphs, the matrix structure is especially convenient to work with. The ith row of the adjacency matrix consists of all ones except for a zero in the ith position. Thus once we construct the first row of the matrix, the rest follow by simply rotating the row i positions. Rotating an existing vector is faster than constructing a new one from scratch.

```
K[0] := Graph[{},{}]
K[1] := Graph[{{0}},{{0,0}}]

K[n_Integer?Positive] := CirculantGraph[n,Range[1,Floor[(n+1)/2]]]

CirculantGraph[n_Integer?Positive,l_List] :=
        Block[{i,r},
                r = Prepend[MapAt[1&,Table[0,{n-1}], Map[List,Join[l,n-l]]], 0];
                Graph[ Table[RotateRight[r,i], {i,0,n-1}], CircularVertices[n] ]
        ]

EmptyGraph[n_Integer?Positive] :=
        Block[{i},
                Graph[ Table[0,{n},{n}], Table[{0,i},{i,(1-n)/2,(n-1)/2}] ]
        ]
```

Constructing Circulant Graphs

A *circulant* graph $C_n(n_1, n_2, ..., n_k)$ is the graph on n vertices with each v_i adjacent to each vertex $v_{i \pm n_j}$ mod n, where $n_1 < n_2 < \cdots < n_k < (n+1)/2$ [BH90]. Circulants are graphs whose adjacency matrix can be constructed by rotating a vector n times, and include complete graphs and cycles as special cases.

The complete graph on five vertices K_5 is famous as being the smallest non-planar graph. Rotated appropriately, it becomes a supposed Satanic symbol, the pentagram.

```
In[21]:= ShowGraph[ RotateVertices[ K[5], 11 Pi/10] ];
```

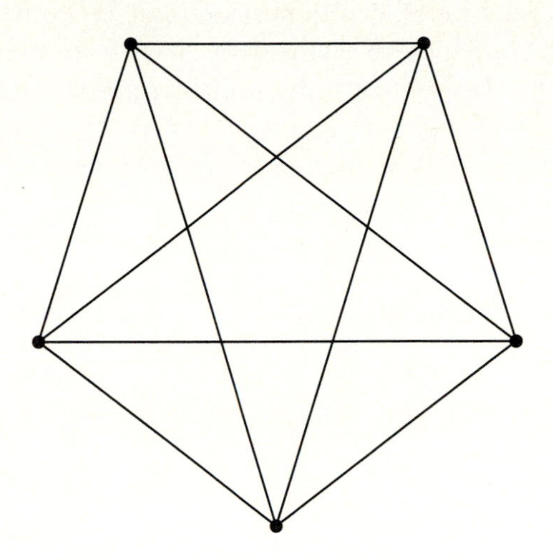

The graph with no vertices and edges is called the *null graph* and has been extensively studied [HR73]. A generalization of the null graph is the *empty graph* on n vertices, the complement of K_n.

```
In[22]:= K[0]
Out[22]= Graph[{}, {}]
```

Even random circulant graphs have an interesting, regular structure.

```
In[23]:= ShowGraph[
CirculantGraph[21,RandomKSubset[Range[10],3]] ];
```

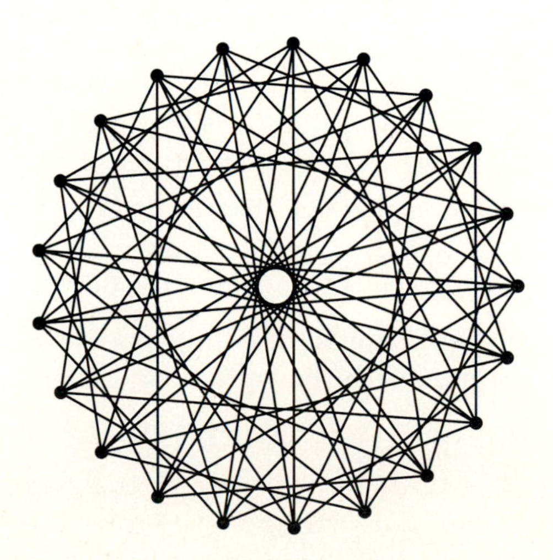

4.2.2 Complete k-Partite Graphs

A graph is *bipartite* when its vertices can be partitioned into sets A and B such that no edge connects an element of A with an element of B. This notion can be generalized into *k-partite* graphs, which contain k stages such that no two vertices of the same stage are related. A *complete* k-partite graph contains k stages and relates all vertices which are not members of the same stage. Thus they are parameterized by the number of vertices per stage.

```
K[l__] :=
        Block[{ll=List[l],t,i,x,row,stages=Length[List[l]]},
                t = Accumulate[Plus,List[0,l]];
                Graph[
                        Apply[
                                Join,
                                Table [
                                        row = Join[
                                                Table[1, {t[[i-1]]}],
                                                Table[0, {t[[i]]-t[[i-1]]}],
                                                Table[1, {t[[stages+1]]-t[[i]]}]
                                        ];
                                        Table[row, {ll[[i-1]]}],
                                        {i,2,stages+1}
                                ]

                        ],
                        Apply [
                                Join,
                                Table[
                                        Table[{x,i-1+(1-ll[[x]])/2},{i,ll[[x]]}],
                                        {x,stages}
                                ]
                        ]
                ]
        ] /; TrueQ[Apply[And, Map[Positive,List[l]]]] && (Length[List[l]]>1)
```

Constructing Complete k-Partite Graphs

The most obvious way to draw a k-partite graph partitions the vertices into equally spaced stages, with the vertices of each stage drawn in a line.

The most famous bipartite graph is $K_{3,3}$, the "other" smallest non-planar graph, although $K_{18,18}$ plays an important role in the novel *Foucault's Pendulum* [Eco89]. A ranked embedding shows the structure of the graph since all edges are between vertices of different ranks.

```
In[24]:= ShowGraph[ K[3,3] ];
```

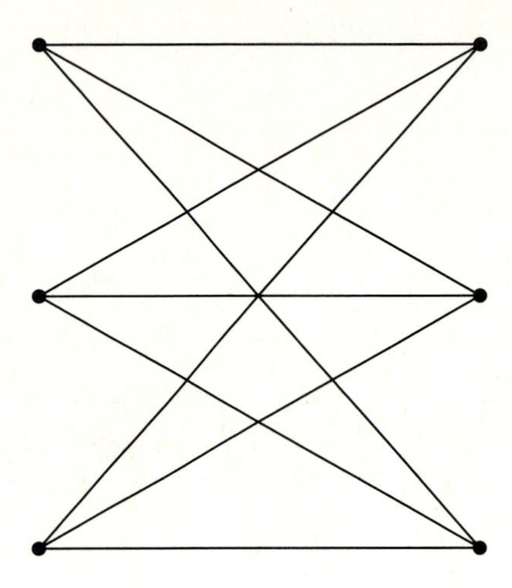

We can construct complete k-partite graphs for any k. These graphs can get very dense when there are many stages, and in fact the complete graph K_n can be defined as $K_{1,1,\ldots,1}$. Whenever three vertices are collinear, there might be edges which overlap. This is unfortunate but costly to prevent.

```
In[25]:= ShowGraph[ K[2,2,2,2] ];
```

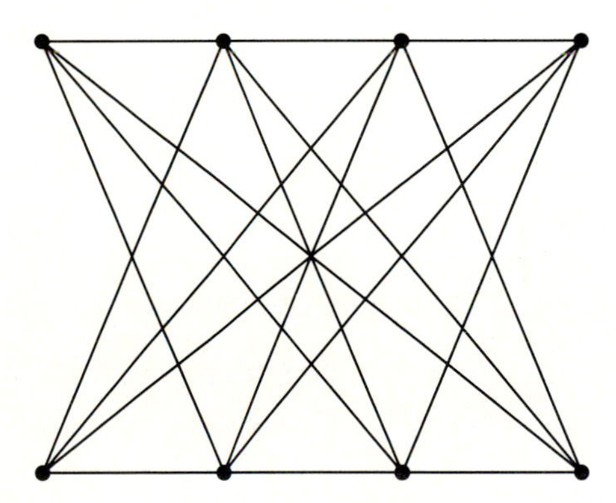

A special case of a complete k-partite graph is the *Turán graph*, which provided the answer to the first problem in what is now known as extremal graph theory [Tur41]. An *extremal graph* $Ex(n, G)$ is the largest graph of order n which does not contain G as a subgraph. Turán was interested in finding the extremal graph which does not contain K_p as a subgraph. Since Turán's paper, extremal graph theory has grown to have a very large literature [Bol78].

```
Turan[n_Integer,p_Integer] :=
        Block[{k = Floor[ n / (p-1) ], r},
                r = n - k (p-1);
                Apply[K, Join[ Table[k,{p-1-r}], Table[k+1,{r}] ] ]
        ] /; (n > 0 && p > 1)
```

Constructing the Turán Graph

Since the Turán graph is $(p-1)$-partite, it cannot contain K_p as a subgraph. The idea behind the construction is to balance the size of each stage as evenly as possible, maximizing the number of edges between every pair of stages.

In[26]:= **ShowGraph[Turan[10,4]];**

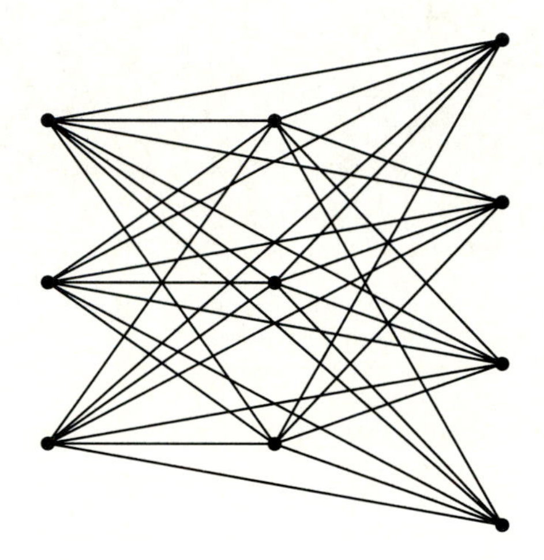

■ 4.2.3 Cycles, Stars, and Wheels

The *cycle* C_n is a connected 2-regular graph of order n. Special cases of interest include the *triangle* K_3 and the *square* C_4. Cycles are a special case of circulant graphs.

```
Cycle[n_Integer] := CirculantGraph[n,{1}]  /; n>=3
```

Constructing a Cycle

Any two-regular connected graph is a cycle. Circular embeddings are the obvious way to represent circulant graphs.

```
In[27]:= ShowGraph[ Cycle[20] ];
```

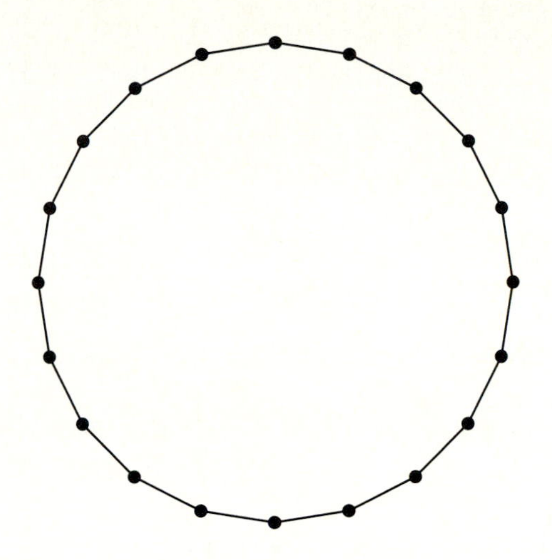

The minimum number of swaps between vertices in a random circular embedding of a cycle to get it into its proper configuration is an interesting combinatorial problem. This is related to the *Bruhat order* of a group [BW82, Sta86].

```
In[28]:= ShowGraph[ Graph[ Edges[Cycle[20]],
Permute[Vertices[Cycle[20]], RandomPermutation[20]] ] ];
```

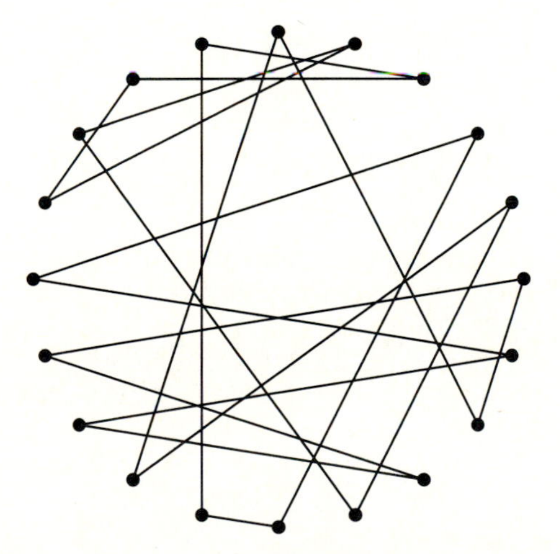

A *star* is a tree with one vertex of degree $n-1$. Since a star is a tree, it is acyclic, but adding any new edge creates a cycle of length three. An interesting property of the star is that the distance between any two vertices is at most two, despite the fact that it contains the minimum number of edges to be connected. This makes it a good topology for computer networks because it minimizes communication time between nodes, at least until the center node goes down.

```
Star[n_Integer?Positive] :=
        Block[{g},
                g = Append [ Table[0,{n-1},{n}], Append[ Table[1,{n-1}], 0] ];
                Graph[
                        g + Transpose[g],
                        Append[ CircularVertices[n-1], {0,0}]
                ]
        ]
```

Constructing a Star

In this construction, the center of the star is the nth vertex.

The complete bipartite graph $K_{1,n-1}$ is a star on n vertices.

```
In[29]:= Isomorphism[ K[1,5], Star[6] ]
Out[29]= {6, 1, 2, 3, 4, 5}
```

A *wheel* is the graph on n vertices obtained by joining a single isolated vertex K_1 to each vertex of a cycle C_{n-1}. The resulting edges which form the star are, logically enough, called the *spokes* of the wheel.

```
Wheel[n_Integer] :=
        Block[{i,row = Join[{0,1}, Table[0,{n-4}], {1}]},
                Graph[
                        Append[
                                Table[ Append[RotateRight[row,i-1],1], {i,n-1}],
                                Append[ Table[1,{n-1}], 0]
                        ],
                        Append[ CircularVertices[n-1], {0,0} ]
                ]
        ] /; n >= 3
```

Constructing a Wheel

The *dual* of a planar embedding of a graph G is a graph with a vertex for each region of G, with edges if the corresponding regions are adjacent. The dual graph of a wheel is a wheel.

```
In[30]:= ShowGraph[ Wheel[20] ];
```

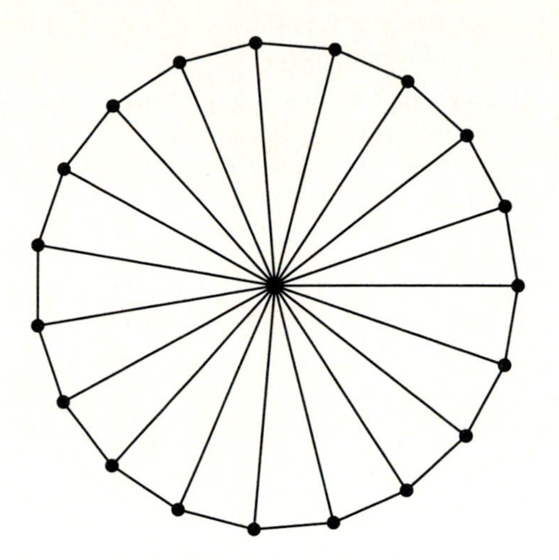

■ 4.2.4 Grid Graphs

Anyone who has used a piece of graph paper is familiar with what a *grid graph* should look like. Each edge corresponds to an instance of the smallest distance in an $m \times n$ integer lattice [RS89]. As discussed in Section 6.1.3, it is a classic combinatorial exercise to count the number of distinct paths between two points in a grid graph.

An $m \times n$ grid graph is the product of paths of length m and n. The structure of paths makes them easy to construct.

```
Path[1] := K[1]
Path[n_Integer?Positive] :=
        FromUnorderedPairs[ Partition[Range[n],2,1], Map[({#,0})&,Range[n]] ]

GridGraph[n_Integer?Positive,m_Integer?Positive] :=
        GraphProduct[
                ChangeVertices[Path[n], Map[({Max[n,m]*#,0})&,Range[n]]],
                Path[m]
        ]
```

Constructing a Grid Graph

The graph consisting of a simple path on n vertices is the fundamental building block of grid graphs, and yet the simple path is also the special case GridGraph[n,1].

```
In[31]:= Isomorphism[Path[10], GridGraph[10,1]]
Out[31]= {1, 2, 3, 4, 5, 6, 7, 8, 9, 10}
```

Grid graphs are Hamiltonian if either the number of rows or columns is even. Such a cycle can be constructed by starting in the lower left-hand corner, going all way to the right, and then zig-zagging up and down until reaching the original corner.

```
In[32]:= ShowLabeledGraph[ GridGraph[4,5] ];
```

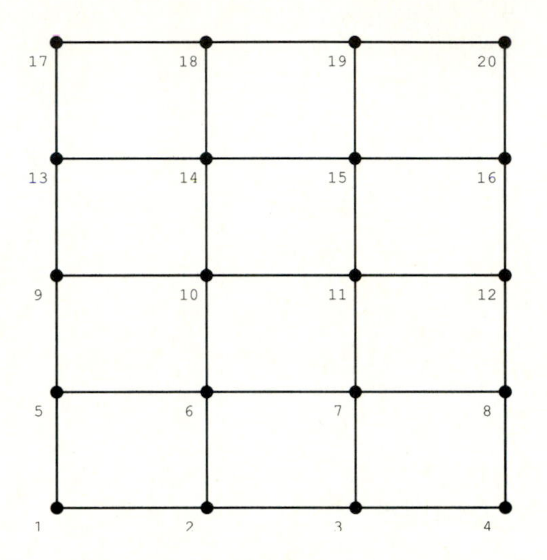

This Hamiltonian cycle doesn't follow this pattern, thus demonstrating there can be distinct Hamiltonian cycles of a graph.

```
In[33]:= HamiltonianCycle[ GridGraph[4,5] ]
Out[33]= {1, 2, 3, 4, 8, 7, 6, 10, 11, 12, 16, 20, 19,
  15, 14, 18, 17, 13, 9, 5, 1}
```

Grid graphs are bipartite, for the vertices can be partitioned like the squares on a chessboard.

```
In[34]:= TwoColoring[ GridGraph[4,5] ]
Out[34]= {1, 2, 1, 2, 2, 1, 2, 1, 1, 2, 1, 2, 2, 1, 2, 1,
  1, 2, 1, 2}
```

■ 4.2.5 Hypercubes

A n-dimensional *hypercube* or *n-cube* is defined as the product of K_2 and an $n-1$ dimensional hypercube, and this recursive definition can be used to construct one.

```
Hypercube[0] := K[1]
Hypercube[1] := Path[2]
Hypercube[2] := Cycle[4]

Hypercube[n_Integer] := Hypercube[n] =
	GraphProduct[
```

```
            RotateVertices[ Hypercube[Floor[n/2]], 2Pi/5],
            Hypercube[Ceiling[n/2]]
        ]
```

Constructing Hypercubes

All hypercubes are Hamiltonian, and this was exploited in Section 1.5.3 to generate the binary reflected Gray code. A Gray code minimizes the number of changes between successive subsets to one element at a time, which corresponds to changing one bit between two binary strings of length n. By labeling the two vertices of `Hypercube[1]` 0 and 1, and appending 0 and 1 to the vertex labels of the different copies of `Hypercube[i-1]` which form `Hypercube[i]`, each vertex will be labeled with a string of length n. The labels of adjacent vertices in the hypercube differ by exactly one bit, so any Hamiltonian cycle of the labeled hypercube defines a Gray code.

Hypercubes represent an important topology for parallel computers, since any processor can communicate with any other processor in $O(\log n)$ time using only $O(n \log n)$ wires.

The three-dimensional hypercube is what we usually think of as a cube, while a square is a two-dimensional hypercube. The maximum number of vertices of a 3-cube which induce a cycle is six, namely $(1, 2, 4, 6, 5, 7)$ [DK67].

In[35]:= `ShowLabeledGraph[ Hypercube[3] ];`

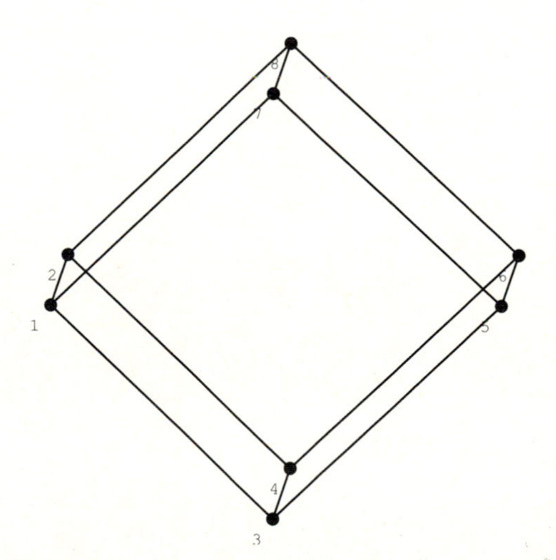

A four-dimensional hypercube is perhaps the easiest four-dimensional object to visualize. Two three-dimensional cubes are connected in a symmetric way.

```
In[36]:= ShowGraph[ Hypercube[4] ];
```

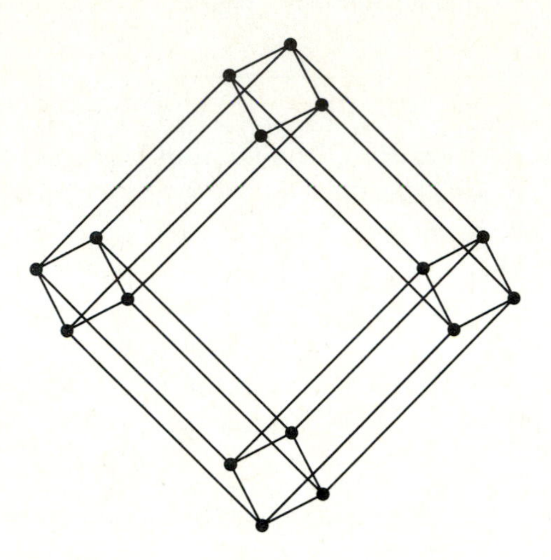

Even in five dimensions, the recursive structure of the hypercube is still apparent. There is a three-dimensional cube at each corner of the square.

```
In[37]:= ShowGraph[ Hypercube[5] ];
```

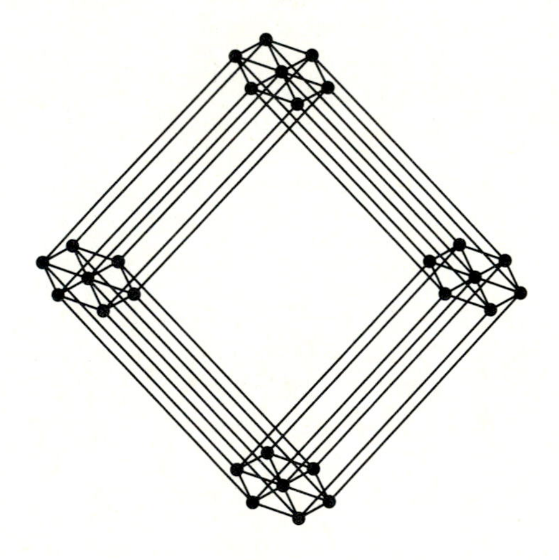

All n-cubes are bipartite, as can be seen from the recursive construction.

```
In[38]:= TwoColoring[ Hypercube[4] ]
Out[38]= {1, 2, 1, 2, 2, 1, 2, 1, 1, 2, 1, 2, 2, 1, 2, 1}
```

4.3 Trees

A *tree* is a connected graph with no cycles. Trees are the simplest interesting class of graphs, so "Can you prove it for trees?" should be the first question asked after formulating a new conjecture.

One of the first theorems in graphical enumeration was Cayley's proof [Cay89] that there are n^{n-2} distinct labeled trees on n vertices. Prüfer [Pru18] established a bijection between such trees and strings of $n-2$ integers between 1 and n, providing a constructive proof of Cayley's result. This bijection can then be exploited to give algorithms for systematically and randomly generating labeled trees.

The key to Prüfer's bijection is the observation that for any tree there are always at least two vertices of degree one. Thus in a labeled tree T, the vertex v incident to the lowest labeled such vertex is well defined, and v is the first symbol in our string, or Prüfer code. After deleting this edge we have a tree on $n-1$ vertices, and repeating this operation until only one edge is left gives us $n-2$ integers between 1 and n.

```
LabeledTreeToCode[g_Graph] :=
        Block[{e=ToAdjacencyLists[g],i,code},
                Table [
                        {i} = First[ Position[ Map[Length,e], 1 ] ];
                        code = e[[i,1]];
                        e[[code]] = Complement[ e[[code]], {i} ];
                        e[[i]] = {};
                        code,
                        {V[g]-2}
                ]
        ]
```

Constructing a Prüfer Code from a Labeled Tree

To reconstruct T from its Prüfer code, we observe that a particular vertex appears in the code exactly one time less than the degree of the vertex in T. Thus we can compute the degree sequence of T, and thereby identify the lowest labeled degree-one vertex in the tree. Since the first symbol in the code is the vertex it is incident upon, we have determined the first edge and, by induction, the entire tree.

```
CodeToLabeledTree[l_List] :=
        Block[{m=Range[Length[l]+2],x,i},
                FromUnorderedPairs[
                        Append[
                                Table[
                                        x = Min[Complement[m,Drop[l,i-1]]];
                                        m = Complement[m,{x}];
```

```
                                        {x,l[[i]]},
                                        {i,Length[l]}
                                 ],
                                 m
                          ]
                   ]
            ]

RandomTree[n_Integer?Positive] :=
       RadialEmbedding[CodeToLabeledTree[ Table[Random[Integer,{1,n}],{n-2}] ], 1]
```

Constructing a Labeled Tree from its Code

A star contains $n-1$ vertices of degree 1, all incident on one center. Since the degree of a particular vertex is one more than its frequency in the Prüfer code, the ith labeled star is defined by a code of $n-2$ i's.

```
In[39]:= ShowGraph[ RadialEmbedding[
CodeToLabeledTree[{10,10,10,10,10,10,10,10}] ] ];
```

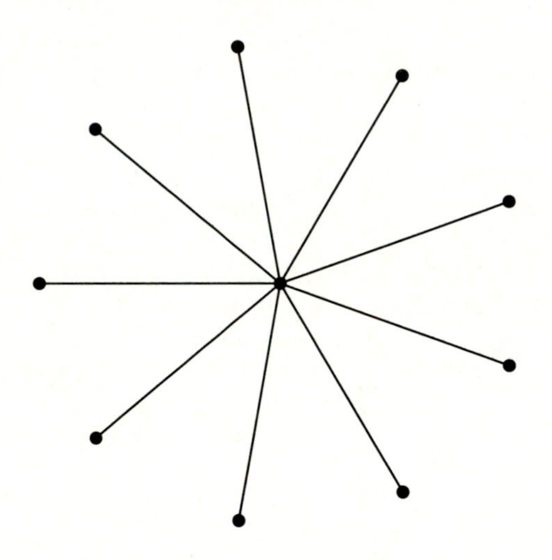

Since there is a bijection between trees and codes, composing the two functions gives an identity operation.

```
In[40]:= LabeledTreeToCode[ CodeToLabeledTree[{3,3,3,2,3}]
]

Out[40]= {3, 3, 3, 2, 3}
```

Each labeled path on n vertices is represented by a pair of permutations of length n, for a total of $n!/2$ distinct labeled paths. The Prüfer codes of paths are exactly the sequences of $n-2$ distinct integers, since the two vertices of degree one do not appear in the code.

```
In[41]:= LabeledTreeToCode[Path[10]]

Out[41]= {2, 3, 4, 5, 6, 7, 8, 9}
```

Cayley's result was the first of many in graphical enumeration [HP73], counting how many distinct graphs of a given type exist. An interesting table of the number of a dozen different types of graphs on up to eight vertices appears in [Wil85b].

With most enumeration problems, counting the number of unlabeled things is harder than counting the number of labeled things. So it is with trees. Algorithms for the systematic generation of free and rooted trees appear in [NW78, Wil89].

4.4 Random Graphs

The easiest way to generate a graph is by tossing a coin for each edge to decide whether it should be included. *Random graph theory* considers questions of the type, "What density of edges in a graph is necessary, on average, to ensure that the graph has monotone graph property X?". A *monotone graph property* is an invariant which is preserved as more edges are added, such as whether the graph is connected or has a cycle.

4.4.1 Constructing Random Graphs

There are two distinct models of random graphs. In the first, a parameter p represents the probability that any given edge is in the graph. When $p = \frac{1}{2}$, this model generates all labeled graphs with equal probability, since the edge probabilities are independent.

```
RandomGraph[n_Integer,p_] := RandomGraph[n,p,{1,1}]

RandomGraph[n_Integer,p_,range_List] :=
        Block[{i,g},
                g = Table[
                        Join[
                                Table[0,{i}],
                                Table[
                                        If[Random[Real]<p, Random[Integer,range], 0],
                                        {n-i}
                                ]
                        ],
                        {i,n}
                ];
                Graph[ g + Transpose[g], CircularVertices[n] ]
        ]
```

Constructing Model 1 Random Graphs

The second model generates random unlabeled graphs with exactly m edges, which is difficult to do both efficiently and correctly, since it implies describing the set of non-isomorphic graphs. Our algorithm produces random *labeled* graphs of the given size, by selecting m integers from 1 to $\binom{n}{2}$ and defining a bijection between such integers and ordered pairs. This does not result in a random unlabeled graph, since there will be n K_{n-1}'s produced for every K_n because of isomorphisms.

```
ExactRandomGraph[n_Integer,e_Integer] :=
        FromUnorderedPairs[
                Map[ NthPair, Take[ RandomPermutation[n(n-1)/2], e] ],
                CircularVertices[n]
        ]

NthPair[0] := {}
NthPair[n_Integer] :=
        Block[{i=2},
                While[ Binomial[i,2] < n, i++];
                {n - Binomial[i-1,2], i}
        ]

RandomVertices[n_Integer] := Table[{Random[], Random[]}, {n}]
RandomVertices[g_Graph] := Graph[ Edges[g], RandomVertices[V[g]] ]
```

Constructing Random Labeled Graphs on m Edges

We exploit a bijection between unordered pairs and the integers with the nice property that all unordered pairs containing n are ranked at most $\binom{n}{2}$, so selecting a random edge is equivalent to selecting a random integer.

```
In[42]:= Table[NthPair[n], {n,1,10}]
Out[42]= {{1, 2}, {1, 3}, {2, 3}, {1, 4}, {2, 4}, {3, 4},
  {1, 5}, {2, 5}, {3, 5}, {4, 5}}
```

This random graph can be expected to have half the number of edges of a complete graph, even though all labeled graphs occur with equal probability.

```
In[43]:= ShowGraph[ RandomGraph[20,0.5] ];
```

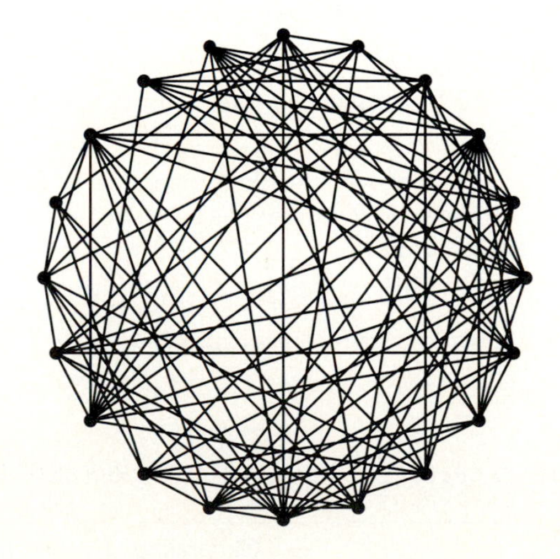

Since there are ten edges in K_5, this generates a random labeled graph with exactly half the number of edges.

```
In[44]:= M[ ExactRandomGraph[5,5] ]
Out[44]= 5
```

The ability to generate random weighted graphs will be important to test algorithms for minimum spanning trees and shortest paths.

```
In[45]:= TableForm[ Edges[ RandomGraph[5,0.5,{1,10}] ] ]
                      0  8  8  9  4
                      8  0  0  7  9
                      8  0  0  0  5
                      9  7  0  0  0
Out[45]//TableForm=   4  9  5  0  0
```

There are six labeled graphs of four vertices and five edges, because any pair of vertices can be the missing edge from the complete graph.

```
In[46]:= Distribution[ Table[ExactRandomGraph[4,5], {200}]
]
Out[46]= {33, 39, 41, 25, 29, 33}
```

There are eight labeled graphs on three vertices. When the edge probability is 1/2, each is equally likely to occur. A table of all unlabeled graphs on up to six vertices, directed graphs up to four vertices, and trees up to nine vertices appears in [Har69].

```
In[47]:= Distribution[ Table[RandomGraph[3,0.5], {200}] ]
Out[47]= {22, 20, 17, 34, 33, 23, 25, 26}
```

The interesting issue in random graph theory is how many edges are needed before the graph can be expected to have some monotone graph property. Random graph theory is discussed in [Bol79, ES74, Pal85]. Through the probabilistic model, it can be proven that almost all graphs are connected and almost all graphs are non-planar.

Random directed graphs can be constructed in a similar manner to random undirected graphs.

```
RandomGraph[n_Integer,p_,range_List,Directed] :=
        RemoveSelfLoops[
                Graph[
                        Table[If[Random[Real]<p,Random[Integer,range],0],{n},{n}],
                        CircularVertices[n]
                ]
        ]

RandomGraph[n_Integer,p_,Directed] := RandomGraph[n,p,{1,1},Directed]
```

Generating Random Directed Graphs

When directed edges are selected with probability p, the expected number of undirected edges in the graph is $n(n-1)/4p$.

```
In[48]:= M[ RandomGraph[10,0.25,Directed], Directed]
Out[48]= 18
```

4.4.2 Realizing Degree Sequences

The *degree sequence* of an undirected graph is the number of edges incident upon each vertex. Perhaps the most elementary theorem in graph theory is that the sum of the degree sequence for any graph must be even, or equivalently that any simple graph has an even number of odd degree vertices. By convention, degree sequences are sorted into decreasing order. The degree of a vertex is sometimes called its *valency*, and the minimum and maximum degrees of graph G are denoted $\delta(G)$ and $\Delta(G)$, respectively.

A degree sequence is *graphic* if there exists a simple graph with that degree sequence. Erdös and Gallai [EG60] proved that a degree sequence is graphic if and only if the sequence observes the following condition for each integer $r < n$:

$$\sum_{i=1}^{r} d_i \leq r(r-1) + \sum_{i=r+1}^{n} \min(r, d_i)$$

The Erdös-Gallai condition also generalizes to directed graphs [Ful65, Rys57].

Alternatively, [Hak62, Hav55] proved that if a degree sequence is graphic, there exists a graph G where the vertex of highest degree is adjacent to the $\Delta(G)$ next highest degree vertices of G. This gives an inductive definition of a graphic degree sequence, as well as a construction of a graph which realizes it.

```
DegreeSequence[g_Graph] := Reverse[ Sort[ Degrees[g] ] ]

Degrees[Graph[g_,_]] := Map[(Apply[Plus,#])&, g]

GraphicQ[s_List] := False /; (Min[s] < 0) || (Max[s] >= Length[s])
GraphicQ[s_List] := (First[s] == 0) /; (Length[s] == 1)
GraphicQ[s_List] :=
        Block[{m,sorted = Reverse[Sort[s]]},
                m = First[sorted];
                GraphicQ[ Join[ Take[sorted,{2,m+1}]-1, Drop[sorted,m+1] ] ]
        ]
```

Identifying Graphic Degree Sequences

The degree sequence shows that K_{30} is a 29-regular graph.

```
In[49]:= DegreeSequence[K[30]]
Out[49]= {29, 29, 29, 29, 29, 29, 29, 29, 29, 29, 29, 29,
  29, 29, 29, 29, 29, 29, 29, 29, 29, 29, 29, 29, 29, 29,
  29, 29, 29, 29}
```

By definition, the degree sequence of any graph is graphic.

```
In[50]:= GraphicQ[DegreeSequence[RandomGraph[10,1/2]]]
Out[50]= True
```

The diagonal of the square of the adjacency matrix of a simple graph gives the unsorted degree sequence of the graph. This is because the square of the adjacency matrix counts the number of paths of length two between any pair of vertices, and the only paths of length two from a vertex to itself are to an adjacent vertex and directly back again.

```
In[51]:= ( g = RealizeDegreeSequence[{3,2,2,1}];
TableForm[Edges[g] . Edges[g]] )
                         3  1  1  0
                         1  2  1  1
                         1  1  2  1
Out[51]//TableForm=      0  1  1  1
```

No degree sequence can be graphic if all the degrees occur with multiplicity one [BC67]. Any sequence of positive integers summing up to an even number can be realized by a multigraph with self-loops [Hak62].

```
In[52]:= GraphicQ[{7,6,5,4,3,2,1}]
Out[52]= False
```

The direct implementation of the inductive definition of graphic degree sequences gives a deterministic algorithm for realizing any such sequence. However, the `GraphicQ` predicate provides a means to construct semi-random graphs of a particular degree sequence. We attempt to connect the vertex of degree Δ to a random set of Δ other vertices. If, after deleting one from the degrees of each of these vertices, the remaining sequence is graphic, then there exists a way to finish off the construction of the graph appropriately. If not, we keep trying other random vertex sets until it does.

```
RealizeDegreeSequence[d_List] :=
        Block[{i,j,v,set,seq,n=Length[d],e},
                seq = Reverse[ Sort[ Table[{d[[i]],i},{i,n}]] ];
                FromUnorderedPairs[
                        Flatten[ Table[
                                {{k,v},seq} = {First[seq],Rest[seq]};
                                While[ !GraphicQ[
                                        MapAt[
                                                (# - 1)&,
                                                Map[First,seq],
                                                set = RandomKSubset[Table[{i},{i,n-j}],k]
                                        ] ],
                                ];
```

```
                                e = Map[(Prepend[seq[[#,2]],v])&,set];
                                seq = Reverse[ Sort[
                                        MapAt[({#[[1]]-1,#[[2]]})&,seq,set]
                                ] ];
                                e,
                                {j,Length[d]-1}
                        ], 1],
                        CircularVertices[n]
                ]
        ] /; GraphicQ[d]

RealizeDegreeSequence[d_List,seed_Integer] :=
        (SeedRandom[seed]; RealizeDegreeSequence[d])
```

Realizing Graphic Degree Sequences

For a deterministic construction, specify a seed of the random number generator as a second argument.

The construction is powerful enough to produce disconnected graphs when there is no other way to realize the degree sequence.

```
In[53]:= ConnectedComponents[
RealizeDegreeSequence[{2,2,2,1,1,1,1}] ]
Out[53]= {{1, 2, 6, 4}, {3, 5, 7}}
```

An important set of degree sequences are defined by *regular graphs*. A graph is regular if all vertices are of equal degree. An algorithm for constructing regular random graphs appears in [Wor84], although here we construct semi-random regular graphs as a special case of RealizeDegreeSequence.

```
RegularQ[Graph[g_,_]] := Apply[ Equal, Map[(Apply[Plus,#])& , g] ]

RegularGraph[k_Integer,n_Integer] := RealizeDegreeSequence[Table[k,{n}]]
```

Constructing and Testing Regular Graphs

Here we construct and test a 4-regular graph on eight vertices. All cycles and complete graphs are regular.

```
In[54]:= RegularQ[ RegularGraph[4,8] ]
Out[54]= True
```

Complete bipartite graphs are regular if and only if the two stages contain equal numbers of vertices.

```
In[55]:= RegularQ[ K[10,9] ]
Out[55]= False
```

The join of two regular graphs is not necessarily regular.

```
In[56]:= DegreeSequence[ GraphJoin[Cycle[4],K[2]] ]
Out[56]= {5, 5, 4, 4, 4, 4}
```

Constructing regular graphs is a special case of realizing arbitrary degree sequences. Here we build a 3-regular graph of order 12.

```
In[57]:= ShowGraph[ RegularGraph[3,12] ];
```

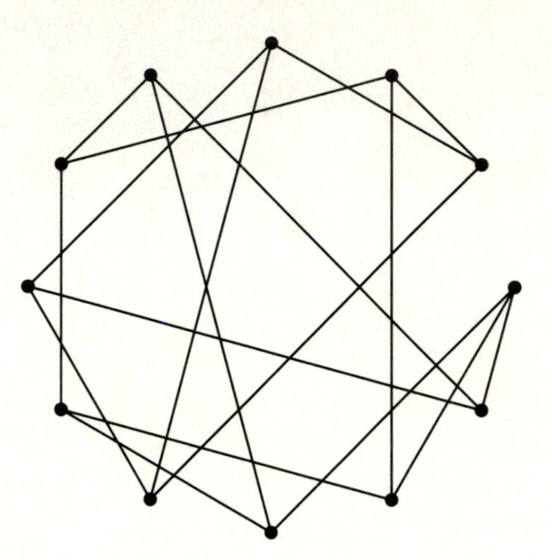

■ 4.5 Relations and Functional Graphs

The reason graphs are so important for modeling structures is that they are representations of binary relations. Graph theory can then be applied to determine properties of the model. Here we provide tools for constructing graphs out of relationships between objects and functions.

■ 4.5.1 Binary Relations

Given a set of objects V, a binary relation is a subset of the Cartesian product $V \times V$. Any boolean predicate which takes two elements of V as arguments defines a binary relation and hence a graph.

```
MakeGraph[v_List,f_] :=
        Block[{n=Length[v],i,j},
                Graph [
                        Table[If [Apply[f,{v[[i]],v[[j]]}], 1, 0],{i,n},{j,n}],
                        CircularVertices[n]
                ]
        ]
```

Making a Graph from a Binary Relation

This binary relation defines the complete graph K_5. `MakeGraph` can be a very powerful tool for gaining insight into binary relations.

```
In[58]:= ShowGraph[ MakeGraph[Range[5], (#1 != #2)&] ];
```

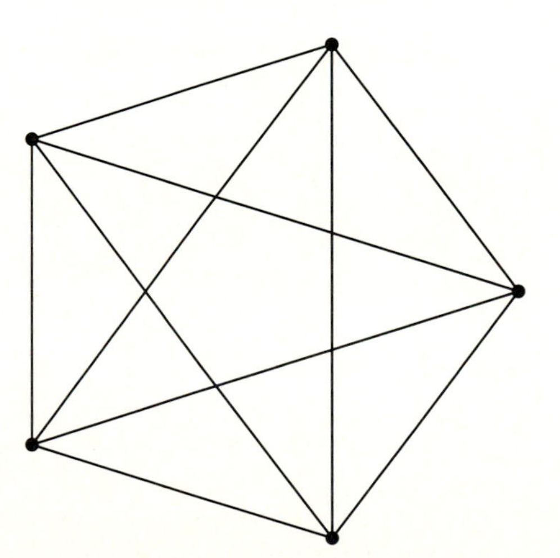

The *odd graphs* O_k [Big74] have vertices corresponding to the $k-1$ subsets of $\{1, ..., 2k-1\}$ where two vertices are connected by an edge iff the associated subsets are disjoint. O_2 gives K_3, and here we prove that O_3 is the Petersen graph.

```
In[59]:= Isomorphism[ ReadGraph["graphs/petersen"],
MakeGraph[KSubsets[Range[5],2],
(SameQ[Intersection[#1,#2], {}])&] ]

Out[59]= {1, 2, 6, 9, 10, 8, 7, 3, 5, 4}
```

Permutations differing by one transposition can be ordered by the number of inversions. The *inversion poset* [SW86] relates permutations p and q if there is a sequence of transpositions from p to q such that each transposition increases the number of inversions. We can begin construction of this graph by linking all permutations which differ by one transposition in the direction of increasing disorder.

```
In[60]:= ShowGraph[ g=MakeGraph[ Permutations[{1,2,3,4}],
(Count[#1-#2,0] == 2 && (Inversions[#1]>Inversions[#2]))&]
];
```

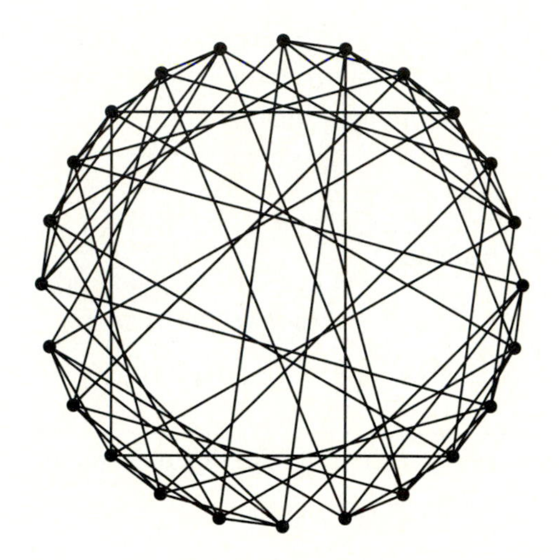

Taking the *transitive closure* of the previous graph adds an edge between any two vertices connected by a path, thus completing the construction of the inversion poset.

```
In[61]:= ShowGraph[ h = TransitiveClosure[g] ];
```

The *Hasse diagram* of a poset (Section 5.4.2) eliminates redundant edges and ranks the vertices according to their position in the defined order. In the inversion poset, the lowest ranked vertex corresponds to the identity permutation and the highest ranked vertex to the reverse of the identity permutation. Each path from bottom to top through this lattice defines a sequence of transpositions which increase the number of inversions. `ShowLabeledGraph` with `Permutations[{1,2,3,4}]` as labels could be used to reveal such sequences.

```
In[62]:= ShowGraph[ HasseDiagram[h] ];
```

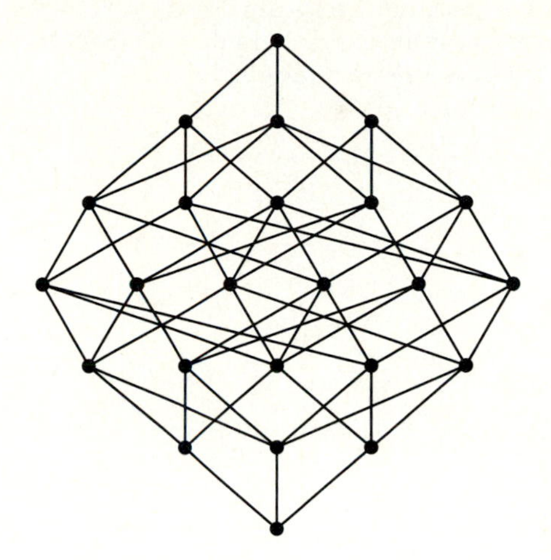

An *interval graph* [BL62, GH64] is defined by a set of intervals on the real line, such that two intervals are related if they overlap. Viewing such graphs as binary relations makes them easy to construct.

```
IntervalGraph[l_List] :=
        MakeGraph[
                l,
                ( ((First[#1] <= First[#2]) && (Last[#1] >= First[#2])) ||
                  ((First[#2] <= First[#1]) && (Last[#2] >= First[#1])) )&
        ]
```

Constructing an Interval Graph

Labeling the vertices of an interval graph appropriately makes it easier to understand. Stars are interval graphs, but cycles are not. Testing if a graph is an interval graph and finding a set of intervals to realize it can be done in linear time [BL76].

```
In[63]:= (l={{1,10},{2,3},{4,5},{6,7},{8,9}};
ShowLabeledGraph[IntervalGraph[l], l];)
```

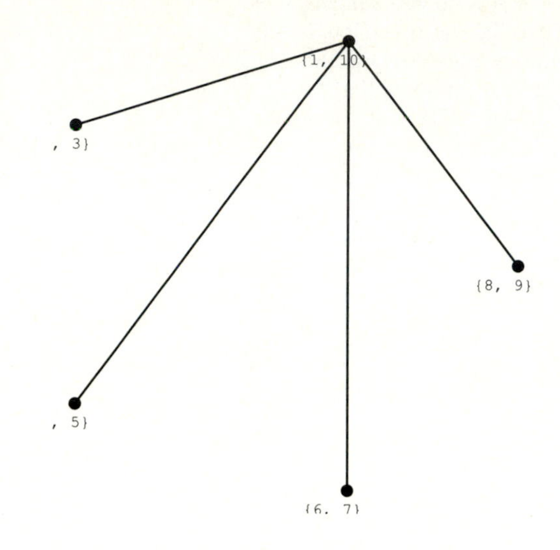

■ 4.5.2 Functional Graphs

In a *functional graph*, each vertex has outdegree one. Thus a functional graph can be constructed with a function that maps each vertex to one other vertex. If the vertices are labeled from 1 to n, this can be made into a function mapping the integers from 1 to n onto themselves.

```
FunctionalGraph[f_,n_] :=
        Block[{i,x},
                FromOrderedPairs[
                        Table[{i, x=Mod[Apply[f,{i}],n]; If[x!=0,x,n]}, {i,n} ],
                        CircularVertices[n]
                ]
        ]
```

Constructing a Functional Graph

An example of a functional graph is $y \equiv x^2 + 2x$ **mod** n. The arrowheads without arrows represent self-loops in the graph, which is about as well as we can do with a straight-line drawing.

```
In[64]:= ShowGraph[ FunctionalGraph[(#^2 + 2*#)&, 20],
Directed];
```

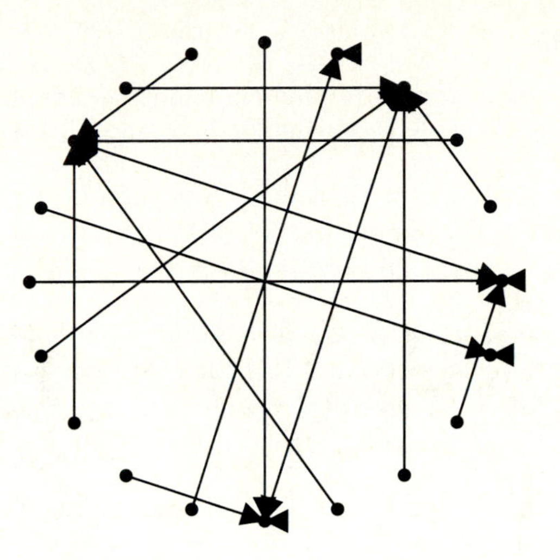

The increment function produces a directed cycle as its functional digraph. Since functional digraphs have exactly n edges, directed cycles are the only strongly connected functional digraphs.

```
In[65]:= ShowGraph[ FunctionalGraph[(#+1)&, 20],
Directed];
```

4.6 Exercises and Research Problems

Exercises

1. Give more direct implementations of `GraphUnion` and `GraphIntersection`. How does their performance compare with the existing versions?

2. Write a function to embed a hypercube in a grid graph, meaning to assign vertices of a hypercube to all the vertices of a grid graph such that the grid graph is contained within an induced subgraph of these hypercube vertices.

3. The *total graph* $T(G)$ of a graph G has a vertex for each edge and vertex of G, and an edge in $T(G)$ for every edge-edge and vertex-edge adjacency in G [CM78]. Thus total graphs are a generalization of line graphs. Write a function to construct the total graph of a graph.

4. Use `MakeGraph` to construct the knight's tour graph for an $n \times n$ chessboard. This graph has a vertex for each square on the chessboard, with an edge connected two squares which are separated by one knight move. Is this graph Hamiltonian for small values of n?

5. Write functions to construct the following graphs from `GraphIntersection`, `GraphJoin`, `GraphUnion`, and `GraphProduct`: complete graphs, complete bipartite graphs, cycles, stars, and wheels.

6. Which of the following graphs are interval graphs: complete graphs, complete bipartite graphs, cycles, stars, or wheels? For those that are, implement functions which use `IntervalGraph` to construct them.

7. What special cases of complete k-partite graphs are circulant graphs? Implement them accordingly.

8. Experiment with `RealizeDegreeSequence` to determine how many random subsets must be selected, in connecting the ith vertex of an order n graph, before finding one which leaves a graphic degree sequence.

9. An alternate non-deterministic `RealizeDegreeSequence` can be based on the result [Egg75, FHM65] that every realization of a given degree sequence can be constructed from any other using a finite number of *edge interchange* operations, where an edge interchange takes a graph with edges (x, y) and (w, z) and replaces them with edges (x, w) and (y, z). Implement a non-deterministic `RealizeDegreeSequence` function based on a sequence of edge-interchange operations.

10. A *degree set* for a graph G is the set of integers which make up the degree sequence. Any set of positive integers is the degree set for some graph. Design and implement an algorithm for constructing a graph which realizes an arbitrary degree set.

Research Problems

1. Assign the vertices of a cycle C_n to a convex set of n points. A *swap* operation exchanges the points associated with two vertices of the graph. How many swaps are required to disentangle the graph and leave a planar embedding of the cycle?

2. What is the probability that a random unlabeled tree has a center of one vertex versus two vertices? For labeled trees, asymptoticially half the trees are central and half bicentral [Sze83].

3. Luiz Goddyn and Tomaž Pisanski ask if there exists a polynomial-time algorithm to test whether the complement of a line graph is Hamiltonian. An equivalent formulation is testing whether there is a cyclic ordering of the edges of a graph in which no two consecutive edges have a vertex in common.

Chapter 5.

Properties of Graphs

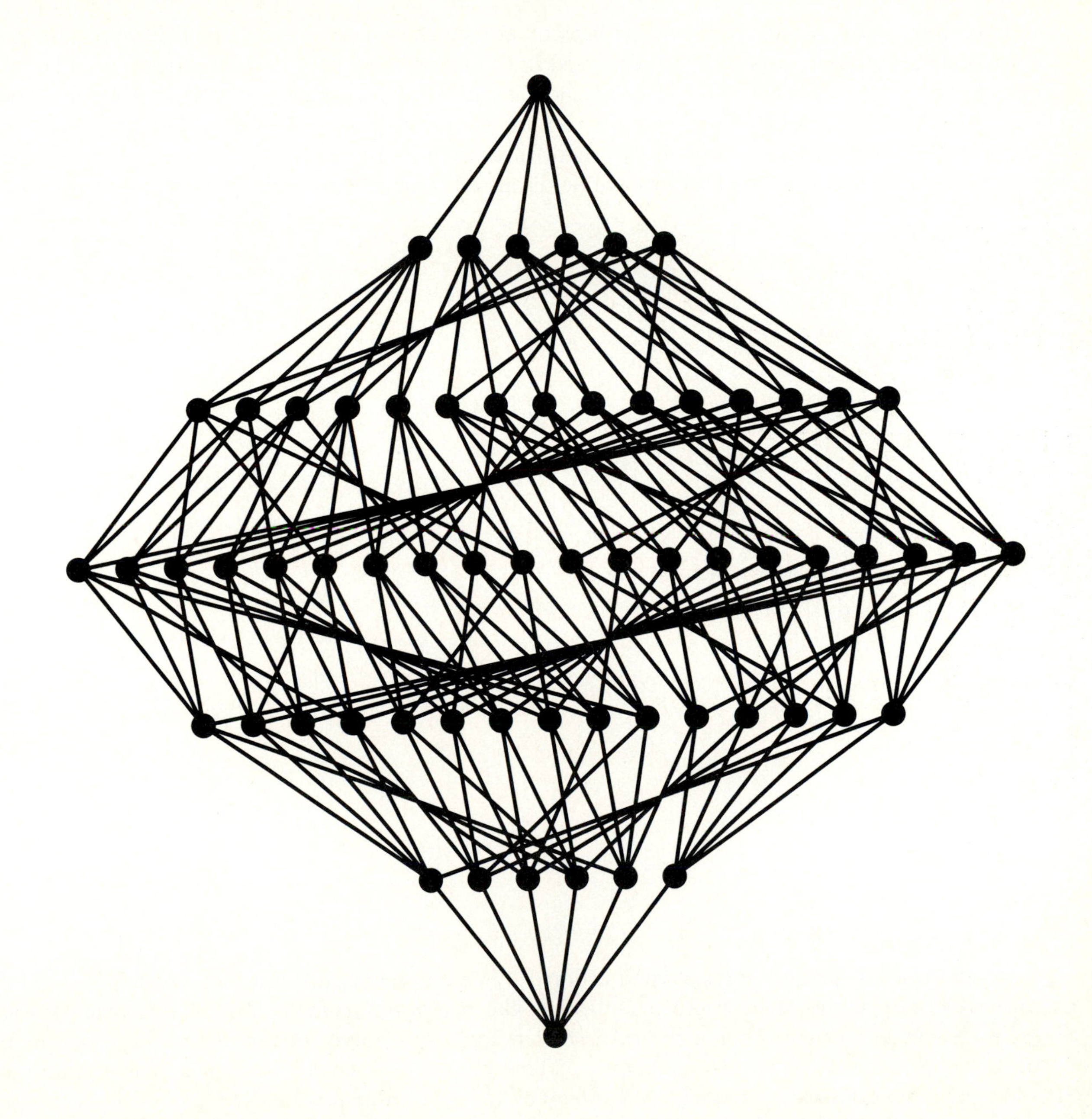

Graph theory is the study of properties or *invariants* of graphs. Among the properties of interest are such things as connectivity, cycle structure, and chromatic number. In this chapter, we discuss invariants of graphs and how to compute them.

We make a somewhat arbitrary distinction between the properties of graphs discussed in this chapter and algorithmic graph theory, discussed in the next. Algorithms are necessary to calculate the properties of graphs, and the study of these properties is necessary for the development of efficient algorithms, so clearly there is a lot of room for overlap between these chapters. Because of this incestuous relationship, some of the algorithms used as subroutines in this chapter, such as `NetworkFlow` and `BipartiteMatching`, will be discussed in detail only after they have been used.

About the illustration overleaf:

The subsets of a set define a lattice when ordered by containment, meaning one set is a subset of the other. The recursive structure of this lattice, the *boolean algebra*, is apparent from its Hasse diagram. There is a neat partition between the left and right halves of the lattice, each of which is the boolean algebra on $n-1$ elements. This picture was constructed with the command:

```
ShowGraph[ HasseDiagram[ MakeGraph[Subsets[6], ((Intersection[#2,#1]===#1)
&& (#1 != #2))&] ] ];
```

■ 5.1 Connectivity

Most interesting graphs tend to be *connected*, meaning they come in one piece. Connectivity can be generalized in two distinct ways to directed graphs, and considering how much of the structure must be removed to disconnect the graph provides a natural measure of whether one graph is "more" connected than another.

■ 5.1.1 Connected Components

An undirected graph is *connected* if there exists a path between any pair of vertices. Performing a depth-first search from any vertex eventually visits all vertices in a connected component. Since the vertices in distinct connected components are disjoint, starting a new traversal from any unvisited vertex results in a new component.

```
ConnectedComponents[g_Graph] :=
        Block[{untraversed=Range[V[g]],traversed,comps={}},
                While[untraversed != {},
                        traversed = DepthFirstTraversal[g,First[untraversed]];
                        AppendTo[comps,traversed];
                        untraversed = Complement[untraversed,traversed]
                ];
                comps
        ]

ConnectedQ[g_Graph] := Length[ DepthFirstTraversal[g,1] ] == V[g]
```

Testing for Connectivity in Undirected Graphs

All the interesting graphs we have defined are connected.

```
In[1]:= ConnectedQ[ K[1,1,3,2,1] ]
Out[1]= True
```

Deleting an edge from a connected graph can disconnect it. Such an edge is called a *bridge*.

```
In[2]:= ConnectedQ[ DeleteEdge[ Star[10], {1,10} ] ]
Out[2]= False
```

`GraphUnion` can be used to create disconnected graphs. Calling `InduceSubgraph` on each of the connected components provides a means to extract them.

```
In[3]:= ConnectedComponents[ GraphUnion[K[3],K[4]] ]
Out[3]= {{1, 2, 3}, {4, 5, 6, 7}}
```

If a graph is disconnected, its complement must be connected.

```
In[4]:= ConnectedComponents[ GraphComplement[
GraphUnion[K[3],K[4]] ] ]
Out[4]= {{1, 4, 2, 5, 3, 6, 7}}
```

Any graph with minimum degree $\delta = (n-1)/2$ is connected.

```
In[5]:= ConnectedQ[
RealizeDegreeSequence[{4,4,4,4,4,4,4,4}] ]

Out[5]= True
```

5.1.2 Strong and Weak Connectivity

There are two distinct notions of connectivity in directed graphs. A directed graph is *weakly connected* if there is an undirected path between any pair of vertices. Thus the graph hangs together in one piece.

```
WeaklyConnectedComponents[g_Graph] := ConnectedComponents[ MakeUndirected[g] ]

ConnectedQ[g_Graph,Undirected] := Length[ WeaklyConnectedComponents[g] ] == 1
```

Finding Weakly Connected Components

However, suppose the directed graph represents the Manhattan traffic grid. If it was weakly connected, only taxi drivers, with their willingness to ignore one-way signs, would be able to travel between any two points in the city. A directed graph is *strongly connected* if there is a directed path between every pair of vertices.

The following linear-time strong connectivity algorithm is based on depth-first search and due to Tarjan [Tar72]. When a depth-first search is performed, after everything reachable from a vertex v has been traversed, the search backs up to the parent of v. If there were no back edges to v or an ancestor of v, there is no directed path from some descendent of v to v, so v is in a strongly connected component separate from its children.

```
StronglyConnectedComponents[g_Graph] :=
        Block[{e=ToAdjacencyLists[g],s,c=1,i,cur={},low=dfs=Table[0,{V[g]}],scc={}},
                While[(s=Select[Range[V[g]],(dfs[[#]]==0)&]) != {},
                        SearchStrongComp[First[s]];
                ];
                scc
        ]

SearchStrongComp[v_Integer] :=
        Block[{r},
                low[[v]]=dfs[[v]]=c++;
                PrependTo[cur,v];
                Scan[
                        (If[dfs[[#]] == 0,
                                SearchStrongComp[#];
                                low[[v]]=Min[low[[v]],low[[#]]],
                                If[(dfs[[#]] < dfs[[v]]) && MemberQ[cur,#],
```

```
                                        low[[v]]=Min[low[[v]],dfs[[#]] ]
                                ];
                        ])&,
                        e[[v]]
                ];
                If[low[[v]] == dfs[[v]],
                        {r} = Flatten[Position[cur,v]];
                        AppendTo[scc,Take[cur,r]];
                        cur = Drop[cur,r];
                ];
        ]

ConnectedQ[g_Graph,Directed] := Length[ StronglyConnectedComponents[g] ] == 1
```

Finding Strongly Connected Components

A sparse directed graph is likely to have more strongly connected components than weakly connected ones.

```
In[6]:= ShowGraph[ g = RandomGraph[20,0.1,Directed],
Directed];
```

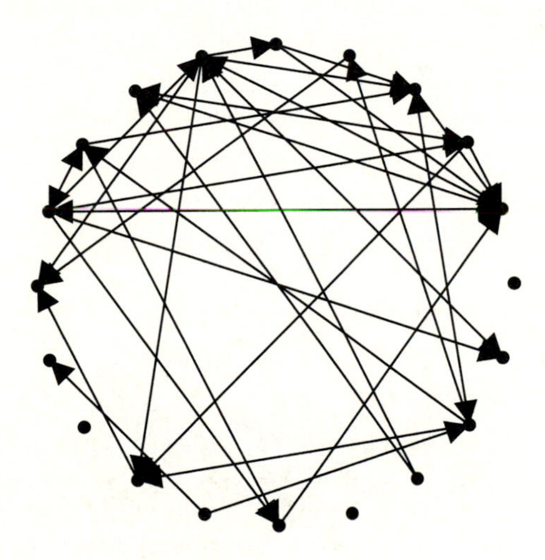

The number of weakly connected components of this graph is greater...

```
In[7]:= WeaklyConnectedComponents[g]
Out[7]= {{1, 2, 7, 17, 6, 3, 5, 8, 9, 15, 19, 18, 4, 10,
   13, 14, 11}, {12}, {16}, {20}}
```

... than the number of strongly connected components.

```
In[8]:= StronglyConnectedComponents[g]
Out[8]= {{10}, {4}, {13}, {19},
  {5, 15, 7, 2, 9, 8, 18, 3, 6, 1}, {11}, {12}, {14},
  {16}, {17}, {20}}
```

An alternative algorithm for identifying strongly connected components uses the *transitive closure* $T(G)$ of the adjacency matrix of graph G, where an element (x, y) of $T(G)$ is 1 if and only if y is reachable from x. The strongly connected component of vertex v_i is exactly the set of vertices with ones in the ith row of $G \cdot G^T$, where G^T is the transpose of R.

```
In[9]:= TableForm[ {m = Edges[GraphUnion[K[2],K[3]]], m . Transpose[m]} ]

                  0  1  0  0  0    1  0  0  0  0
                  1  0  0  0  0    0  1  0  0  0
                  0  0  0  1  1    0  0  2  1  1
                  0  0  1  0  1    0  0  1  2  1
Out[9]//TableForm= 0  0  1  1  0    0  0  1  1  2
```

■ 5.1.3 Orienting Graphs

An *orientation* of an undirected graph G is an assignment of exactly one direction to each of the edges of G. For transportation applications, it is often necessary to find a strongly connected orientation of a graph, if one exists.

Only connected, bridgeless graphs can have a strongly connected orientation, since deleting a bridge leaves two disconnected components, and so assigning the bridge an orientation permits travel to proceed in only one direction between them. Robbins [Rob39] showed that this condition is sufficient, by presenting an algorithm to find such an orientation. Extract all the cycles from the graph, and orient each as a directed cycle. If two or more directed cycles share a vertex, there exists a directed path between each pair of vertices in the component. If the union of these cycles is disconnected, there are at least two edges connecting them, since the graph is bridgeless. Therefore, orienting them in opposite directions yields a strongly connected graph.

```
OrientGraph[g_Graph] :=
        Block[{pairs,newg,rest,cc,c,i,e},
                pairs = Flatten[Map[(Partition[#,2,1])&,ExtractCycles[g]],1];
                newg = FromUnorderedPairs[pairs,Vertices[g]];
                rest = ToOrderedPairs[ GraphDifference[ g, newg ] ];
                cc = Sort[ConnectedComponents[newg], (Length[#1]>=Length[#2])&];
                c = First[cc];
                Do[
                        e = Select[rest,(MemberQ[c,#[[1]]] &&
                                         MemberQ[cc[[i]],#[[2]]])&];
                        rest = Complement[rest,e,Map[Reverse,e]];
                        c = Union[c,cc[[i]]];
                        pairs = Join[pairs, Prepend[ Rest[e],Reverse[e[[1]]] ] ],
                        {i,2,Length[cc]}
                ];
                FromOrderedPairs[
                        Join[pairs, Select[rest,(#[[1]] > #[[2]])&] ],
                        Vertices[g]
```

```
            ]
    ] /; SameQ[Bridges[g],{}]
```

Orienting a Directed Graph

This orientation of a wheel directs each edge in the outer cycle in the same direction, and completes it by giving the center an in-degree of $n-1$ and out-degree of 1.

```
In[10]:= ShowGraph[ OrientGraph[Wheel[10]], Directed];
```

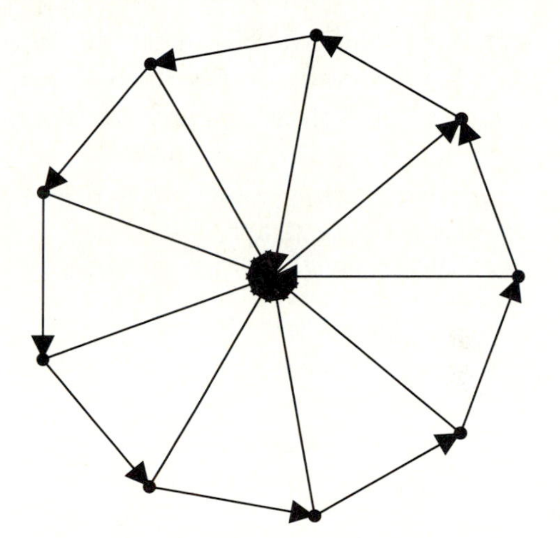

A *tournament* is an oriented complete graph. Tournaments model the results of a game where each pair of n players competes against each other, determining a winner and a loser. The direction of each edge in the tournament can be interpreted as pointing at the loser. A *Hamiltonian path* is a path between two vertices in a graph which visits every vertex exactly once. Interestingly, every tournament has an odd number of Hamiltonian paths [Red34, Sze43], meaning every tournament contains at least one of them. Tournaments have directed Hamiltonian cycles if and only if they are strongly connected [HM66, Fou60]. Moon [Moo68] presents other topics on tournaments, and optimal orientations under special conditions are discussed in [BT80, CT78, Rob78].

■ 5.1.4 Biconnected Components

An *articulation vertex* of a graph G is a vertex whose deletion disconnects G. Any graph with no articulation vertices is said to be *biconnected* or a *block*. Biconnectivity is an important property, for a graph of order $n > 2$ is biconnected if and only if there are at least two vertex disjoint paths between any pair of points. From the perspective of communications, biconnected networks are more reliable, since blowing away any single node doesn't cut off communication for any other node.

An undirected graph can be tested for biconnectivity using a similar algorithm to that which tested a directed graph for strong connectivity. In each case, a depth-first search is performed, with an articulation vertex occurring when there are no back edges, implying no cycles. The same function cannot be used to compute both quantities only because an undirected edge in a directed graph implies a cycle, which isn't really there when the graph is perceived as undirected.

This implementation finds the biconnected components and articulation vertices during the same traversal and returns both. Special care must be taken to test whether the root of the depth-first search tree is an articulation vertex, since no back edge can originate from the root [AHU74].

```
FindBiconnectedComponents[g_Graph] :=
        Block[{e=ToAdjacencyLists[g],n=V[g],par,c=0,act={},back,dfs,ap=bcc={}},
                back=dfs=Table[0,{n}];
                par = Table[n+1,{n}];
                Map[(SearchBiConComp[First[#]])&, ConnectedComponents[g]];
                {bcc,Drop[ap, -1]}
        ]

SearchBiConComp[v_Integer] :=
        Block[{r},
                back[[v]]=dfs[[v]]=++c;
                Scan[
                        (If[ dfs[[#]] == 0,
                                If[!MemberQ[act,{v,#}], PrependTo[act,{v,#}]];
                                par[[#]] = v;
                                SearchBiConComp[#];
                                If[ back[[#]] >= dfs[[v]],
                                        {r} = Flatten[Position[act,{v,#}]];
                                        AppendTo[bcc,Union[Flatten[Take[act,r]]]];
                                        AppendTo[ap,v];
                                        act = Drop[act,r]
                                ];
                                back[[v]] = Min[ back[[v]],back[[#]] ],
                                If[# != par[[v]],back[[v]]=Min[dfs[[#]],back[[v]]]]
                        ])&,
                        e[[v]]
                ];
        ]
```

Finding Biconnected Components

A *bridge* is an edge whose deletion disconnects the graph. An edge is a bridge if and only if it does not lie on any cycle. Thus the bridges of a graph are exactly the biconnected components of length two, because if they were involved with a cycle they would form part of a larger component.

```
ArticulationVertices[g_Graph]  := Union[Last[FindBiconnectedComponents[g]]];

Bridges[g_Graph] := Select[BiconnectedComponents[g],(Length[#] == 2)&]

BiconnectedComponents[g_Graph] := First[FindBiconnectedComponents[g]];

BiconnectedQ[g_Graph] := Length[ BiconnectedComponents[g] ] == 1
```

Applications of the Biconnectivity Algorithm

Any graph with a vertex of degree 1 cannot be biconnected, since deleting the other vertex which defines its only edge disconnects the graph.

```
In[11]:= BiconnectedComponents[
RealizeDegreeSequence[{4,4,3,3,3,2,1}] ]

Out[11]= {{2, 7}, {1, 2, 3, 4, 5, 6}}
```

All Hamiltonian graphs are biconnected.

```
In[12]:= HamiltonianQ[ Cycle[10] ]

Out[12]= True
```

The only articulation vertex of a star is its center, even though its deletion leaves $n-1$ connected components. Deleting a leaf leaves a connected tree.

```
In[13]:= ArticulationVertices[ Star[10] ]

Out[13]= {10}
```

Every edge in a tree is a bridge

```
In[14]:= Bridges[ RandomTree[10] ]

Out[14]= {{2, 6}, {3, 4}, {4, 5}, {4, 9}, {7, 10},
  {4, 10}, {4, 8}, {2, 8}, {1, 2}}
```

A *cubic* or 3-regular graph contains a bridge if and only if it contains an articulation vertex.

```
In[15]:= (g = RegularGraph[3,10];
{ArticulationVertices[g], Bridges[g]} )

Out[15]= {{}, {}}
```

■ 5.1.5 k-Connectivity

A graph is said to be *k-connected* if there does not exist a set of $k-1$ vertices whose removal disconnects the graph. Thus a connected graph is 1-connected and a biconnected graph 2-connected. A graph is *k-edge-connected* if there does not exist a set of $k-1$ edges whose removal disconnects the graph.

Both edge and vertex connectivity can be found using network flow techniques. The network flow problem, to be further discussed in Section 6.3, interprets a weighted graph as a network of pipes where the maximum capacity of an edge is its weight and seeks to maximize the flow between two given vertices of the graph. The maximum-flow, minimum-cut theorem [FF62] states that the maximum flow between v_i, v_j in G is exactly the weight of the smallest set of edges to disconnect G with v_i and v_j in different components. Thus the edge connectivity can be found by minimizing the flow in an unweighted graph between v_i and each of the $n-1$ other vertices in G since, after deleting the minimum cut set, v_i will be in a different component from some other vertex.

```
EdgeConnectivity[g_Graph] :=
        Block[{i},
                Apply[Min, Table[NetworkFlow[g,1,i], {i,2,V[g]}]]
        ]
```

Finding the Edge Connectivity of a Graph

The edge connectivity of a graph is at most the minimum degree δ, since deleting those edges disconnects the graph. Complete bipartite graphs realize this bound.

```
In[16]:= EdgeConnectivity[K[3,4]]
Out[16]= 3
```

Deleting any edge from a tree disconnects it, so the edge connectivity of a tree is 1.

```
In[17]:= EdgeConnectivity[ RandomTree[10] ]
Out[17]= 1
```

Vertex connectivity is characterized by *Menger's theorem* [Men27, Whi32], which states that a graph is k-connected if and only if every pair of vertices is joined by at least k vertex-disjoint paths. Network flow can again be used to perform this calculation, since in an unweighted graph G a flow of k between a pair of vertices implies k *edge*-disjoint paths. We must construct a graph G' with the property that any set of edge disjoint paths in G' corresponds to vertex disjoint paths in G. This can be done by replacing each vertex v_i of G with two vertices $v_{i,1}$ and $v_{i,2}$, such that edge $(v_{i,1}, v_{i,2}) \in G'$ for all $v_i \in G$, and for every edge $(v_i, x) \in G$ adding edges $(v_{i,j}, x_k)$, $j \neq k \in \{0,1\}$ to G. Thus two edge-disjoint paths in G' correspond to vertex-disjoint paths in G, and so the minimum flow between a pair of vertices in G' gives the vertex connectivity of G.

```
VertexConnectivityGraph[g_Graph] :=
        Block[{n=V[g],e},
                e=Table[0,{2 n},{2 n}];
                Scan[ (e[[#-1,#]] = 1)&, 2 Range[n] ];
                Scan[
```

```
                    (e[[#[[1]], #[[2]]-1]] = e[[#[[2]],#[[1]]-1]] = Infinity)&,
                    2 ToUnorderedPairs[g]
                ];
                Graph[e,Apply[Join,Map[({#,#})&,Vertices[g]]]]
        ]

VertexConnectivity[g_Graph] :=
        Block[{p=VertexConnectivityGraph[g],k=V[g],i=0,notedges},
                notedges = ToUnorderedPairs[ GraphComplement[g] ];
                While[ i++ <= k,
                        k = Min[
                                Map[
                                        (NetworkFlow[p,2 #[[1]],2 #[[2]]-1])&,
                                        Select[notedges,(First[#]==i)&]
                                ],
                                k
                        ]
                ];
                k
        ]
```

Finding the Vertex Connectivity of a Graph

The vertex connectivity of a complete bipartite graph is the number of vertices in the smaller stage, since all of them must be deleted to disconnect the graph.

```
In[18]:= VertexConnectivity[K[3,4]]
Out[18]= 3
```

The wheel is the basic triconnected graph [Tut61].

```
In[19]:= VertexConnectivity[Wheel[5]]
Out[19]= 3
```

5.1.6 Harary Graphs

In general, the higher the connectivity of a graph, the more edges it must contain. Every vertex in a k-connected graph must have degree at least k, so a k-connected graph on n vertices must have at least $\lceil kn/2 \rceil$ edges. The *Harary graph* $H_{k,n}$ [Har62] realizes this bound and hence is the smallest k-connected graph with n vertices.

There are three different cases, depending upon the parity of n and k. Two are circulant graphs, but when n and k are odd the graph is not symmetrical.

```
Harary[k_?EvenQ, n_Integer] := CirculantGraph[n,Range[k/2]]

Harary[k_?OddQ, n_?EvenQ] := CirculantGraph[n,Append[Range[k/2],n/2]]

Harary[k_?OddQ, n_?OddQ] :=
```

```
Block[{g=Harary[k-1,n],i},
        FromUnorderedPairs[
                Join[
                        ToUnorderedPairs[g],
                        { {1,(n+1)/2}, {1,(n+3)/2} },
                        Table [ {i,i+(n+1)/2}, {i,2,(n-1)/2} ]
                ],
                Vertices[g]
        ]
]
```

Constructing the Harary Graph

A cycle is the minimum biconnected graph.

In[20]:= **IdenticalQ[Cycle[12],Harary[2,12]]**

Out[20]= True

When n or k is even, the Harary graph is a circulant graph, which implies that it is regular. In other cases, there must be one distinguished vertex of higher degree. With at least one vertex of degree k, the Harary graph $H_{k,n}$ is clearly at most k-connected. A proof that it is in fact k-connected appears in [BM76].

In[21]:= **ShowGraph[Harary[7,13]];**

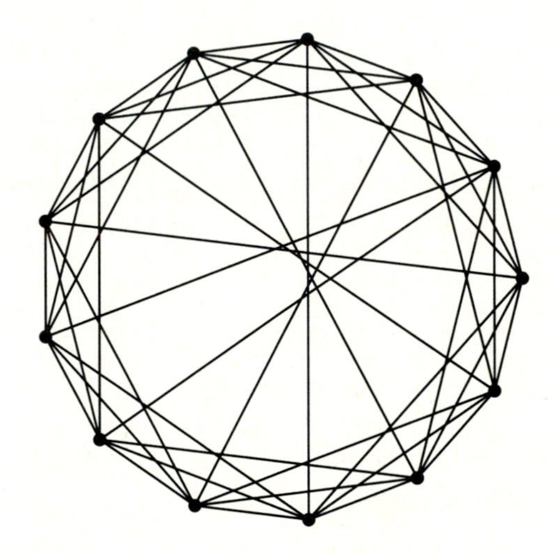

As advertised, $H_{4,6}$ is indeed 4-connected.

In[22]:= **VertexConnectivity[Harary[4,6]]**

Out[22]= 4

The Harary graph $H_{n-1,n}$ gives the complete graph K_n, although this function runs like a dog relative to `K[9]`.

In[23]:= **CompleteQ[Harary[8,9]]**

Out[23]= True

5.2 Graph Isomorphism

Two graphs are *isomorphic* if there exists a renaming of the vertices of the graphs such that they are identical. Two graphs are isomorphic when they have identical structure, although they may be represented differently.

Unfortunately, this notion of equality doesn't come cheap. All the algorithms we have seen thus far are efficient, even if they may not have *seemed* so efficient, for they took time polynomial in the size of the problem to return the answer. Unfortunately, many interesting and important problems have no known polynomial-time algorithms. Since they are interesting, we will provide functions for computing these quantities; and since they are important, care must be taken to make them as efficient as possible, through heuristics or careful programming.

Most of the problems discussed in this chapter which have known no polynomial-time algorithm are *NP-complete*, meaning that if there exists a polynomial time algorithm for solving any one of them (and several hundred other problems), there will exist a polynomial time algorithm for solving them all. Since researchers have been beating their heads against these problems for many years and been coming up empty, the conjecture is that no such polynomial algorithms exist.

Garey and Johnson [GJ79] provides a comprehensive treatment of the theory of NP-completeness, and contains a compilation of all NP-complete problems known when it was published in 1979. Johnson's occasional column in the *Journal of Algorithms* surveys more recent developments in the field.

There exists no known polynomial-time algorithm for isomorphism testing, despite the fact that isomorphism has not been shown to be NP-complete. It seems to fall in the crack somewhere between P and NP-complete (if such a crack exists), although a polynomial-time algorithm is known when the maximum degree vertex is bounded by a constant [Luk80]. Ironically, an original interest of Leonid Levin, one of two independent developers of the notion of NP-completeness [Lev73], in defining this complexity class was proving that graph isomorphism is hard!

In the next section, we will present an algorithm for finding isomorphisms which, although usually efficient in practice, is not guaranteed to take polynomial time.

5.2.1 Finding Isomorphisms

An isomorphism between two graphs is described by a one-to-one mapping between the two sets of vertices. Two labeled graphs are *identical* if their current labelings represent an isomorphism.

```
IdenticalQ[g_Graph,h_Graph] := Edges[g] === Edges[h]

IsomorphismQ[g_Graph,h_Graph,p_List] := False        /;
            (V[g]!=V[h]) || !PermutationQ[p] || (Length[p] != V[g])

IsomorphismQ[g_Graph,h_Graph,p_List] := IdenticalQ[g, InduceSubgraph[h,p] ]
```

Testing if Two Graphs are Identical

Every permutation of a complete graph represents a complete graph.

```
In[24]:= IsomorphismQ[K[5],K[5],RandomPermutation[5]]
Out[24]= True
```

These graphs are clearly isomorphic but not identical.

```
In[25]:= IdenticalQ[K[3,2],K[2,3]]
Out[25]= False
```

To identify isomorphisms between two graphs efficiently, we must avoid testing as many of the $n!$ permutations as possible. This can be done by using graph invariants to partition the vertices into equivalence classes. For example, no two vertices of different degrees can be mapped to each other in an isomorphism, so the set of degrees defines equivalence classes. Observe that since any simple graph except for K_1 contains at least two vertices of equal degree, the degree sequence is not sufficient to completely partition the vertices of any graph. Further, the degree invariant is of no help at all on regular graphs.

A stronger heuristic due to [SD76] uses the all-pairs shortest path matrix to partition the vertices, since the set of shortest distances between one vertex and all the others is invariant for G. For most graphs, this is sufficient to eliminate most non-isomorphic vertex pairs, although there exist non-isomorphic graphs which realize the same set of distances [BH90]. Another approach to finding isomorphisms appears in [CG70].

From these equivalence classes, we can perform a backtrack search, since our partial solution will consist of a list of vertices with the property that the ith element belongs to the ith equivalence class. The predicates for testing partial and complete solutions make certain the induced subgraphs are identical to this point. We use the general purpose backtrack routine discussed in Section 1.1.5.

```
Isomorphism[g_Graph,h_Graph,flag_:One] := {}        /; (V[g] != V[h])

Isomorphism[g_Graph,h_Graph,flag_:One] :=
        Block[{eg=Edges[g],eh=Edges[h]},
                Backtrack[
```

```
                        Equivalences[g,h],
                        (IdenticalQ[InduceSubgraph[g,Range[Length[#]]],
                                 InduceSubgraph[h,#] ] &&
                         !MemberQ[Drop[#,-1],Last[#]])&,
                        (IsomorphismQ[g,h,#])&,
                        flag
                ]
        ]

IsomorphicQ[g_Graph,h_Graph] := True /; IdenticalQ[g,h]
IsomorphicQ[g_Graph,h_Graph] := ! SameQ[ Isomorphism[g,h], {}]

Equivalences[g_Graph,h_Graph] :=
        Equivalences[ AllPairsShortestPath[g], AllPairsShortestPath[h]]

Equivalences[g_List,h_List] :=
        Block[{dg=Map[Sort,g],dh=Map[Sort,h],s,i},
                Table[
                        Flatten[Position[dh,_?(Function[s,SameQ[s,dg[[i]] ]])]],
                        {i,Length[dg]}
                ]
        ] /; Length[g] == Length[h]
```

Finding Isomorphisms

The shortest path information is sufficient to make clear that there are exactly two isomorphisms between paths of the same length. The ith equivalence class contains only the vertices v_i and v_{n-i}.

In[26]:= **Equivalences[Path[5],Path[5]]**

Out[26]= {{1, 5}, {2, 4}, {3}, {2, 4}, {1, 5}}

From the reduced equivalence classes, the backtrack search quickly proceedes to find both isomorphisms.

In[27]:= **Isomorphism[Path[5],Path[5],All]**

Out[27]= {{1, 2, 3, 4, 5}, {5, 4, 3, 2, 1}}

These two complete bipartite graphs are isomorphic, since the order of the two stages is simply reversed. However, in general, isomorphism is as difficult for bipartite graphs as for arbitrary graphs. Any graph can be made bipartite by replacing each edge by two edges connected with a new vertex. Clearly, the original graphs are isomorphic if and only if the transformed graphs are.

In[28]:= **Isomorphism[K[3,2],K[2,3],All]**

Out[28]= {{3, 4, 5, 1, 2}, {3, 4, 5, 2, 1}, {3, 5, 4, 1, 2}, {3, 5, 4, 2, 1}, {4, 3, 5, 1, 2}, {4, 3, 5, 2, 1}, {4, 5, 3, 1, 2}, {4, 5, 3, 2, 1}, {5, 3, 4, 1, 2}, {5, 3, 4, 2, 1}, {5, 4, 3, 1, 2}, {5, 4, 3, 2, 1}}

Two graphs are not necessarily isomorphic if they share the same degree sequences, although it is a prerequisite. Because the shortest path vectors are so different for both of these graphs, the reduced equivalence classes leave no possible isomorphisms.

```
In[29]:= Equivalences[Cycle[6],GraphUnion[2,K[3]]]
Out[29]= {{}, {}, {}, {}, {}, {}}
```

■ 5.2.2 Automorphism Groups

An *automorphism* of a graph is an isomorphism with itself. The set of automorphisms defines a permutations group, as discussed in Section 1.2. The automorphism group provides a tool for understanding symmetries in a graph and has been used as a foundation for graph drawing algorithms which seek to display these symmetries [LNS85]. The function Polya developed in Section 1.2.6 uses automorphism groups for combinatorial enumeration.

```
Automorphisms[g_Graph,flag_:All] :=
        Block[{s=AllPairsShortestPath[g]},
                Backtrack[
                        Equivalences[s,s],
                        (IdenticalQ[InduceSubgraph[g,Range[Length[#]]],
                                    InduceSubgraph[g,#] ] &&
                         !MemberQ[Drop[#,-1],Last[#]])&,
                        (IsomorphismQ[g,g,#])&,
                        flag
                ]
        ]
```

Finding the Automorphisms of a Graph

For every group Γ, there exists a graph whose automorphism group is isomorphic to Γ [Fru39]. The Frucht graph is the smallest 3-regular graph whose group consists only of the identity element.

```
In[30]:= ShowGraph[ f = ReadGraph["graphs/frucht"] ];
```

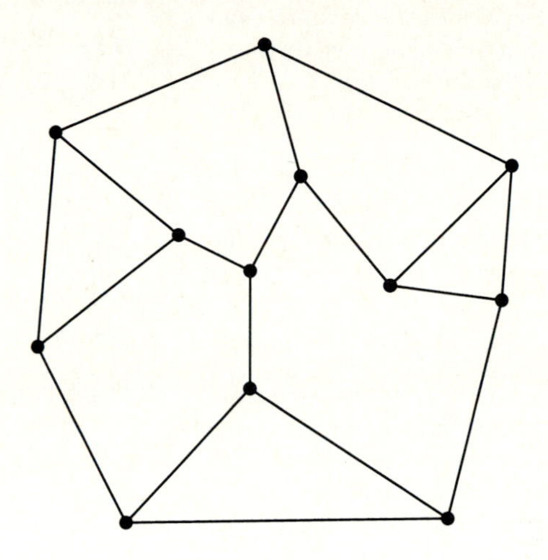

Every graph has at least one automorphism, the identity element.

```
In[31]:= Automorphisms[ f ]

Out[31]= {{1, 2, 3, 4, 5, 6, 7, 8, 9, 10, 11, 12}}
```

It is clear that the automorphism group of this graph contains a cycle, by repeatedly performing a rotation of $\pi/3$ radians about its physical center. In fact, it is the smallest non-trivial graph [HP66] whose group is cyclic.

```
In[32]:= ShowGraph[ c = ReadGraph["graphs/cyclegroup"] ];
```

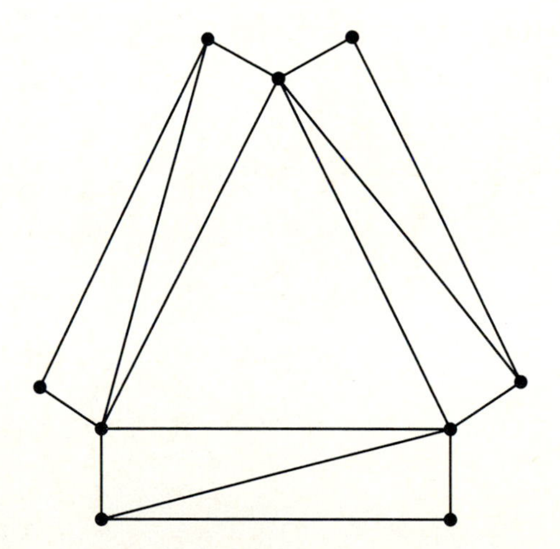

There are three distinct automorphisms of this graph, the structure of which is reflected by the cycles of the permutations. Beyond the identity element, the vertices are partitioned into three cycles, corresponding to vertices of degree 5, 3, and 2, respectively.

```
In[33]:= Map[ ToCycles, Automorphisms[ c ] ]
Out[33]= {{{1}, {2}, {3}, {4}, {5}, {6}, {7}, {8}, {9}},
  {{2, 3, 1}, {8, 6, 4}, {9, 7, 5}},
  {{3, 2, 1}, {6, 8, 4}, {7, 9, 5}}}
```

The automorphism group of the complement of a graph is the same as that of the original graph.

```
In[34]:= Automorphisms[ GraphComplement[ c ] ]
Out[34]= {{1, 2, 3, 4, 5, 6, 7, 8, 9},
  {2, 3, 1, 8, 9, 4, 5, 6, 7},
  {3, 1, 2, 6, 7, 8, 9, 4, 5}}
```

A graph is *vertex-transitive* if every pair of vertices is equivalent under some element of its automorphism group. A graph is *edge-transitive* if any two edges are equivalent under some element of its group. Every non-trivial edge-transitive but not vertex-transitive graph, such as the Folkman graph [Fol67] contains at least twenty vertices.

```
In[35]:= ShowGraph[ f = ReadGraph["graphs/folkman"] ];
```

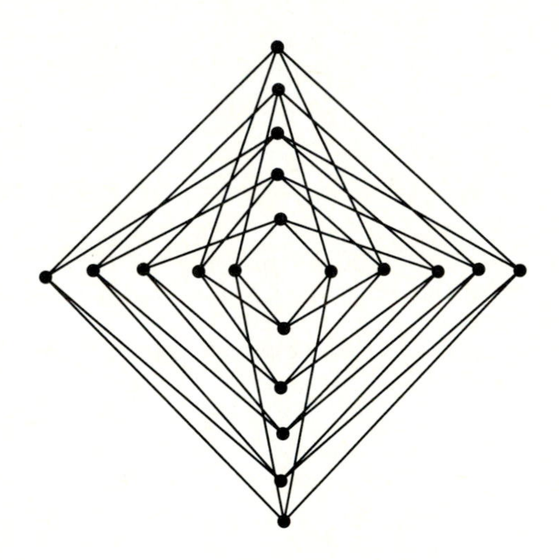

The all-pairs shortest path equivalence classes show that the Folkman graph contains a lot of symmetry, and that the vertices are neatly partitioned into two groups of ten similar vertices.

```
In[36]:= Equivalences[f,f]
Out[36]= {{1, 2, 3, 4, 5, 16, 17, 18, 19, 20},
  {1, 2, 3, 4, 5, 16, 17, 18, 19, 20},
  {1, 2, 3, 4, 5, 16, 17, 18, 19, 20},
  {1, 2, 3, 4, 5, 16, 17, 18, 19, 20},
  {1, 2, 3, 4, 5, 16, 17, 18, 19, 20},
  {6, 7, 8, 9, 10, 11, 12, 13, 14, 15},
  {6, 7, 8, 9, 10, 11, 12, 13, 14, 15},
  {6, 7, 8, 9, 10, 11, 12, 13, 14, 15},
  {6, 7, 8, 9, 10, 11, 12, 13, 14, 15},
  {6, 7, 8, 9, 10, 11, 12, 13, 14, 15},
  {6, 7, 8, 9, 10, 11, 12, 13, 14, 15},
  {6, 7, 8, 9, 10, 11, 12, 13, 14, 15},
  {6, 7, 8, 9, 10, 11, 12, 13, 14, 15},
  {6, 7, 8, 9, 10, 11, 12, 13, 14, 15},
  {6, 7, 8, 9, 10, 11, 12, 13, 14, 15},
  {1, 2, 3, 4, 5, 16, 17, 18, 19, 20},
  {1, 2, 3, 4, 5, 16, 17, 18, 19, 20},
  {1, 2, 3, 4, 5, 16, 17, 18, 19, 20},
  {1, 2, 3, 4, 5, 16, 17, 18, 19, 20},
  {1, 2, 3, 4, 5, 16, 17, 18, 19, 20}}
```

5.2.3 Self-Complementary Graphs

A graph is *self-complementary* if it is isomorphic to its complement.

```
SelfComplementaryQ[g_Graph] := IsomorphicQ[g, GraphComplement[g]]
```

Testing Self-Complementary Graphs

The smallest non-trivial self-complementary graphs are the path on four vertices and the cycle on five.

```
In[37]:= SelfComplementaryQ[ Cycle[5] ] &&
SelfComplementaryQ[ Path[4] ]
Out[37]= True
```

A simple parity argument shows that every self-complementary graph contains $4k$ or $4k+1$ vertices.

```
In[38]:= SelfComplementaryQ[ K[3,3] ]
Out[38]= False
```

All self-complementary graphs have diameter 2 or 3 [Sac62].

```
In[39]:= Diameter[ Cycle[5] ]
Out[39]= 2
```

5.3 Cycles in Graphs

A *cycle* in a graph is simply a closed path. In our representation, a cycle in G will consist of a list of vertices $C = \{v_1, v_2, ..., v_1\}$ such that there is an edge of G for each pair of neighbors in C. Cycles play a role in any connected graph which isn't a tree.

5.3.1 Acyclic Graphs

A graph is *acyclic* if it contains no cycles. Testing whether a graph is acyclic is a simple matter of finding a cycle, which can be done by constructing the depth-first search tree and testing whether there are any back edges.

This implementation shows that the algorithms for finding directed and undirected cycles are virtually identical. One complication is that since the graph is not necessarily connected, on failing to find a cycle in our search from the start vertex, we try again if there is a higher numbered vertex which has never been visited.

```
FindCycle[g_Graph,flag_:Undirected] := FindCycle[g,flag,1] /; V[g] >= 2

FindCycle[g_Graph,flag_,start_Integer] :=
        Block[{edge,n=V[g],x,queue={start},ex,cycle={},parent,next},
                edge=ToAdjacencyLists[g];
                parent=Table[n+1,{n}];
                parent[[start]] = 0;
                While[ queue != {},
                        {x, queue} = {First[queue], Rest[queue]};
                        If [x === {},
                                cycle = Drop[cycle,-1],
                                ex = If[ SameQ[flag,Undirected],
                                        Select[edge[[x]],(parent[[x]] != #)&],
                                        edge[[x]]
                                ];
                                Scan[ (parent[[#]]=x)&, ex];
                                queue = Join[ex,{{}},queue];
                                If[MemberQ[cycle,x], Return[FromParent[parent,x]]];
                                AppendTo[cycle,x]
                        ]
                ];
                If [(next = Position[Drop[parent,start],n+1]) == {},
                        {},
                        FindCycle[g,flag,start+next[[1,1]]]
                ]
        ]

FromParent[parent_List,s_Integer] :=
```

```
        Block[{i=s,lst={s}},
                While[(i=parent[[i]]) != s, PrependTo[lst,i] ];
                PrependTo[lst,i]
        ]
```

Finding Cycles in Directed and Undirected Graphs

A cycle is detected the moment there is a back edge. Determining the cycle involves backing up the chain of parents in the depth-first search tree until the other vertex of the back edge is reached.

Once we can find a cycle, we can use this function as a subroutine to perform other useful operations:

```
AcyclicQ[g_Graph,flag_:Undirected] := SameQ[FindCycle[g,flag],{}]

TreeQ[g_Graph] := ConnectedQ[g] && (M[g] == V[g]-1)

ExtractCycles[gi_Graph,flag_:Undirected] :=
        Block[{g=gi,cycles={},c},
                While[!SameQ[{}, c=FindCycle[g,flag]],
                        PrependTo[cycles,c];
                        g = DeleteCycle[g,c,flag];
                ];
                cycles
        ]

DeleteCycle[g_Graph,cycle_List,flag_:Undirected] :=
        Block[{newg=g},
                Scan[(newg=DeleteEdge[newg,#,flag])&, Partition[cycle,2,1] ];
                newg
        ]
```

Cycle Operations on Graphs

Any simple graph with minimum degree δ contains a cycle of length at least $\delta + 1$.

```
In[40]:= ExtractCycles[
RealizeDegreeSequence[{3,3,3,3,3,3,3,3}] ]

Out[40]= {{1, 3, 4, 2, 6, 1}}
```

A directed cycle in an undirected graph can consist of only two edges.

```
In[41]:= FindCycle[K[5], Directed]

Out[41]= {1, 2, 1}
```

The first cycle found is not necessarily the shortest in the graph.

```
In[42]:= FindCycle[ GraphUnion[Cycle[5],Cycle[3]] ]

Out[42]= {1, 2, 3, 4, 5, 1}
```

A tree can be identified without explicitly showing the graph is acyclic, since this is implied if the graph is connected with exactly $n-1$ edges.

```
In[43]:= TreeQ[ RandomTree[10] ]
Out[43]= True
```

A directed graph with half the edges is almost certain to contain a cycle. Directed acyclic graphs are often called *DAGs*.

```
In[44]:= AcyclicQ[ RandomGraph[7,0.5,Directed], Directed]
Out[44]= False
```

5.3.2 Girth

The *girth* of a graph is the length of its shortest cycle. Graphs of minimal girth can be constructed by simply including a triangle, but achieving large girth in a graph with many edges is difficult.

```
Girth[g_Graph] :=
        Block[{v,dist,queue,n=V[g],girth=Infinity,parent,e=ToAdjacencyLists[g],x},
                Do [
                        dist = parent = Table[Infinity, {n}];
                        dist[[v]] = parent[[v]] = 0;
                        queue = {v};
                        While [queue != {},
                                {x,queue} = {First[queue],Rest[queue]};
                                Scan[
                                        (If [ (dist[[#]]+dist[[x]]<girth) &&
                                                   (parent[[x]] != #),
                                                girth=dist[[#]]+dist[[x]] + 1,
                                          If [dist[[#]]==Infinity,
                                                dist[[#]] = dist[[x]] + 1;
                                                parent[[#]] = x;
                                                If [2 dist[[#]] < girth-2,
                                                        AppendTo[queue,#] ]
                                        ]])&,
                                        e[[ x ]]
                                ];
                        ],
                        {v,n}
                ];
                girth
        ] /; SimpleQ[g]
```

Finding the Girth of a Graph

This algorithm, for simple graphs only, searches for the smallest cycle which originates from each vertex v_i, since the smallest of these gives the girth of the graph. A breadth-first search starting from v_i finds the shortest distance to all expanded vertices. Any edge connecting two vertices in the tree creates a cycle. We continue the search until it is clear that no shorter cycle can be found, which happens when twice the distance from v_i to the closest unexpanded vertex is greater than the shortest cycle thus far identified. The proof of correctness of this procedure is that the shortest cycle through a vertex v and containing an edge (x, y) must consist of the shortest vertex-disjoint paths from v to x and v to y. If the shortest paths are not vertex disjoint, then there will be a shorter cycle associated with another vertex.

The girth of a complete graph is three, since it contains a triangle, the smallest possible cycle.

```
In[45]:= Girth[ K[5] ]
Out[45]= 3
```

Since all cycles in bipartite graphs are of even length, the girth of any such graph is even. The girth of any complete bipartite graph which is not a star is four since it contains $K_{2,2}$ as a subgraph.

```
In[46]:= Girth[ K[4,4] ]
Out[46]= 4
```

The girth of a tree is ∞, since there is no shortest cycle in the graph.

```
In[47]:= Girth[RandomTree[10]]
Out[47]= Infinity
```

A (g, k)-*cage* is the smallest k-regular graph of girth g. K_4 is trivially the smallest 3-regular graph of girth 3, since it is the smallest 3-regular graph. $K_{3,3}$ is a (4,3)-cage, a 3-regular graph of girth 4. The Petersen graph is the unique (5,3)-cage.

```
In[48]:= ShowGraph[ p=ReadGraph["graphs/petersen"] ];
```

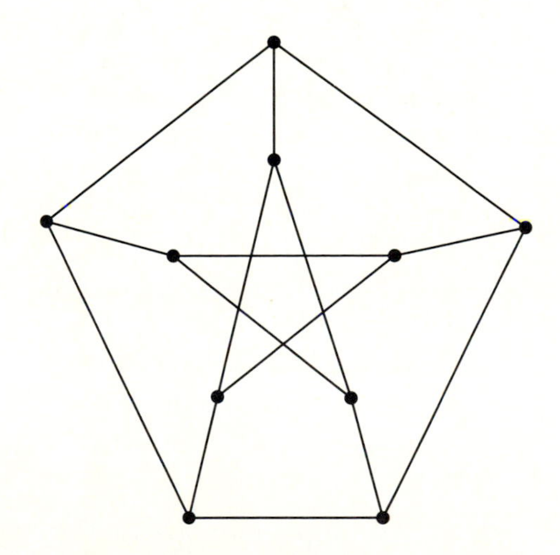

The girth of the Petersen graph is 5 since it is the (5,3)-cage.

```
In[49]:= Girth[p]
Out[49]= 5
```

The Heawood graph is a (6,3)-cage and can be embedded on a torus.

```
In[50]:= ShowGraph[ h=ReadGraph["graphs/heawood"] ];
```

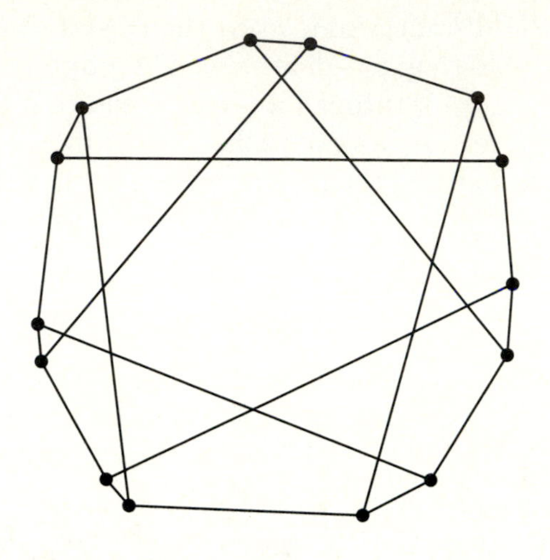

Since search is cut off promptly once the shortest cycle is discovered, `Girth` is reasonably efficient even on graphs of large girth. The time complexity of `Girth` is $O(mn)$.

```
In[51]:= Girth[h]
Out[51]= 6
```

The length of the longest cycle in a graph is its *circumference.* Since testing whether a graph is Hamiltonian is equivalent to asking if its circumference is n, computing the circumference is a computationally expensive operation.

■ 5.3.3 Eulerian Cycles

Euler initiated the study of graph theory in 1736 with the famous Seven Bridges of Königsberg problem. The town of Königsberg straddled the Pregel River with a total of seven bridges connecting the two shores and two islands. Euler proved that a tour of all edges in a connected, undirected graph without repetition is possible if and only if the degree of each vertex is even. Such graphs are known as *Eulerian graphs.*

An *Eulerian cycle* is a complete tour of all the edges of a graph. The term *circuit* is often used instead of cycle, since each vertex can be visited more than once.

Directed graphs have directed Eulerian cycles if and only if they are *weakly* connected and each vertex has the same in-degree as out-degree. The strong connectivity necessary for an Eulerian cycle falls out of the equality between in- and out-degrees.

```
EulerianQ[g_Graph,Directed] :=
        ConnectedQ[g,Undirected] && (InDegree[g] === OutDegree[g])

EulerianQ[g_Graph,flag_:Undirected] := ConnectedQ[g,Undirected] &&
        UndirectedQ[g] && Apply[And,Map[EvenQ,DegreeSequence[g]]]

OutDegree[Graph[e_List,_],n_Integer] := Length[ Select[ e[[n]], (# != 0)& ] ]
OutDegree[g_Graph] := Map[ (OutDegree[g,#])&, Range[V[g]] ]

InDegree[g_Graph,n_Integer] := OutDegree[ TransposeGraph[g], n ];
InDegree[g_Graph] := Map[ (InDegree[g,#])&, Range[V[g]] ]

TransposeGraph[Graph[g_List,v_List]] := Graph[ Transpose[g], v ]
```

Identifying Eulerian Graphs

Since the original system of bridges formed a multigraph, we cannot display it properly with the straight line embeddings we have at our disposal. An English translation of Euler's original paper appears in [BLW76].

```
In[52]:= TableForm[ Edges[ k=Graph[
{{0,1,1,1},{1,0,2,0},{1,2,0,2},{1,0,2,0}},
RandomVertices[4] ] ] ]
                    0  1  1  1

                    1  0  2  0

                    1  2  0  2

Out[52]//TableForm= 1  0  2  0
```

The people of Königsberg never could find a way to tour the bridges, and it is even more difficult today since they were destroyed in World War II. They were located in the present day city of Kalingrad in the Soviet Union [Wil86].

```
In[53]:= EulerianQ[k]

Out[53]= False
```

The complete graph on an odd number of vertices is Eulerian.

```
In[54]:= EulerianQ[ K[9] ]

Out[54]= True
```

It is appropriate that Fleury's algorithm [Luc91], an elegant way to construct an Eulerian cycle, uses bridges for a solution. Start walking from any vertex and erase any edge which has been traversed. The only criterion in picking an edge is that we do not select a bridge unless there are no other alternatives. Of course, no Eulerian graph contains a bridge, but what remains after traversing some edges eventually ceases to be biconnected.

Fleury's algorithm is more elegant than efficient, since it is expensive to recompute which edges are bridges after each deletion. Thus we use a different algorithm, attributed to [Hie73], which extracts all the cycles and then concatenates them together by finding vertices in common.

```
EulerianCycle[g_Graph,flag_:Undirected] :=
        Block[{euler,c,cycles,v},
                cycles = Map[(Drop[#,-1])&, ExtractCycles[g,flag]];
                {euler, cycles} = {First[cycles], Rest[cycles]};
                Do [
                        c = First[ Select[cycles, (Intersection[euler,#]=!={})&] ];
                        v = First[Intersection[euler,c]];
                        euler = Join[
                                RotateLeft[c, Position[c,v] [[1,1]] ],
                                RotateLeft[euler, Position[euler,v] [[1,1]] ]
                        ];
                        cycles = Complement[cycles,{c}],
                        {Length[cycles]}
                ];
                Append[euler, First[euler]]
        ] /; EulerianQ[g,flag]
```

Finding an Eulerian Tour

An Eulerian cycle of a bipartite graph bounces back and forth between the stages.

```
In[55]:= EulerianCycle[ K[4,4] ]
Out[55]= {7, 2, 8, 1, 5, 4, 6, 3, 7, 4, 8, 3, 5, 2, 6, 1,
  7}
```

Every edge is traversed exactly once in an Eulerian cycle.

```
In[56]:= EmptyQ[ DeleteCycle[K[5], EulerianCycle[K[5]]] ]
Out[56]= True
```

The Chinese postman problem [Kwa62] asks for the shortest tour of a graph which visits each edge at least once. Clearly, if the graph is Eulerian, an Eulerian cycle is the optimal postman tour since each edge is visited exactly once. In a tree, the Chinese postman tour traverses each edge twice. It can be shown that no edge need ever be traversed more than twice, and a polynomial-time algorithm for finding the optimal tour appears in [EJ73]. A different generalization of the Eulerian cycle problem for odd vertices is finding the largest subgraph of a graph which is Eulerian. This problem can be shown to be NP-complete by a simple reduction from Hamiltonian cycle on a 3-regular graph.

de Bruijn Sequences

Suppose a safecracker is faced with a combination lock where the possible combinations consist of strings of n numbers from an alphabet of size σ, for a total of σ^n combinations. Further, the lock does not have a reset mechanism, so that it opens when the last n numbers entered are the combination. By exploiting the overlap between successive combinations, all combinations can be tried by entering fewer than $n\sigma^n$ numbers.

A *de Bruijn sequence* [dB46] is a circular sequence of length σ^n with the property that each string of length n over an alphabet of size σ occurs as a contiguous part of the sequence. Since the length of a de Bruijn sequence equals the number of strings of this size, and different strings cannot share the same endpoint in a sequence, a de Bruijn sequence gives the optimal solution to the safecracker's problem.

De Bruijn sequences can be constructed by finding an Eulerian cycle of a specially constructed graph [Goo46]. The vertices of this directed graph will consist of all the strings of length $n-1$, with outgoing edges for vertex $s_1 s_2 ... s_{n-1}$ going to vertices $s_2 ... s_{n-1}\sigma_i$, for each symbol σ_i in the alphabet. An Eulerian cycle of this graph describes a de Bruijn sequence, with the first symbol of each visited string composing the sequence.

```
DeBruijnSequence[alph_List,n_Integer?Positive] :=
        Block[{states = Strings[alph,n-1]},
                Rest[ Map[
                        (First[ states[[#]] ])&,
                        EulerianCycle[
                                MakeGraph[
                                        states,
                                        (Block[{i},
                                         MemberQ[
                                                Table[
                                                        Append[Rest[#1],alph[[i]]],
                                                        {i,Length[alph]}
                                                ],
                                                #2
                                         ]
                                        ])&
                                ],
                                Directed
                        ]
                ] ]
        ]
```

Constructing de Bruijn Sequences

De Bruijn sequences can also be generated by feedback shift registers [Gol66, Ron84] and have properties which are useful for random number generation. For example, each substring occurs an equal number of times.

```
In[57]:= s = DeBruijnSequence[{0,1},4]
Out[57]= {1, 1, 0, 0, 1, 0, 1, 1, 1, 1, 0, 1, 0, 0, 0, 0}
```

Taking every four consecutive symbols of this de Bruijn sequence gives all the strings of length four.

```
In[58]:= Table[Take[RotateLeft[s,q],4], {q,0,15}]
Out[58]= {{1, 1, 0, 0}, {1, 0, 0, 1}, {0, 0, 1, 0},
  {0, 1, 0, 1}, {1, 0, 1, 1}, {0, 1, 1, 1}, {1, 1, 1, 1},
  {1, 1, 1, 0}, {1, 1, 0, 1}, {1, 0, 1, 0}, {0, 1, 0, 0},
  {1, 0, 0, 0}, {0, 0, 0, 0}, {0, 0, 0, 1}, {0, 0, 1, 1},
  {0, 1, 1, 0}}
```

Since the in-degree and out-degree of each vertex is σ, the graph is Eulerian, and hence the sequence exists for arbitrary-sized alphabet. The number of de Bruijn sequences on n symbols of a σ symbol alphabet is determined in [Knu67].

```
In[59]:= DeBruijnSequence[{X,Y,Z},2]
Out[59]= {X, Y, Z, Z, Y, Y, X, X, Z}
```

■ 5.3.4 Hamiltonian Cycles

A *Hamiltonian cycle* of a graph G is a cycle which visits every vertex in G exactly once, as opposed to an Eulerian cycle which visits each edge exactly once. Despite this similarity, testing whether a graph is Hamiltonian is NP-complete, so all known algorithms involve exhaustive search.

We can use backtracking to find one or all the Hamiltonian cycles of a graph, advancing when we reach a new vertex and retreating when all edges return to previously visited vertices. By starting all cycles at vertex 1, each Hamiltonian cycle is generated only once. For better performance on sparse graphs, a custom implementation of backtracking is used instead of `Backtrack`.

```
HamiltonianQ[g_Graph] := False /; !BiconnectedQ[g]
HamiltonianQ[g_Graph] := HamiltonianCycle[g] != {}

HamiltonianCycle[g_Graph,flag_:One] :=
        Block[{s={1},all={},done,adj=Edges[g],e=ToAdjacencyLists[g],x,v,ind,n=V[g]},
                ind=Table[1,{n}];
                While[ Length[s] > 0,
                        v = Last[s];
                        done = False;
                        While[ ind[[v]] <= Length[e[[v]]] && !done,
                                If[!MemberQ[s,(x = e[[v,ind[[v]]++]])], done=True]
```

```
                    ];
                    If[ done, AppendTo[s,x], s=Drop[s,-1]; ind[[v]] = 1];
                    If[(Length[s] == n),
                            If [(adj[[x,1]]>0),
                                    AppendTo[all,Append[s,First[s]]];
                                    If [SameQ[flag,All],
                                            s=Drop[s,-1],
                                            all = Flatten[all]; s={}
                                    ],
                                    s = Drop[s,-1]
                            ]
                    ]
            ];
            all
    ]
```

Finding Hamiltonian Cycles

$K_{n,n}$ for $n > 1$ are the only Hamiltonian complete bipartite graphs.

```
In[60]:= HamiltonianCycle[K[3,3],All]
Out[60]= {{1, 4, 2, 5, 3, 6, 1}, {1, 4, 2, 6, 3, 5, 1},
  {1, 4, 3, 5, 2, 6, 1}, {1, 4, 3, 6, 2, 5, 1},
  {1, 5, 2, 4, 3, 6, 1}, {1, 5, 2, 6, 3, 4, 1},
  {1, 5, 3, 4, 2, 6, 1}, {1, 5, 3, 6, 2, 4, 1},
  {1, 6, 2, 4, 3, 5, 1}, {1, 6, 2, 5, 3, 4, 1},
  {1, 6, 3, 4, 2, 5, 1}, {1, 6, 3, 5, 2, 4, 1}}
```

All Hamiltonian graphs are biconnected, but not vise versa. This provides a useful heuristic in quickly identifying non-Hamiltonian graphs.

```
In[61]:= HamiltonianQ[ K[3,4] ]
Out[61]= False
```

Ore [Ore60] showed that a graph is Hamiltonian if the sum of degrees of non-adjacent vertices is always greater than n. A Hamiltonian cycle in such an *Ore graph* can be constructed in polynomial time [BC76].

```
In[62]:= HamiltonianCycle[ RegularGraph[4,7] ]
Out[62]= {1, 2, 3, 4, 5, 7, 6, 1}
```

Tait [Tai80] "proved" the four-color theorem by using the assumption that all cubic 3-connected graphs are Hamiltonian. Tutte [Tut46, Tut72] gave this non-Hamiltonian graph as a counterexample to Tait's assumption. A graph this size could take longer to test than you will care to wait.

```
In[63]:= ShowGraph[ ReadGraph["graphs/tutte"] ];
```

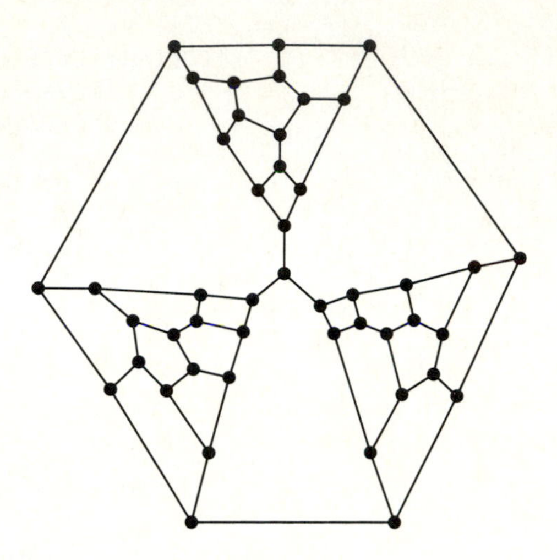

The dodecahedron holds a special place in the history of graph theory. In 1857 [BC87] Hamilton developed the "around the world game" which challenged the player to find a cycle which visits each of the twenty vertices of the dodecahedron exactly once. Unfortunately, the game did not sell well enough to make a profit.

```
In[64]:= ShowLabeledGraph[ d =
ReadGraph["graphs/dodecahedron"] ];
```

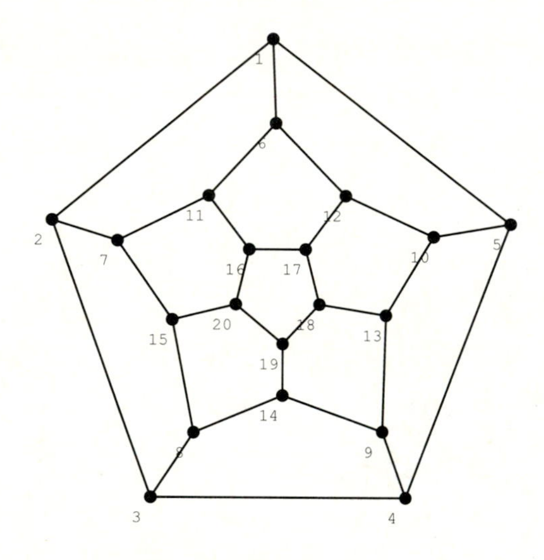

There are thirty possible Hamiltonian cycles in the dodecahedron. The twelve regular faces of the dodecahedron make it of crucial importance to the calendar paperweight industry [Cha85].

```
In[65]:= HamiltonianCycle[ d ]
Out[65]= {1, 2, 3, 4, 5, 10, 12, 17, 16, 20, 19, 18, 13,
  9, 14, 8, 15, 7, 11, 6, 1}
```

■ 5.3.5 Traveling Salesman Tours

The traveling salesman problem is perhaps the most notorious NP-complete problem [LLKS85]. Given a collection of cities connected by roads, find the *shortest* Hamiltonian cycle, or route which visits each city exactly once. The appealing title and the existence of traveling salesman jokes have made this the favorite hard problem of the popular press.

A backtracking algorithm similar to that used in `HamiltonianCycle` serves to find the optimal tour. The best tour found to date is saved, and the search backtracks unless the partial solution is cheaper than the cost of the best tour. In a graph with positive edge weights, such a partial solution cannot expand into a better tour.

```
TravelingSalesman[g_Graph] :=
        Block[{v,s={1},sol={},done,cost,g1,e=ToAdjacencyLists[g],x,ind,best,n=V[g]},
                ind=Table[1,{n}];
                g1 = PathConditionGraph[g];
                best = Infinity;
                While[ Length[s] > 0,
                        v = Last[s];
                        done = False;
                        While[ ind[[v]] <= Length[e[[v]]] && !done,
                                x = e[[v,ind[[v]]++]];
                                done = (best > CostOfPath[g1,Append[s,x]]) &&
                                        !MemberQ[s,x]
                        ];
                        If[done, AppendTo[s,x], s=Drop[s,-1]; ind[[v]] = 1];
                        If[(Length[s] == n),
                                cost = CostOfPath[g1, Append[s,First[s]]];
                                If [(cost < best), sol = s; best = cost ];
                                s = Drop[s,-1]
                        ]
                ];
                Append[sol,First[sol]]
        ]

CostOfPath[Graph[g_,_],p_List] := Apply[Plus, Map[(Element[g,#])&,Partition[p,2,1]] ]

Element[a_List,{index___}] := a[[ index ]]
```

Solving the Traveling Salesman Problem

The function `Element` is a kludge, providing a way to assign a part of a structure a new value without specifying the index in advance.

In simple graphs, the cost of any Hamiltonian cycle is n.

```
In[66]:= CostOfPath[K[3,3],HamiltonianCycle[K[3,3]]]
Out[66]= 6
```

Here we construct a weighted complete graph, so the traveling salesman solution is not just an arbitrary Hamiltonian cycle.

```
In[67]:= TableForm[ Edges[ r = RandomGraph[7,1,{1,10}] ] ]
                    0   2   3   6   6   3   4
                    2   0   7   5   9   5   8
                    3   7   0   10  2   7   2
                    6   5   10  0   7   9   8
                    6   9   2   7   0   10  1
                    3   5   7   9   10  0   6
Out[67]//TableForm= 4   8   2   8   1   6   0
```

Since r is a complete graph, it is Hamiltonian and thus contains a non-infinite solution.

```
In[68]:= TravelingSalesman[r]
Out[68]= {1, 3, 7, 5, 4, 2, 6, 1}
```

The optimal tour avoids the heaviest edges.

```
In[69]:= CostOfPath[r,%]
Out[69]= 26
```

Since the traveling salesman problem is so difficult to solve exactly, it is useful to be able to find good but non-optimal tours quickly as well as establish a decent lower bound to assess how far from optimality this tour is.

The *Euclidean* traveling salesman problem assumes that the distances in the graph correspond to distances for an embedding in the Euclidean plane. Thus the triangle inequality is satisfied between all distances. A function d obeys the *triangle inequality* if and only if for all x, y, and z, $d(x,y) + d(y,z) \geq d(x,z)$. A *metric* is a function that obeys the triangle inequality, is non-negative, and $d(x,y) = 0$ if and only if $x = y$.

```
TriangleInequalityQ[e_?SquareMatrixQ] :=
        Block[{i,j,k,n=Length[e],flag=True},
                Do [

                        If[(e[[i,k]]!=0) && (e[[k,j]]!=0) && (e[[i,j]]!=0),
                                If[e[[i,k]]+e[[k,j]]<e[[i,j]],
                                        flag = False;
                                ]
                        ],
                        {i,n},{j,n},{k,n}
                ];
                flag
        ]
```

```
TriangleInequalityQ[g_Graph] := TriangleInequalityQ[Edges[g]]
```

Testing the Triangle Inequality

Even if the weights of a particular graph are not metric, the all-pairs shortest path function of it is.

```
In[70]:= TriangleInequalityQ[ AllPairsShortestPath[
RandomGraph[6,1,{1,10}] ] ]

Out[70]= True
```

An upper bound on the length of a Euclidean traveling salesman tour can be obtained by doubling the length of the minimum spanning tree, to be discussed in Section 6.2. An actual tour is found by performing a depth-first search on the spanning tree and eliminating duplicate vertices. For Euclidean graphs, this tour is guaranteed to be at most twice the cost of the optimal tour [RSL77].

Another heuristic for the traveling salesman problem which has proven effective takes a Hamiltonian cycle and mutates it by deleting a small number of edges and connecting them up in a different way. The performance of such a heuristic depends on the mutation schedule and selection procedure [Lin65].

A lower bound on the cost of a traveling salesman tour follows from the observation that every vertex must be entered and exited, so summing up the smallest costs associated with each vertex provides a lower bound.

```
TravelingSalesmanBounds[g_Graph] := {LowerBoundTSP[g], UpperBoundTSP[g]}

UpperBoundTSP[g_Graph] :=
        CostOfPath[g, Append[DepthFirstTraversal[MinimumSpanningTree[g],1],1]]

LowerBoundTSP[g_Graph] := Apply[Plus, Map[Min,ReplaceAll[Edges[g],0->Infinity]]]
```

Determining Bounds for the Traveling Salesman Tour

Since both the heuristic and the lower-bound calculations are fairly simple and crude, a gap between them can be expected. The optimal tour for r lies within these bounds.

```
In[71]:= {TravelingSalesmanBounds[r], CostOfPath[r,
TravelingSalesman[r]]}

Out[71]= {{16, 54}, 26}
```

Because r is not Euclidean, the upper bound can be off by more than a factor of two. The shortest-path graph satisfies the triangle inequality.

```
In[72]:= TableForm[ Edges[ r1 =
Graph[AllPairsShortestPath[r],Vertices[r]] ] ]
                      0  2  3  6  5  3  4
                      2  0  5  5  7  5  6
                      3  5  0  9  2  6  2
                      6  5  9  0  7  9  8
                      5  7  2  7  0  7  1
                      3  5  6  9  7  0  6
Out[72]//TableForm=   4  6  2  8  1  6  0
```

On the metric graph, the upper bound is within a factor of two of the optimal tour.

```
In[73]:= {TravelingSalesmanBounds[r1],
CostOfPath[r1,TravelingSalesman[r1]]}
Out[73]= {{16, 39}, 26}
```

5.4 Partial Orders

A *partially ordered set*, or *poset*, is a set with a consistent ordering relation over its elements. More formally, a partial order is a binary relation which is reflexive, transitive, and anti-symmetric. Interesting binary relations between combinatorial objects, such as those constructed with MakeGraph, often turn out to be partial orders.

```
PartialOrderQ[g_Graph] := ReflexiveQ[g] && AntiSymmetricQ[g] && TransitiveQ[g]

TransitiveQ[g_Graph] := IdenticalQ[g,TransitiveClosure[g]]

ReflexiveQ[Graph[g_List,_]] :=
        Block[{i},
                Apply[And, Table[(g[[i,i]]!=0),{i,Length[g]}] ]
        ]

AntiSymmetricQ[g_Graph] :=
        Block[{e = Edges[g], g1 = RemoveSelfLoops[g]},
                Apply[And, Map[(Element[e,Reverse[#]]==0)&,ToOrderedPairs[g1]] ]
        ]
```

Partial Order Predicates

5.4.1 Transitive Closure and Reduction

A graph G is *transitive* if any three vertices x, y, z such that edges $\{x, y\}, \{y, z\} \in G$ imply $\{x, z\} \in G$. Any graph can be made transitive by adding enough extra edges. The *transitive closure* $C(G)$ of a graph G contains an edge $\{u, v\}$ whenever there is a directed path from u to v. The *transitive reduction* of a graph G is the smallest graph $R(G)$ such that $C(G) = C(R(G))$.

The definition of transitive closure yields a straightforward $O(n^3)$ time algorithm for computing it. Determining the transitive reduction of a binary relation is more difficult, since arcs which were not part of the relation can be in the reduction. Although it is difficult to find the smallest subset of the *arcs* determining the relation, [AGU72] give an efficient algorithm to determine the smallest subset of *vertex pairs* determining the relation.

```
TransitiveClosure[g_Graph] :=
        Block[{i,j,k,e=Edges[g],n=V[g]},
                Do [
                        If[ e[[j,i]] != 0,
                                Do [
                                        If[ e[[i,k]] != 0, e[[j,k]]=1],
```

```
                                        {k,n}
                                ]
                        ],
                        {i,n},{j,n}
                ];
                Graph[e,Vertices[g]]
        ]
```

Finding the Transitive Closure of a Graph

All of these complications go away when the graph is acyclic, as is true with a partial order. In this case, any edge of G which would have been added in finding the transitive closure cannot appear in the transitive reduction. For directed cyclic graphs, this policy could delete every edge in the cycle, and so we delete any edge which is implied by two edges in the current transitive reduction. This gives *a* reduction, but one which is not necessarily minimal.

```
TransitiveReduction[g_Graph] :=
        Block[{closure=reduction=Edges[g],i,j,k,n=V[g]},
                Do[
                        If[ closure[[i,j]]!=0 && closure[[j,k]]!=0 &&
                                reduction[[i,k]]!=0 && (i!=j) && (j!=k) && (i!=k),
                                        reduction[[i,k]] = 0
                        ],
                        {i,n},{j,n},{k,n}
                ];
                Graph[reduction,Vertices[g]]
        ] /; AcyclicQ[RemoveSelfLoops[g],Directed]

TransitiveReduction[g_Graph] :=
        Block[{reduction=Edges[g],i,j,k,n=V[g]},
                Do[
                        If[ reduction[[i,j]]!=0 && reduction[[j,k]]!=0 &&
                                reduction[[i,k]]!=0 && (i!=j) && (j!=k) && (i!=k),
                                        reduction[[i,k]] = 0
                        ],
                        {i,n},{j,n},{k,n}
                ];
                Graph[reduction,Vertices[g]]
        ]
```

Finding the Transitive Reduction of an Acyclic Graph

The divisibility relation between integers is reflexive since each integer divides itself and anti-symmetric, since x cannot divide y if $x > y$. Finally, it is transitive, as $x \setminus y$ implies $y = cx$ for some integer c, so $y \setminus z$ implies $x \setminus z$.

```
In[74]:= ShowLabeledGraph[ g =
MakeGraph[Range[8],(Mod[#1,#2]==0)&] ];
```

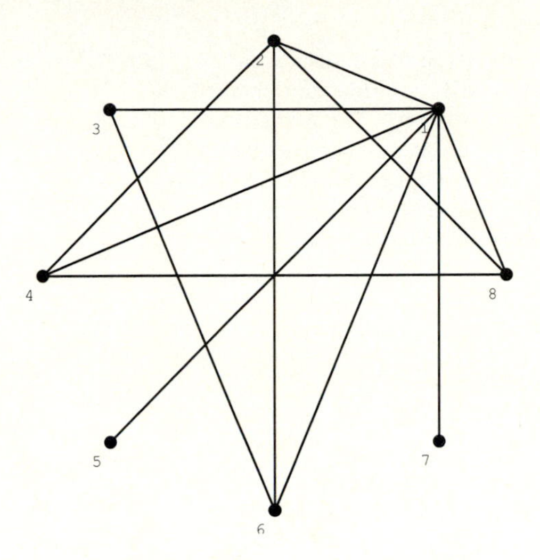

Since the divisibility relation is reflexive, transitive, and anti-symmetric, it is a partial order.

```
In[75]:= PartialOrderQ[g]
Out[75]= True
```

The transitive reduction eliminates all implied edges in the divisibility relation, such as $4 \setminus 8$, $1 \setminus 4$, $1 \setminus 6$, and $1 \setminus 8$.

```
In[76]:= ShowLabeledGraph[ TransitiveReduction[g] ];
```

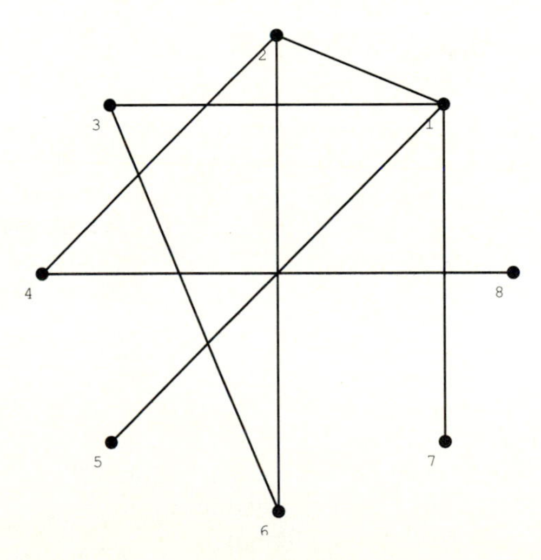

The transitive closure of the transitive reduction of a transitive graph is an identity operation.

```
In[77]:= IdenticalQ[g,
TransitiveClosure[TransitiveReduction[g]] ]
Out[77]= True
```

The transitive reduction of a non-acyclic graph is not necessarily minimal under our simple reduction algorithm. The algorithm works under the assumption that the graph is transitive in the first place, which isn't true for complete graphs, since by our convention they contain no self-loops.

```
In[78]:= ShowGraph[TransitiveReduction[K[10]]];
```

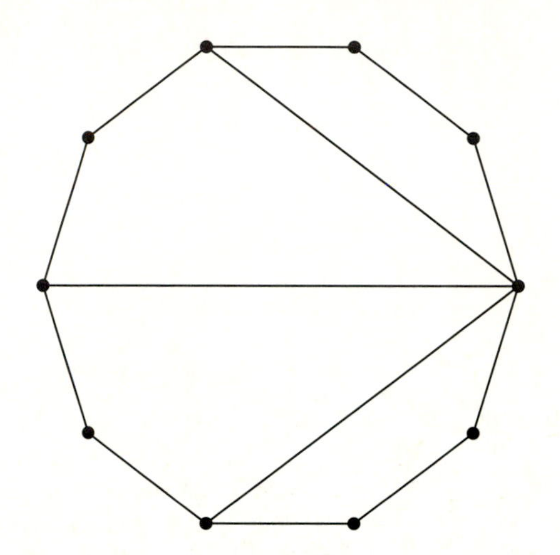

■ 5.4.2 Hasse Diagrams

What is the best embedding for a partial order? It should clearly identify the hierarchy imposed by the order and contain as few edges as possible to avoid cluttering up the picture. Such a drawing is a *Hasse diagram*, the pictorial representation of a partial order.

We have seen how ranked embeddings (Section 3.3.4) can be used to represent a hierarchy. At the bottom of the hierarchy are the vertices which have no ancestors or, equivalently, have in-degree 0. The vertices which have out-degree 0 all represent maxima of the partial order and thus are ranked together at the top. Performing a transitive reduction minimizes the number of edges in the graph.

```
HasseDiagram[g_Graph] :=
        Block[{r,rank,m,stages,freq=Table[0,{V[g]}]},
                r = TransitiveReduction[ RemoveSelfLoops[g] ];
                rank = RankGraph[
                                MakeUndirected[r],
                                Select[Range[V[g]],(InDegree[r,#]==0)&]
                ];
                m = Max[rank];
```

```
            rank = MapAt[(m)&,rank,Position[OutDegree[r],0]];
            stages = Distribution[ rank ];
            Graph[
                    Edges[r],
                    Table[
                            m = ++ freq[[ rank[[i]] ]];
                            {(m-1) + (1-stages[[rank[[i]] ]])/2, rank[[i]]},
                            {i,V[g]}
                    ]
            ]
    ] /; AcyclicQ[RemoveSelfLoops[g],Directed]
```

Constructing the Hasse Diagram of a Partial Order

It is impossible to understand the *boolean algebra*, the partial order on subsets defined by inclusion, from this circular embedding. Beyond the arbitrary position of the vertices, this graph contains too many edges to understand.

```
In[79]:= ShowGraph[ s = MakeGraph[Subsets[4],
((Intersection[#2,#1]===#1) && (#1 != #2))&] ];
```

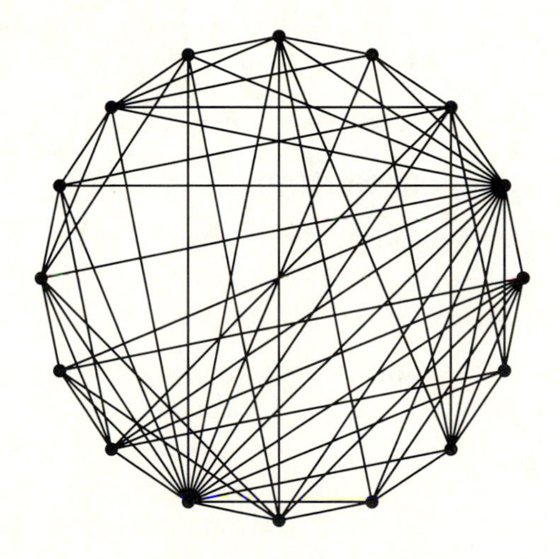

The Hasse diagram clearly shows the lattice structure, with top and bottom elements. All implied edges have been removed. The boolean algebra defined by the subsets of a six-element set appears as the frontispiece for this chapter.

```
In[80]:= ShowLabeledGraph[ HasseDiagram[s], Subsets[4] ];
```

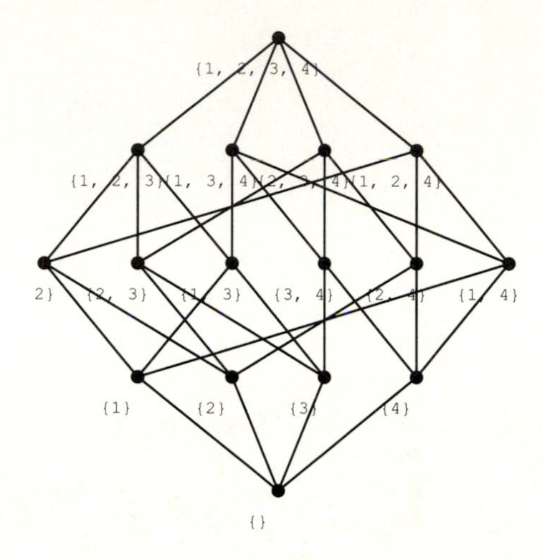

■ 5.4.3 Topological Sorting

A directed acyclic graph defines a precedence relation upon the vertices, if edge $\{i, j\}$ means vertex i must occur before vertex j. A *topological sort* is a permutation p of the vertices of a graph such that an edge $\{i, j\}$ implies i appears before j in p.

It is apparent that only directed, acyclic graphs can be topologically sorted, as no vertex in a directed cycle can take precedence over all the rest. Every acyclic graph contains at least one vertex v of out-degree zero. Clearly, v can appear last in the topological ordering. Deleting v leaves a graph with at least one other vertex of outdegree zero, and repeating this argument gives an algorithm for topologically sorting any directed acyclic graph.

```
TopologicalSort[g_Graph] :=
        Block[{g1 = RemoveSelfLoops[g],e,indeg,zeros,v},
                e=ToAdjacencyLists[g1];
                indeg=InDegree[g1];
                zeros = Flatten[ Position[indeg, 0] ];
                Table [
                        {v,zeros}={First[zeros],Rest[zeros]};
                        Scan[
                                ( indeg[[#]]--;
                                  If[indeg[[#]]==0, AppendTo[zeros,#]] )&,
                                e[[ v ]]
                        ];
```

```
                        v,
                        {V[g]}
               ]
        ] /; AcyclicQ[RemoveSelfLoops[g],Directed]
```

Topologically Sorting a Graph

Any topological ordering of the relation graph must begin with one. In this ordering, the vertices are ranked in terms of distance from a source and hence the primes all immediately follow the identity.

```
In[81]:= TopologicalSort[
MakeGraph[Range[12],(Mod[#2,#1]==0)&] ]

Out[81]= {1, 2, 3, 5, 7, 11, 4, 6, 9, 10, 8, 12}
```

Any permutation of the vertices of an empty graph defines a topological order.

```
In[82]:= TopologicalSort[ EmptyGraph[10] ]

Out[82]= {1, 2, 3, 4, 5, 6, 7, 8, 9, 10}
```

On the other hand, a complete directed acyclic graph defines a total order, so there is only one possible output from `TopologicalSort`.

```
In[83]:= TopologicalSort[ MakeGraph[Range[10],(#1 > #2)&]
]

Out[83]= {10, 9, 8, 7, 6, 5, 4, 3, 2, 1}
```

■ 5.5 Graph Coloring

A *vertex coloring* is an assignment of labels or colors to each vertex of a graph such that no edge connects two identically colored vertices. It is easy to color a graph with n colors by coloring each vertex uniquely, but the problem gets harder when the goal is minimizing the number of colors. The *chromatic number* of a graph $\chi(G)$ is the smallest number of colors with which it is possible to color all vertices of G.

The only one-colorable graphs are totally disconnected, and two-colorability is a synonym for bipartiteness. The situation gets more interesting when at least three colors are necessary.

Perhaps the most notorious problem in graph theory was whether all planar graphs are four-colorable. The history of the four-color problem dates back to 1852 [SK86], culminating with the computer-aided proof of Appel and Haken [AHK77, AH77]. The unusual nature of this proof has made it the focus of some nasty rumors, which are debunked in [AH86].

Applications of graph coloring include scheduling and compiler optimization [OR81].

■ 5.5.1 Chromatic Polynomials

Any labeled graph G can be colored in a certain number of ways with exactly k colors $1, \ldots, k$. Thus G defines a particular chromatic function $\pi_G(z)$ such that $\pi_G(k)$ is the number of ways G can be k-colored. Birkhoff [Bir12, BL46] presented a nice way to compute this *chromatic polynomial* of a graph, for the function is indeed always a polynomial.

Observe that for an empty graph E on n vertices, $\pi_E(z) = z^n$, since every vertex can be assigned any color. Now consider the chromatic polynomial for an arbitrary graph G. If we delete an edge $\{x, y\}$ of G, yielding G', we get a graph which can be colored in more ways than G, since x and y in G' can be assigned the same color, which was prevented by $\{x, y\}$. Contracting $\{x, y\}$ in G' gives a graph which can be colored each way G' can, if x and y recieve the same color. Thus the chromatic polynomial of G is the chromatic polynomial of $G - \{x, y\}$ minus the chromatic polynomial of the contraction of $\{x, y\}$ in G, and since both graphs are smaller than G, this computation eventually ends with an empty graph.

```
ChromaticPolynomial[g_Graph,z_] := 0 /; Identical[g,K[0]]

ChromaticPolynomial[g_Graph,z_] :=
        Block[{i}, Product[z-i, {i,0,V[g]-1}] ] /; CompleteQ[g]
```

```
ChromaticPolynomial[g_Graph,z_] := z ( z - 1 ) ^ (V[g]-1) /; TreeQ[g]

ChromaticPolynomial[g_Graph,z_] :=
        If [M[g]>Binomial[V[g],2]/2, ChromaticDense[g,z], ChromaticSparse[g,z]]

ChromaticSparse[g_Graph,z_] := z^V[g] /; EmptyQ[g]
ChromaticSparse[g_Graph,z_] :=
        Block[{i=1, v, e=Edges[g], none=Table[0,{V[g]}]},
                While[e[[i]] === none, i++];
                v = Position[e[[i]],1] [[1,1]];
                ChromaticSparse[ DeleteEdge[g,{i,v}], z ] -
                        ChromaticSparse[ Contract[g,{i,v}], z ]
        ]

ChromaticDense[g_Graph,z_] := ChromaticPolynomial[g,z] /; CompleteQ[g]
ChromaticDense[g_Graph,z_] :=
        Block[
                {i=1, v, e=Edges[g], all=Join[Table[1,{V[g]-1}],{0}] },
                While[e[[i]] === RotateRight[all,i], i++];
                v = Last[ Position[e[[i]],0] ] [[1]];
                ChromaticDense[ AddEdge[g,{i,v}], z ] +
                        ChromaticDense[ Contract[g,{i,v}], z ]
        ]
```

Computing the Chromatic Polynomial of a Graph

By rearranging the recurrence relation, we can determine the chromatic polynomial of a graph by adding edges and terminating instead when we reach a complete graph. This is more efficient on dense graphs. Further, we can exploit the fact that the chromatic polynomial is known for trees. Except for such special cases, the computation is exponential in the minimum of the number of edges in the graph and its complement.

```
ChromaticNumber[g_Graph] :=
        Block[{ways, z},
                ways[z_] = ChromaticPolynomial[g,z];
                For [z=0, z<=V[g], z++,
                        If [ways[z] > 0, Return[z]]
                ]
        ]
```

Computing the Chromatic Number

Since the chromatic polynomial of a graph determines the number of ways to color the graph with z colors, the chromatic number is the smallest positive integer z such that the value of the polynomial is greater than zero. As the chromatic number decision problem is NP-complete, calculating the chromatic polynomial must be at least as hard.

On dense graphs, the computation terminates on complete graphs, which have a chromatic polynomial defined in terms of the falling factorial function.

```
In[84]:= ChromaticPolynomial[ DeleteEdge[K[5],{1,2}], z ]
Out[84]= (-3 + z) (-2 + z) (-1 + z) z +
  (-4 + z) (-3 + z) (-2 + z) (-1 + z) z
```

The computed polynomial is not always in the simplest possible form.

```
In[85]:= Simplify[ % ]
                 2
Out[85]= (-3 + z)  (-2 + z) (-1 + z) z
```

Finding the chromatic polynomial of a cycle would proceed more quickly if the special case of the tree were tested in each iteration. However, repeatedly testing this special case slows down the computation for most graphs.

```
In[86]:= ChromaticPolynomial[Cycle[5], z]
                  2       3      4    5
Out[86]= 4 z - 10 z  + 10 z  - 5 z  + z
```

The chromatic polynomial of a disconnected graph is the product of the chromatic polynomials of the connected components.

```
In[87]:= ChromaticPolynomial[ GraphUnion[K[2,2],Cycle[3]],
z ]
             2       3       4       5      6    7
Out[87]= -6 z  + 21 z  - 29 z  + 20 z  - 7 z  + z
```

The chromatic polynomial of a graph of order n has degree $n+1$. The coefficients alternate in sign and the coefficient of the $n-1$st term is $-m$, where m is the number of edges.

```
In[88]:= Expand[ wheelpoly[z_] =
ChromaticPolynomial[Wheel[5],z] ]
                   2       3      4    5
Out[88]= 14 z - 31 z  + 24 z  - 8 z  + z
```

To three-color the wheel, one color must be selected for the center and the other two alternate red-black or black-red around the cycle, for a total of six three-colorings. Observe that the chromatic polynomial is still defined when the number of colors exceeds the degree.

```
In[89]:= Table[ wheelpoly[z], {z,1,7} ]
Out[89]= {0, 0, 6, 72, 420, 1560, 4410}
```

This wheel has chromatic number 3, since the values of the chromatic polynomial were zero for 1 and 2.

```
In[90]:= ChromaticNumber[ Wheel[5] ]
Out[90]= 3
```

There is no known NP-complete problem which can have at most only a polynomial number of solutions, since certain models of reduction are solution preserving and certain NP-complete problems have a possibly exponential number of solutions. And yet the number of k-colorings of a graph is given by a polynomial! The catch is that the polynomial is of degree $n+1$, so $O(k^{n+1})$ k-colorings are possible.

■ 5.5.2 Coloring Bipartite Graphs

A graph is bipartite if the vertices can be partitioned into two disjoint sets such that there exists no edge between vertices of different sets. Thus 'two-colorable' and 'bipartite' are exactly the same property. An alternate definition is that a graph is bipartite if and only if all cycles are of even length. A greedy strategy is sufficient to two-color a bipartite graph, since colors must alternate along any path.

```
TwoColoring[g_Graph] :=
        Block[{queue,elem,edges,col,flag=True,colored=Table[0,{V[g]}]},
                edges = ToAdjacencyLists[g];
                While[ MemberQ[colored,0],
                        queue = First[ Position[colored,0] ];
                        colored[[ First[queue] ]] = 1;
                        While[ queue != {},
                                elem = First[queue];
                                col = colored[[elem]];
                                Scan[
                                        (Switch[colored[[ # ]],
                                                col, flag = False,
                                                0, AppendTo[queue, # ];
                                                   colored[[#]] = Mod[col,2]+1
                                        ])&,
                                        edges[[elem]]
                                ];
                                queue = Rest[queue];
                        ]
                ];
                If [!flag, colored[[1]] = 0];
                colored
        ]

BipartiteQ[g_Graph] := ! MemberQ[ TwoColoring[g], 0 ]
```

Finding a Two-Coloring of a Graph

The two-coloring of a complete bipartite graph partitions it by vertex stages.

```
In[91]:= TwoColoring[ K[5,4] ]
Out[91]= {1, 1, 1, 1, 1, 2, 2, 2, 2}
```

All trees are bipartite.

```
In[92]:= BipartiteQ[ RandomTree[10] ]
Out[92]= True
```

Odd cycles are not bipartite, although even cycles are.

```
In[93]:= BipartiteQ[ Cycle[11] ]
Out[93]= False
```

5.5.3 Finding a Vertex Coloring

Finding a minimum coloring can be done using brute-force search [Chr71, Wil84], but to effectively color large graphs heuristics are faster and usually quite effective. In this section we consider a heuristic approach to graph coloring. Since it is not necessarily minimal, to determine the chromatic number of a particular graph, we can compute the chromatic polynomial as discussed in the previous section.

Brelaz [Bre79] recommends the following heuristic algorithm, which relies on defining a total order on the colors. This is easy to achieve by numbering the colors from 1 to k. First, color the vertex of largest degree with color 1. Then repeatedly select the vertex with highest *color degree*, where the color degree is the number of adjacent vertices which have already been colored, and color it with the smallest possible color.

```
VertexColoring[g_Graph] :=
        Block[{v,l,n=V[g],e=ToAdjacencyLists[g],x,color=Table[0,{V[g]}]},
                v = Map[(Apply[Plus,#])&, Edges[g]];
                Do[
                        l = MaximumColorDegreeVertices[e,color];
                        x = First[l];
                        Scan[(If[ v[[#]] > v[[x]], x = #])&, l];
                        color[[x]] = Min[
                                Complement[ Range[n], color[[ e[[x]] ]] ]
                        ],
                        {V[g]}
                ];
                color
        ]

MaximumColorDegreeVertices[e_List,color_List] :=
        Block[{n=Length[color],l,i,x},
                l = Table[ Count[e[[i]], _?(Function[x,color[[x]]!=0])], {i,n}];
                Do [
                        If [color[[i]]!=0, l[[i]] = 0],
                        {i,n}
                ];
                Flatten[ Position[ l, Max[l] ] ]
        ]
```

Brelaz's Heuristic Graph-Coloring Algorithm

A complete k-partite graph is k-colorable. Further, Brelaz's algorithm colors such a graph with the minimum number of colors.

```
In[94]:= ShowLabeledGraph[ K[1,2,3,1],
VertexColoring[K[1,2,3,1]] ]
```

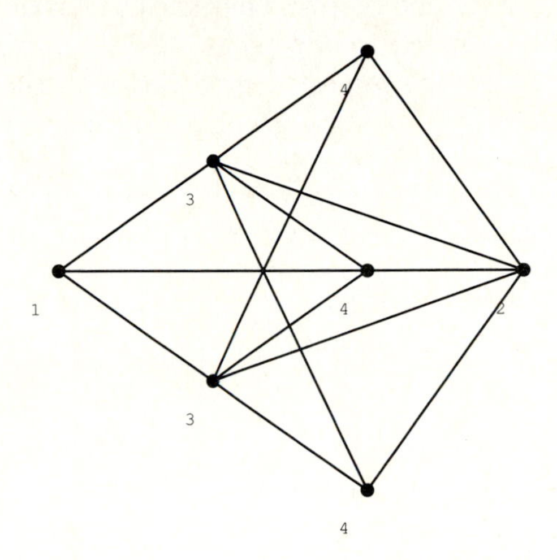

Brooks' theorem [Bro41, Lov75] states that the chromatic number of a graph is at most the maximum vertex degree Δ unless the graph is complete or an odd cycle. `RealizeDegreeSequence` generates semirandom graphs, so the colorings are not necessarily identical each time.

```
In[95]:= TableForm[ Table[ VertexColoring[
RealizeDegreeSequence[{4,4,4,4,4,4}] ], {5}] ]

                     1  2  2  1  3  3
                     1  2  1  3  3  2
                     1  2  3  1  3  2
                     1  2  3  1  2  3
Out[95]//TableForm=  1  2  1  2  3  3
```

The size of the largest clique, or complete subgraph, represents a lower bound on the chromatic number, since each vertex in the clique must be colored differently. However, large cliques (or even short cycles) are not necessary for high chromatic numbers, as [Erd61, Lov68] showed that for any two positive integers g and k there exists a graph of girth at least g with chromatic number at least k.

```
In[96]:= VertexColoring[ K[10] ]
Out[96]= {1, 2, 3, 4, 5, 6, 7, 8, 9, 10}
```

Like any heuristic, Brelaz's can be improved by making it more sophisticated, at a cost of speed and complexity. Other studies of coloring heuristics include [Man85b, MMI72], and empirical results on coloring heuristics appear in [Gou88].

■ 5.5.4 Edge Colorings

Finding a minimum edge-coloring of a graph is equivalent to finding the minimum vertex coloring of its line graph. Thus the chromatic polynomial and vertex coloring functions can be applied to edge colorings. The *edge chromatic number* of a graph must be at least Δ, the largest degree vertex of the graph, but Vizing [Viz64] and Gupta [Gup66] proved that any graph can be edge-colored with at most $\Delta+1$ colors.

```
EdgeColoring[g_Graph] := VertexColoring[ LineGraph[g] ]

EdgeChromaticNumber[g_Graph] := ChromaticNumber[ LineGraph[g] ]
```

Finding an Edge Coloring of a Graph

The edge chromatic number of a complete bipartite graph is Δ.

```
In[97]:= EdgeColoring[ K[3,4] ]
Out[97]= {1, 2, 3, 4, 2, 1, 4, 3, 3, 4, 1, 2}
```

Since the line graph of a cycle is a cycle, the edge and vertex chromatic numbers are identical. The vertex labelings for the line graph of a cycle are not the same as the cycle, hence the two colorings are not identical.

```
In[98]:= {EdgeColoring[Cycle[10]],
VertexColoring[Cycle[10]]}
Out[98]= {{1, 2, 2, 1, 2, 1, 2, 1, 2, 1},
  {1, 2, 1, 2, 1, 2, 1, 2, 1, 2}}
```

Despite Vizing's tight bounds on the edge chromatic number of a graph, determining the exact value is NP-complete [Hol81].

```
In[99]:= EdgeChromaticNumber[ Cycle[5] ]
Out[99]= 3
```

Finding the minimum edge-coloring occurs in a variety of scheduling problems. For example, consider a group of processors each capable of performing a group of jobs. This defines a bipartite graph, and the minimum edge-coloring of this graph is the minimum amount of time it will take to complete the job.

5.6 Cliques, Vertex Covers, and Independent Sets

NP-completeness proofs consist of reductions from a known NP-complete problem to the problem in question. Usually these constructions are too baroque to be particularly useful for computation. The most notable exception is the relationship between maximum clique, minimum vertex cover, and maximum independent set, which are different manifestions of the same underlying structure.

5.6.1 Maximum Clique

When a group of people form a clique, everyone knows everyone else. Likewise, a *clique* in a graph G is a subset of the vertices which induce a complete graph. Each vertex and edge define cliques of size one and two, respectively, but the most interesting cliques are the largest ones.

Finding the largest clique can be done by testing all subsets of the vertices of G to see if they induce a complete graph, but this is frightfully expensive. If we consider the subsets in order of decreasing size, the first clique which is found must be the largest. Further, we do not have to consider all k-subsets of the vertices in testing for a clique of size k, merely all k-subsets of the vertices which are of degree $k-1$ or greater, since they cannot be in any clique of that size.

```
CliqueQ[g_Graph,clique_List] :=
        IdenticalQ[ K[Length[clique]], InduceSubgraph[g,clique] ] /; SimpleQ[g]

MaximumClique[g_Graph] := {} /; g === K[0]

MaximumClique[g_Graph] :=
        Block[{d = Degrees[g],i,clique=Null,k},
                i = Max[d];
                While[(SameQ[clique,Null]),
                        k = K[i+1];
                        clique = FirstExample[
                                KSubsets[Flatten[Position[d,_?((#>=i)&)]], i+1],
                                (IdenticalQ[k,InduceSubgraph[g,#]])&
                        ];
                        i--;
                ];
                clique
        ]

FirstExample[list_List, predicate_] := Scan[(If [predicate[#],Return[#]])&,list]
```

Finding the Maximum Clique

Any subset of the vertices of a complete graph forms a clique.

```
In[100]:= CliqueQ[ K[10], RandomSubset[Range[10]] ]
Out[100]= True
```

No bipartite graph contains a clique larger than two. Since the search algorithm looks for big cliques first, it can be slow when the graph contains only small cliques.

```
In[101]:= MaximumClique[K[3,3]]
Out[101]= {1, 4}
```

The Turán graph $T_{n,p}$ is the largest order n graph which does not contain K_p as a subgraph. Any complete k-partite graph has a maximum clique of size k.

```
In[102]:= MaximumClique[Turan[6,4]]
Out[102]= {1, 3, 5}
```

■ 5.6.2 Minimum Vertex Cover

A *vertex cover* V' of a graph G is a subset of the vertices such that every edge in E is incident upon a member of V'. Two vertices not in the cover cannot be connected by an edge, or else V' isn't a cover. Thus in the complement graph of G, the vertices not in a clique define a cover. This observation permits us to use `MaximumClique` to find the minimum vertex cover.

```
VertexCoverQ[g_Graph,vc_List] :=
        CliqueQ[ GraphComplement[g], Complement[Range[V[g]], vc] ]

MinimumVertexCover[g_Graph] :=
        Complement[ Range[V[g]], MaximumClique[ GraphComplement[g] ] ]
```

Finding the Smallest Vertex Cover

Any subset of $n-1$ vertices is a minimum vertex cover of K_n.

```
In[103]:= MinimumVertexCover[K[5]]
Out[103]= {2, 3, 4, 5}
```

In a complete k-partite graph, any vertex cover contains vertices from at least $k-1$ different stages.

```
In[104]:= MinimumVertexCover[K[2,2,2]]
Out[104]= {3, 4, 5, 6}
```

■ 5.6.3 Maximum Independent Set

An *independent set* of a graph G is a subset of the vertices such that no two vertices in the subset represent an edge of G. Given any vertex cover of a graph, all vertices not in the cover define an independent set, since all edges are incident upon the cover.

The *independence number* of a graph is the cardinality of the largest independent set. By definition, the independence number of a graph plus the cover number equals the number of vertices.

```
IndependentSetQ[g_Graph,indep_List] :=
        VertexCoverQ[ g, Complement[ Range[V[g]], indep] ]

MaximumIndependentSet[g_Graph] := Complement[Range[V[g]], MinimumVertexCover[g]]
```

Finding the Largest Independent Set

The largest independent set in a grid graph consists of the white squares of a chessboard, every other point on every other row.

```
In[105]:= MaximumIndependentSet[GridGraph[3,3]]
Out[105]= {1, 3, 5, 7, 9}
```

Covers and independent sets can be defined in a similar way for edges. Gallai [Gal59] showed that the size of the minimum edge cover plus the size of the maximum number of independent edges adds up to the number of vertices n.

■ 5.6.4 Perfect Graphs

A graph G is *perfect* if for each induced subgraph of G, the size of the largest clique equals its chromatic number. The chromatic number and maximum independent set of a perfect graph can be computed in polynomial time; hence they have an important connection to optimization problems.

```
PerfectQ[g_Graph] :=
        Apply[
                And,
                Map[(ChromaticNumber[#] == Length[MaximumClique[#]])&,
                        Map[(InduceSubgraph[g,#])&, Subsets[Range[V[g]]] ] ]
        ]
```

Identifying Perfect Graphs

This definition requires solving two NP-complete problems on each of an exponential number of subgraphs. No matter how fast your machine is, this function can be practical only on very small graphs.

The complement of any bipartite graph is perfect. This is a special case of the perfect graph theorem [Ful71, Lov72] which states that the complement of a perfect graph is perfect.

```
In[106]:= PerfectQ[ GraphComplement[ RandomTree[5] ] ]
Out[106]= True
```

■ 5.7 Exercises and Research Problems

Exercises

1. Write a function to find the strongly connected components of a graph using `TransitiveClosure`. How does your function's performance compare to `StronglyConnectedComponents`?

2. Improve the efficiency of the k-connectivity functions `EdgeConnectivity` and `VertexConnectivity` by using heuristics to minimize the number of times `NetworkFlow` is called.

3. Write a function to compute the circumference of a graph by finding the longest cycle.

4. Implement a version of `MaximumClique` that deletes all vertices of degree less than the desired clique size from the graph, possibly creating more low-degree vertices which can be deleted. How much does this speed up the algorithm?

5. Use the generalized backtrack algorithm `Backtrack` to find the Hamiltonian cycles of a graph.

6. Use backtracking to find an optimal vertex coloring of a graph [Wil84]. Compare the quality of the resulting colorings with what is constructed by Brelaz's algorithm.

7. Are there planar graphs which Brelaz's algorithm colors with at least five colors? If so, find an example of such a graph. Similarly, does Brelaz's algorithm always two-color bipartite graphs? Prove it does or find a counter-example.

8. Use *Mathematica* color graphics to develop a version of `ShowGraph` which colors each of the vertices appropriately.

9. Modify `Girth` to return the shortest cycle in a graph. Also, make `Girth` work correctly for weighted graphs.

10. Experiment with connectivity in random graphs. Estimate the expected connectivity of a graph on n vertices with m edges, for small n and m. How many edges are usually necessary before the graph is connected?

11. Modify `TopologicalSort` to return all the topological orderings of a directed acyclic graph.

Research Problems

1. Find a (6,5)-cage [BM76], meaning the smallest 6-regular graph of girth 5. The basic tools for a computer search exist in `Girth` and `RegularGraph`. Gould [Gou88] contains a table of the order of known cages. Other open problems include finding the (3,9)-, (4,7)-, and (5,8)-cages.

2. Read [Rea68] conjectured that for any chromatic polynomial $c_n z^n + \cdots + c_1 z$, there does not exist a $1 \leq p \leq q \leq r \leq n$ such that $|c_p| > |c_q|$ and $|c_q| < |c_r|$. Other interesting aspects of the chromatic polynomial are discussed in [Chv70, Tut70].

3. Does the integral or derivative of a chromatic polynomial have any interesting combinatorial interpretation? What about generalizing chromatic polynomials to real numbers?

4. Shift register sequences are used to generate random numbers in the poker slot machines which are popular in casinos. Is sufficient information about the state of the register provided by the display to permit a positive return on investment?

5. One of the most important open problems in graph theory is the strong perfect graph conjecture [Gol80], which states that a graph is perfect if and only if neither the graph nor its complement contains an odd cycle of length at least five as an induced subgraph.

6. We define the most *interesting* Eulerian cycle to be one which maximizes the minimum time between visiting the same city twice. Is it NP-complete to find the most interesting Eulerian cycle of a graph?

Chapter 6.

Algorithmic Graph Theory

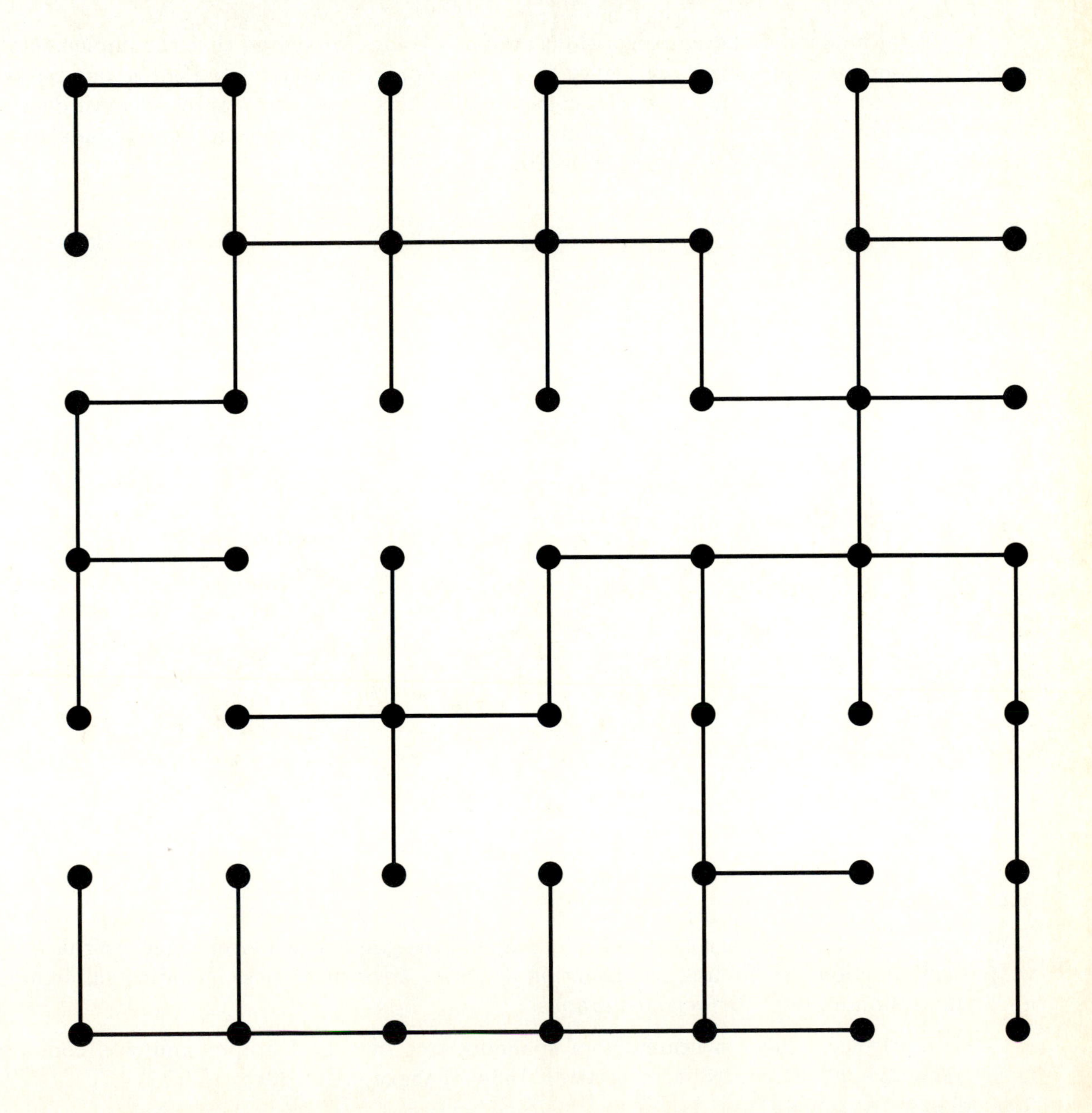

Algorithmic graph theory is one of the best examples of mathematics inspired and nurtured by computer science. As we have seen, there is a natural division into problems which can be solved efficiently and problems which are hard. The problems discussed in this chapter all have polynomial time algorithms, and are of practical importance.

Several fundamental graph algorithms are fairly complicated, and thus the implementations in the chapter are often longer and uglier than in the previous chapters. That is one reason for hiding them in the back of the book. Not coincidentally, this is also the chapter we most dearly pay the price for not having access to a RAM model of computation, because most of these algorithms make frequent modification to adjacency matrices.

About the illustration overleaf:

A spanning tree of a graph is a subset of $n-1$ edges of the graph which form a tree. In this chapter, we present functions for finding and counting spanning trees in graphs, including minimum and maximum spanning trees for weighted graphs.

The frontispiece shows a semirandom spanning tree of a grid graph, and was constructed using: `ShowGraph[ MinimumSpanningTree[ InduceSubgraph[GridGraph[7,7], RandomPermutation[49]] ] ];`

6.1 Shortest Paths

The typical example of a graph for the layperson is a network of roads connecting cities on a map. The most natural algorithmic problem on such a graph is finding the shortest path between a pair of vertices.

The shortest path between two vertices in an unweighted graph, called a *geodesic*, can be found by a simple breadth-first search [Moo59], as used in Section 5.3.2 in computing the girth of a graph. Things get more complicated with weighted graphs, since a shortest path between two nodes might not be a direct edge between them, but might instead involve a detour through other nodes.

6.1.1 Dijkstra's Algorithm

Dijkstra's algorithm [Dij59], independently discovered by [WH60], finds the shortest path between two vertices by constructing a shortest-path tree from the start vertex s to every other vertex in the graph. This is done by performing a best-first search of the graph starting from s, where at the ith iteration we expand the closest unexpanded vertex v_i by considering each edge $\{v_i, w\}$ incident to v_i. If w is a previously expanded vertex, this edge has already been considered. If not, the current shortest path from s to w may be longer than the shortest path from s to v_i plus the cost of the edge $\{v_i, w\}$. If so, this gives us a shorter path and so we update the shortest distance from s to w accordingly. At each iteration, we identify the shortest path from s to exactly one new vertex.

This implementation returns both the parent relation of the shortest-path tree as well as the distance to each vertex from s. The structure of the shortest-path spanning tree tells a lot about the structure of the underlying graph.

```
Dijkstra[g_Graph,start_Integer] := First[ Dijkstra[g,{start}] ]

Dijkstra[g_Graph, l_List] :=
        Block[{x,start,e=ToAdjacencyLists[g],i,p,parent,untraversed},
                p=Edges[PathConditionGraph[g]];
                Table[
                        start = l[[i]];
                        parent=untraversed=Range[V[g]];
                        dist = p[[start]]; dist[[start]] = 0;
                        Scan[ (parent[[#]] = start)&, e[[start]] ];
                        While[ untraversed != {} ,
                                x = First[untraversed];
                                Scan[(If [dist[[#]]<dist[[x]],x=#])&, untraversed];
                                untraversed = Complement[untraversed,{x}];
                                Scan[
                                        (If[dist[[#]] > dist[[x]]+p[[x,#]],
```

```
                                        dist[[#]] = dist[[x]]+p[[x,#]];
                                        parent[[#]] = x ])&,
                                e[[x]]
                        ];
                ];
                {parent, dist},
                {i,Length[l]}
        ]
]

ShortestPath[g_Graph,s_Integer,e_Integer] :=
        Block[{parent=First[Dijkstra[g,s]],i=e,lst={e}},
                While[ (i != s) && (i != parent[[i]]),
                        PrependTo[lst,parent[[i]]];
                        i = parent[[i]]
                ];
                If[ i == s, lst, {}]
        ]

ShortestPathSpanningTree[g_Graph,s_Integer] :=
        Block[{parent=First[Dijkstra[g,s]],i},
                FromUnorderedPairs[
                        Map[({#,parent[[#]]})&, Complement[Range[V[g]],{s}]],
                        Vertices[g]
                ]
        ]
```

Dijkstra's Shortest Path Algorithm

The first list contains the parent relation in the shortest-path spanning tree, and shows that in this instance all edges go from lower- to higher-numbered vertices. The second list gives the distance information for the graph, and illustrates that the shortest path across the main diagonal of an $n \times m$ grid graph is $n + m - 2$.

```
In[1]:= Dijkstra[ g = GridGraph[5,5], 1]
Out[1]= {{1, 1, 2, 3, 4, 1, 2, 3, 4, 5, 6, 7, 8, 9, 10,
   11, 12, 13, 14, 15, 16, 17, 18, 19, 20},
  {0, 1, 2, 3, 4, 1, 2, 3, 4, 5, 2, 3, 4, 5, 6, 3, 4, 5,
   6, 7, 4, 5, 6, 7, 8}}
```

In an unweighted graph there can be many different shortest paths between any pair of vertices. This path between two opposing corners goes all the way to the right, then all the way to top.

```
In[2]:= ShortestPath[g,1,25]
Out[2]= {1, 2, 3, 4, 5, 10, 15, 20, 25}
```

The shortest-path spanning tree of a grid graph is defined in terms of Manhattan distance, where the distance between points with coordinates (x, y) and (u, v) is $|x - u| + |y - v|$.

```
In[3]:= ShowGraph[ ShortestPathSpanningTree[g,1] ];
```

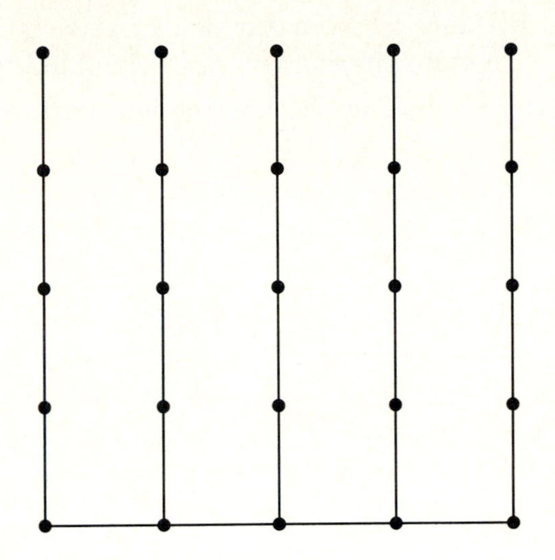

Starting in the center of a graph minimizes the diameter of the shortest-path spanning tree. I see a candelabra. What do *you* see?

```
In[4]:= ShowGraph[ ShortestPathSpanningTree[g,13] ];
```

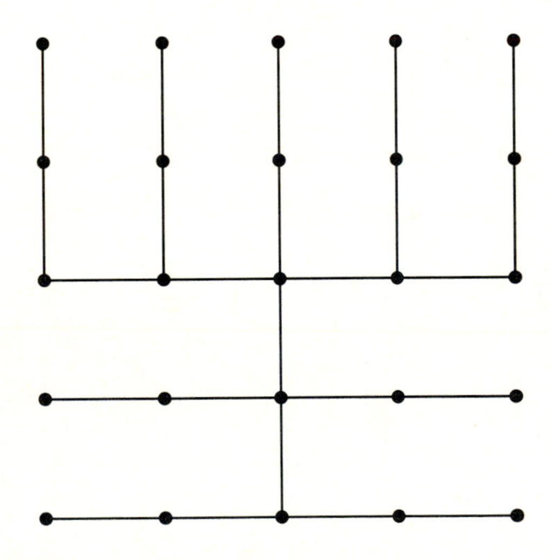

Dijkstra's algorithm works correctly on disconnected graphs as well.

```
In[5]:= Dijkstra[GraphUnion[K[3],K[3]],1]
Out[5]= {{1, 1, 1, 4, 5, 6},
  {0, 1, 1, Infinity, Infinity, Infinity}}
```

6.1.2 All Pairs Shortest Paths

The shortest distance between any pair of vertices in a graph G is defined by the path between them in the shortest-path spanning tree of G, *if* this path does not go through the root. Thus if this tree is a path, all the $\binom{n}{2}$ shortest paths are described, but if it is a star, only $n-1$ shortest paths are determined. Therefore, even in the worst case, running Dijkstra's algorithm n times, once with each of the vertices as the root, provides an algorithm which solves the *all-pairs* shortest-path problem for G.

Dijkstra's algorithm will not necessarily give the correct shortest path when some of the weights are negative, because with negative arcs it is impossible to proclaim with certainty the shortest path between any two vertices before considering all of them. The shortest path is not even defined when negative cost cycles exist in the graph, since the cost of any tour can be made arbitrarily small by repeatedly looping around the negative cost cycle. Note that an undirected edge of negative weight automatically gives rise to a negative cycle.

```
AllPairsShortestPath[g_Graph] :=
        Block[{p=Edges[ PathConditionGraph[g] ],i,j,k,n=V[g]},
                Do [
                        p = Table[Min[p[[i,k]]+p[[k,j]],p[[i,j]]],{i,n},{j,n}],
                        {k,n}
                ];
                p
        ] /; Min[Edges[g]] < 0

AllPairsShortestPath[g_Graph] := Map[ Last, Dijkstra[g, Range[V[g]]]]

PathConditionGraph[Graph[e_,v_]] := RemoveSelfLoops[Graph[ReplaceAll[e,0->Infinity],v]]
```

Floyd's Shortest Path Algorithm

Floyd's algorithm [Flo62] correctly handles negative edges and is an application of dynamic programming. The weights of edges in the initial graph provide the shortest way to get between pairs of vertices with no intermediate stops. The algorithm successively computes the shortest path between all pairs of vertices using the first k vertices as possible intermediaries, as k goes from v_1 to v_n. The best way to get from s to t with $v_1, ..., v_k$ as possible intermediaries is the minimum of the best way using the first $k-1$ vertices and the best known way from s to t through v_k. This gives an $O(n^3)$ time algorithm which is essentially the same as the transitive closure algorithm of Section 5.4. For dense graphs and conventional machines, Floyd's algorithm is usually faster than n calls to Dijkstra's algorithm, although their asymptotic complexities are identical. This isn't necessarily true in

Mathematica, but Floyd's algorithm remains of interest because it handles negative weight edges correctly.

This random complete graph has a distance associated with each edge. For the larger distances, the shortest path will involve more than one edge.

```
In[6]:= TableForm[ Edges[r=RandomGraph[8,1,{1,100}]] ]
                    0   52  90  83  9   16  24  67
                    52  0   6   34  56  35  93  77
                    90  6   0   91  86  39  45  52
                    83  34  91  0   8   22  45  88
                    9   56  86  8   0   79  48  91
                    16  35  39  22  79  0   49  42
                    24  93  45  45  48  49  0   96
Out[6]//TableForm=  67  77  52  88  91  42  96  0
```

The matrix of shortest distances is much more uniform, with the largest distances removed but the smallest ones unchanged.

```
In[7]:= TableForm[ AllPairsShortestPath[r] ]
                    0   51  55  17  9   16  24  58
                    51  0   6   34  42  35  51  58
                    55  6   0   40  48  39  45  52
                    17  34  40  0   8   22  41  64
                    9   42  48  8   0   25  33  67
                    16  35  39  22  25  0   40  42
                    24  51  45  41  33  40  0   82
Out[7]//TableForm=  58  58  52  64  67  42  82  0
```

Subtracting the two distance matrices shows which entries have changed and by how much.

```
In[8]:= TableForm[ Edges[r] - % ]
                    0   1   35  66  0   0   0   9
                    1   0   0   0   14  0   42  19
                    35  0   0   51  38  0   0   0
                    66  0   51  0   0   0   4   24
                    0   14  38  0   0   54  15  24
                    0   0   0   0   54  0   9   0
                    0   42  0   4   15  9   0   14
Out[8]//TableForm=  9   19  0   24  24  0   14  0
```

■ 6.1.3 Number of Paths

The kth *power* of a graph G is a graph with the same set of vertices as G and an edge between two vertices if and only if there is a path of length at most k between them.

Since a path of length two between u and v exists for every vertex v' such that $\{u, v'\}$ and $\{v', v\}$ are edges in G, the square of the adjacency matrix of G counts the number of such paths. By induction, the (u, v)th element of the kth power of the adjacency matrix of G gives the number of paths of length k between vertices u and v. Summing up the first k powers of the adjacency matrix therefore counts all paths of length up to k, the kth power of G.

```
GraphPower[g_Graph,1] := g

GraphPower[g_Graph,n_Integer] :=
        Block[{prod=power=p=Edges[g]},
                Do [
                        prod = prod . p;
                        power = prod + power,
                        {n-1}
                ];
                Graph[power, Vertices[g]]
        ]
```

Computing the Power of a Graph

The diameter of a cycle on seven vertices is three, so the cube of such a graph is complete. Note that there are four distinct paths of length at most three between adjacent vertices: $\{v_i, v_{i+1}, v_{i+2}, v_{i+1}\}$, $\{v_i, v_{i+1}, v_i, v_{i+1}\}$, $\{v_i, v_{i-1}, v_i, v_{i+1}\}$, and $\{v_i, v_{i+1}\}$.

```
In[9]:= TableForm[ Edges[ GraphPower[Cycle[7],3] ] ]
                    2  4  1  1  1  1  4
                    4  2  4  1  1  1  1
                    1  4  2  4  1  1  1
                    1  1  4  2  4  1  1
                    1  1  1  4  2  4  1
                    1  1  1  1  4  2  4
Out[9]//TableForm=  4  1  1  1  1  4  2
```

The number of shortest paths between opposite corners of a $m \times n$ grid graph is $\binom{n+m-2}{m-1}$.

```
In[10]:= Edges[g=GraphPower[GridGraph[5,5],8]] [[1,25]]
Out[10]= 70
```

Raising any graph to the power of its diameter yields the complete graph.

```
In[11]:= ShowGraph[g];
```

Fleischner [Fle74] showed that the square of any biconnected graph is Hamiltonian.

```
In[12]:= HamiltonianCycle[ GraphPower[ K[10,2], 2] ]
Out[12]= {1, 2, 3, 4, 5, 6, 7, 8, 9, 10, 11, 12, 1}
```

6.2 Minimum Spanning Trees

A *minimum spanning tree* [GH85] of a weighted graph is a set of $n - 1$ edges of minimum total weight which form a spanning tree of the graph. Two classical algorithms for finding such a tree are Prim's algorithm [Pri57], which greedily adds edges which extend the existing tree and do not create cycles, and Kruskal's algorithm [Kru56], which greedily adds edges which connect components. We implement Kruskal's algorithm.

Since every course in algorithms includes a discussion at least one of these algorithms, and a greedy approach suffices for finding the optimal solution, students cast a jaded eye at the problem. However, it is quite remarkable that the minimum spanning tree can be found in polynomial time. Many simple variants of the problem, such as finding a spanning tree with maximum degree k or the spanning tree minimizing the total length between all pairs of vertices, are NP-complete [GJ79]. The minimum spanning tree problem can be formulated as a *matroid* [PS82], a system of independent sets whose largest weighted independent set can be found using the greedy algorithm.

6.2.1 Union-Find

Kruskal's algorithm requires a data structure for set-union operations and membership queries, since we need a fast way to test whether the lowest-cost edge remaining connects two components, meaning it is in the spanning tree. If the edge connects two vertices of the same component, it causes a cycle and therefore cannot be in the tree. The *union-find* data structure supports two operations: `FindSet`, which for any element returns the name of the set containing it, and `UnionSet`, which given elements of two sets merges them into one. This implementation maintains each set as a tree such that the name of each set is its root, and `UnionSet` makes the bigger tree the root of the shorter one, thus giving the sets the same name while minimizing the height of the tree for efficiency.

Because the smaller height tree becomes a child of the taller one on each union, `FindSet` can be shown to take $O(\log n)$ time per operation. Adding *path compression* [Tar75] would make the data structure even more efficient.

```
InitializeUnionFind[n_Integer] := Block[{i}, Table[{i,1},{i,n}] ]

FindSet[n_Integer,s_List] := If [n == s[[n,1]], n, FindSet[s[[n,1]],s] ]

UnionSet[a_Integer,b_Integer,s_List] :=
        Block[{sa=FindSet[a,s], sb=FindSet[b,s], set=s},
                If[ set[[sa,2]] < set[[sb,2]], {sa,sb} = {sb,sa} ];
                set[[sa]] = {sa, Max[ set[[sa,2]], set[[sb,2]]+1 ]};
```

```
            set[[sb]] = {sa, set[[sb,2]]};
            set
    ]
```

The Union-Find Data Structure

In this example, all the elements belong to the same set. The second element of each record is the height of the subtree rooted in the node, so this tree has a height of two.

```
In[13]:= UnionSet[1,2, UnionSet[2,3, UnionSet[3,4,
UnionSet[4,5, InitializeUnionFind[5] ] ] ] ]

Out[13]= {{4, 1}, {4, 1}, {4, 1}, {4, 2}, {4, 1}}
```

The name of the root is not particularly significant. For the previous set, the root of all elements is the same.

```
In[14]:= FindSet[1,%]

Out[14]= 4
```

■ 6.2.2 Kruskal's Algorithm

Kruskal's algorithm sorts the edges in order of increasing cost, then repeatedly adds edges which don't create a cycle until the graph is fully connected. The union-find data structure maintains the connected components in the forest, and so an edge between vertices in different sets cannot create a cycle. A proof that this algorithm gives the minimum spanning tree is that in any cycle, the most expensive edge will be the last one considered, and so cannot be in the minimum spanning tree.

```
MinimumSpanningTree[g_Graph] :=
        Block[{edges=Edges[g],set=InitializeUnionFind[V[g]]},
                FromUnorderedPairs[
                        Select [
                                Sort[
                                        ToUnorderedPairs[g],
                                        (Element[edges,#1]>=Element[edges,#2])&
                                ],
                                (If [FindSet[#[[1]],set] != FindSet[#[[2]],set],
                                        set=UnionSet[#[[1]],#[[2]],set]; True,
                                        False
                                ])&
                        ],
                        Vertices[g]
                ]
        ] /; UndirectedQ[g]

MaximumSpanningTree[g_Graph] := MinimumSpanningTree[Graph[-Edges[g],Vertices[g]]]
```

Finding the Minimum Spanning Tree

Any spanning tree is a minimum spanning tree when the graphs are unweighted.

```
In[15]:= ShowGraph[ MinimumSpanningTree[ K[6,6,6] ] ];
```

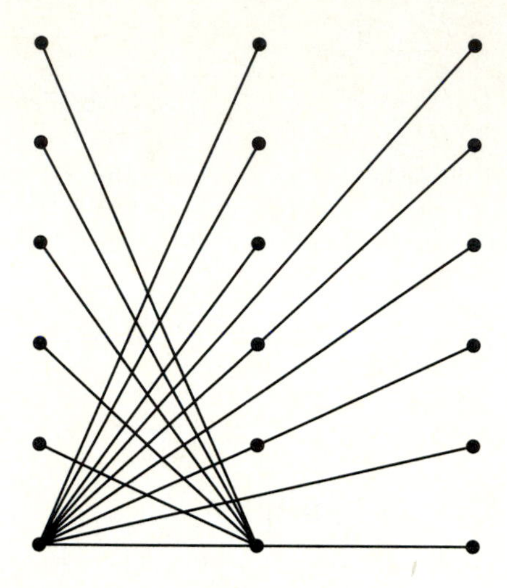

If the weights of the edges are all different, the minimum spanning tree is unique.

```
In[16]:= ShowGraph[ MinimumSpanningTree[
g=RandomGraph[10,0.5,{1,100}] ] ];
```

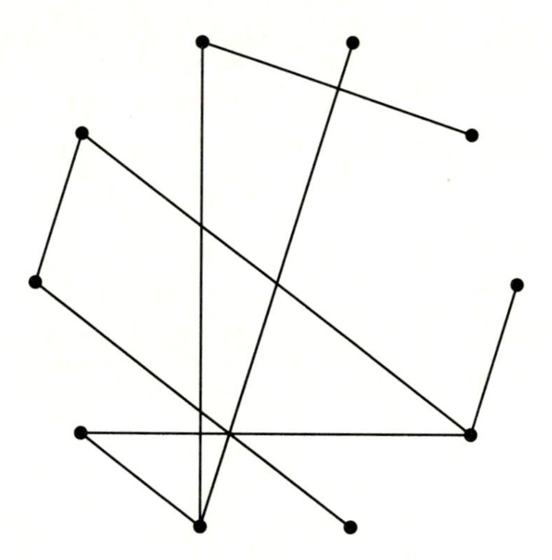

By negating the weights for each edge, the same implementation can also be used to find the maximum-weight spanning tree. With a wide range of weights in the graph, it is unlikely that the spanning trees of maximum and minimum weight will have any edges in common.

```
In[17]:= ShowGraph[ MaximumSpanningTree[g] ];
```

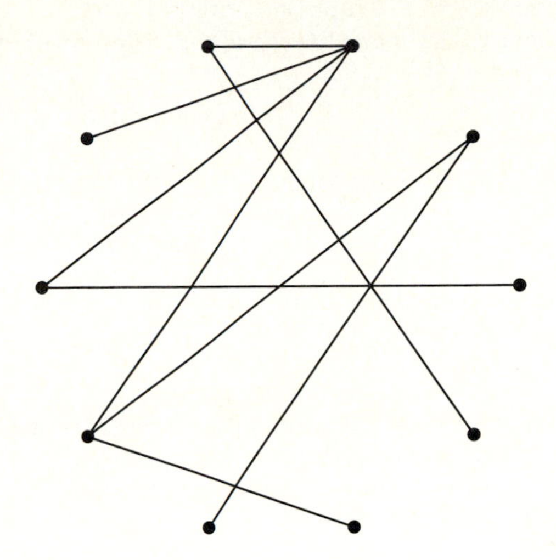

Prim's algorithm consists of $n-1$ iterations, each of which takes $O(n)$ time to select the smallest edge connecting a vertex in the connected component to one which isn't, giving a time complexity of $O(n^2)$. Thus it is optimal on dense graphs. The trick is maintaining the shortest edge to each outside vertex in an array, so after adding the ith edge to the tree only $n-i-1$ edges must be tested to update this array. More efficient algorithms rely on sophisticated data structures [FT87].

■ 6.2.3 Counting Spanning Trees

The number of non-identical spanning trees of a graph was determined by Kirchhoff [Kir47] using some simple computations on the adjacency matrix.

The (i,j) *minor* of a matrix M is the determinant of M with the ith row and jth column deleted. The (i,j) *cofactor* of a matrix M is $(-1)^{i+j}$ times the (i,j) minor of M.

```
Cofactor[m_List,{i_Integer,j_Integer}] :=
        (-1)^(i+j) * Det[ Drop[ Transpose[ Drop[Transpose[m],{j,j}] ], {i,i}] ]
```

Computing the Cofactor of a Matrix

The number of non-identical spanning trees of a graph G is equal to any cofactor of the degree matrix of G minus the adjacency matrix of G, where the degree matrix of a graph is a diagonal matrix with the degree of v_i in the ith position of the matrix. A combinatorial proof of the matrix-tree theorem is due to [Cha82].

```
NumberOfSpanningTrees[Graph[g_List,_]] :=
        Cofactor[ DiagonalMatrix[Map[(Apply[Plus,#])&,g]] - g, {1,1}]
```

Counting Spanning Trees of a Graph

Any tree contains exactly one spanning tree.

```
In[18]:= NumberOfSpanningTrees[Star[20]]
Out[18]= 1
```

A cycle on n vertices contains exactly n non-identical spanning trees, since deleting any edge creates a tree.

```
In[19]:= NumberOfSpanningTrees[Cycle[20]]
Out[19]= 20
```

The number of spanning trees of a complete graph is n^{n-2}, as was shown by the bijection between labeled trees and strings of integers discussed in Section 4.3.

```
In[20]:= NumberOfSpanningTrees[K[10]]
Out[20]= 100000000
```

The set of spanning trees of a graph can be ranked and unranked and thus can be generated systematically or randomly. See [CDN89] for $O(n^3)$ ranking and unranking algorithms.

■ 6.3 Network Flow

Graphs are usually introduced by comparing them to a network of roads connecting cities. When we are interested in pushing sewerage through the network instead of cars, we have a *network flow* problem.

More formally, a network flow problem consists of a weighted graph G and *source* and *sink* vertices s and t. The weight of each edge signifies its capacity, the maximum amount of stuff which can be pumped through it. A path from s to t has as its capacity the minimum of the weights of its edges. We seek the maximum flow from s to t through the entire network, which will be composed of a set of these paths.

The importance of the network flow problem is that many other problems can be easily reduced to it, such as k-connectivity and bipartite matching, so a function to find the maximum flow through a network results in functions to solve several other problems.

Most network flow algorithms use the *augmenting path* idea of Ford and Fulkerson [FF62], which repeatedly finds a path of positive capacity from source to sink and adds it to the flow. It can be shown that the flow through a network of rational capacities is optimal if and only if it contains no augmenting path, and since each augmentation adds to the flow, we will eventually find the maximum. If we are not careful, however, each augmenting path will add but a little to the total flow, and so the algorithm might take a long time to converge. Further, with networks of irrational capacities, by consistently picking the wrong augmenting path, it is possible to converge on a suboptimal flow.

The difference between network flow algorithms is *how* they select the augmenting path. Edmonds and Karp [EK72] proved that by always selecting the *shortest* geodesic augmenting path, at most $O(n^3)$ augmentations will be performed, giving a polynomial-time algorithm for the problem. Edmonds-Karp is fairly easy to implement, since a breadth-first search from the source can find the shortest path in linear time.

```
NetworkFlow[g_Graph,source_Integer,sink_Integer] :=
        Block[{flow=NetworkFlowEdges[g,source,sink], i},
                Sum[flow[[i,sink]], {i,V[g]}]
        ]


NetworkFlowEdges[g_Graph,source_Integer,sink_Integer] :=
        Block[{e=Edges[g], x, y, flow=Table[0,{V[g]},{V[g]}], p, m},
                While[ !SameQ[p=AugmentingPath[g,source,sink], {}],
                        m = Min[Map[({x,y}=#[[1]];
```

```
                    If[SameQ[#[[2]],f],e[[x,y]]-flow[[x,y]],
                            flow[[x,y]]]])&,p]];
            Scan[
                    ({x,y}=#[[1]];
                     If[ SameQ[#[[2]],f],
                            flow[[x,y]]+=m,flow[[x,y]]-=m])&,
                     p
            ]
        ];
        flow
    ]

AugmentingPath[g_Graph,src_Integer,sink_Integer] :=
    Block[{l={src},lab=Table[0,{V[g]}],v,c=Edges[g],e=ToAdjacencyLists[g]},
        lab[[src]] = start;
        While[l != {} && (lab[[sink]]==0),
            {v,l} = {First[l],Rest[l]};
            Scan[ (If[ c[[v,#]] - flow[[v,#]] > 0 && lab[[#]] == 0,
                    lab[[#]] = {v,f}; AppendTo[l,#]])&,
                    e[[v]]
            ];
            Scan[ (If[ flow[[#,v]] > 0 && lab[[#]] == 0,
                    lab[[#]] = {v,b}; AppendTo[l,#]] )&,
                    Select[Range[V[g]],(c[[#,v]] > 0)&]
            ];
        ];
        FindPath[lab,src,sink]
    ]

FindPath[l_List,v1_Integer,v2_Integer] :=
    Block[{x=l[[v2]],y,z=v2,lst={}},
        If[SameQ[x,0], Return[{}]];
        While[!SameQ[x, start],
            If[ SameQ[x[[2]],f],
                    PrependTo[lst,{{ x[[1]], z }, f}],
                    PrependTo[lst,{{ z, x[[1]] }, b}]
            ];
            z = x[[1]]; x = l[[z]];
        ];
        lst
    ]
```

Finding the Maximum Flow through a Network

The maximum flow through an unweighted complete bipartite graph G is the minimum degree $\delta(G)$.

```
In[21]:= NetworkFlow[ K[4,4], 1, 8]
Out[21]= 4
```

For certain applications, knowing the actual flow is less important than knowing which edges are involved and how much actually goes through them. The maximum flow through a complete graph involves the directed edge between source and sink and $n-2$ paths of length two, through each of the other vertices.

```
In[22]:= TableForm[ NetworkFlowEdges[K[5],1,2] ]
                    0  1  1  1  1
                    0  0  0  0  0
                    0  1  0  0  0
                    0  1  0  0  0
Out[22]//TableForm= 0  1  0  0  0
```

A fast algorithm for 0-1 network flow appears in [ET75]. This is useful for problems which use network flow as a subroutine. Such algorithms often require solving flow problems between every pair of vertices in the graph. For undirected graphs, $n-1$ applications of the maximum flow algorithm suffice to determine all $\binom{n}{2}$ pairwise flows [GH61].

Active research continues in finding better network flow algorithms, with [Tar83] an excellent text and [Orl88] reflective of the state of the art.

6.4 Matching

A *matching* in a graph G is a set of edges of G such that no two of them share a vertex in common. Clearly, the largest possible matching consists of $n/2$ edges, and such a matching is called *perfect*. Not all graphs have perfect matchings, but every graph has a *maximum* or largest matching, and finding it reveals a lot about the structure of the graph.

This problem can be naturally generalized to weighted graphs, where the goal is to find the matching such that the sum of the weights of the edges in the matching is maximized. An excellent reference on matching theory is [LP86].

6.4.1 Bipartite Matching

The study of matching in graphs arose from the *marriage problem*, where each of b boys knows some subset of g girls. The marriage problem asks under what conditions the boys can be married so each of them gets a girl that he knows. In graph theoretic terms, under what condition is there a complete matching in a bipartite graph?

Hall's marriage theorem [Hal35] states that there is a complete matching if and only if every subset of boys between them know a subset of girls at least as large. This criterion provides a way to test whether such a marriage is possible and also to construct one, albeit in exponential time.

Network flow techniques can be used to determine the maximum matching in a bipartite graph [HK75] in $O(n^{1/2}m)$. Add a source connected to each vertex in the first stage by edges of weight one and a sink connected to each vertex in the second stage by edges of weight one. The maximum flow in this graph must correspond to the maximum matching, since it will find the largest set of vertex disjoint paths, which in a bipartite graph consists only of edges.

```
BipartiteMatching[g_Graph] :=
        Block[{p,v1,v2,coloring=TwoColoring[g],n=V[g]},
                v1 = Flatten[Position[coloring,1]];
                v2 = Flatten[Position[coloring,2]];
                p = BipartiteMatchingFlowGraph[g,v1,v2];
                flow = NetworkFlowEdges[p,V[g]+1,V[g]+2];
                Select[ToOrderedPairs[Graph[flow,Vertices[p]]], (Max[#]<=n)&]
        ] /; BipartiteQ[g]

BipartiteMatchingFlowGraph[g_Graph,v1_List,v2_List] :=
        Block[{edges = Table[0,{V[g]+2},{V[g]+2}],i,e=ToAdjacencyLists[g]},
                Do[
                            Scan[ (edges[[v1[[i]],#]] = 1)&, e[[ v1[[i]] ]] ],
```

```
                {i,Length[v1]}
            ];
            Scan[(edges[[V[g] + 1, #]] = 1)&, v1];
            Scan[(edges[[#, V[g] + 2]] = 1)&, v2];
            Graph[edges,RandomVertices[V[g] + 2] ]
    ]
```

Finding a Maximum Bipartite Matching

Hypercubes are bipartite and always have perfect matchings.

```
In[23]:= ShowGraph[ FromUnorderedPairs[ BipartiteMatching[
Hypercube[4] ], Vertices[Hypercube[4]] ] ];
```

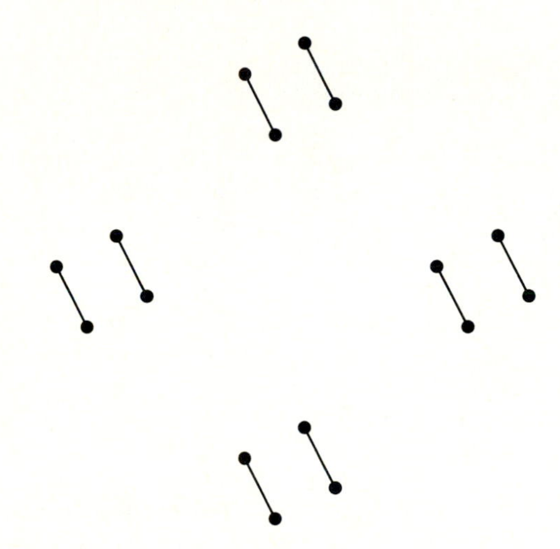

A perfect matching of an even cycle consists of alternating edges in the cycle.

```
In[24]:= BipartiteMatching[ Cycle[8] ]

Out[24]= {{1, 2}, {3, 4}, {5, 6}, {7, 8}}
```

A maximum matching of a star consists of any single edge.

```
In[25]:= BipartiteMatching[ Star[8] ]

Out[25]= {{1, 8}}
```

■ 6.4.2 Dilworth's Theorem

A *chain* in a partially ordered set is a collection of elements $v_1, v_2, ..., v_k$ such that v_i is related to v_{i+1}, $i < k$. An *antichain* is a collection of elements no pair of which are related. Dilworth's theorem [Dil50] states that for any partial order, the maximum size of an antichain equals the minimum number of chains which partition the elements.

To compute the maximum antichain, observe that there is an edge between any two related elements in any transitive relation, such as a partial order. Thus the largest antichain is described by the maximum independent set in the order. To compute the minimum chain partition of the transitive reduction of a partial order G, of order n, we construct a bipartite graph $D(G)$ with two stages of n vertices each, with each vertex v_i of G now associated with vertices $v_{i'}$ and $v_{i''}$. Now each directed edge $\{x, y\}$ of G defines an undirected edge $\{x', y''\}$ of $D(G)$. Each matching of $D(G)$ defines a chain partition of G, since edges $\{x', y''\}$ and $\{y', z''\}$ in the matching represent the chain $\{x, y, z\}$ in G. Further, the maximum matching describes the minimum chain partition.

```
MinimumChainPartition[g_Graph] :=
        ConnectedComponents[
                FromUnorderedPairs[
                        Map[(#-{0,V[g]})&, BipartiteMatching[DilworthGraph[g]]],
                        Vertices[g]
                ]
        ]

MaximumAntichain[g_Graph] := MaximumIndependentSet[TransitiveClosure[g]]

DilworthGraph[g_Graph] :=
        FromUnorderedPairs[
                Map[
                        (#+{0,V[g]})&,
                        ToOrderedPairs[RemoveSelfLoops[TransitiveReduction[g]]]
                ]
        ]
```

Partitioning a Partial Order into a Minimum Number of Chains

As we have seen before, divisibility among the integers defines a partial order. The Hasse diagram represents this order, once it is realized that there is an edge from three to six, not three to four or four to six.

```
In[26]:= ShowLabeledGraph[ HasseDiagram[
d=MakeGraph[Range[8], (Mod[#2,#1]==0 && #1<#2)&] ] ];
```

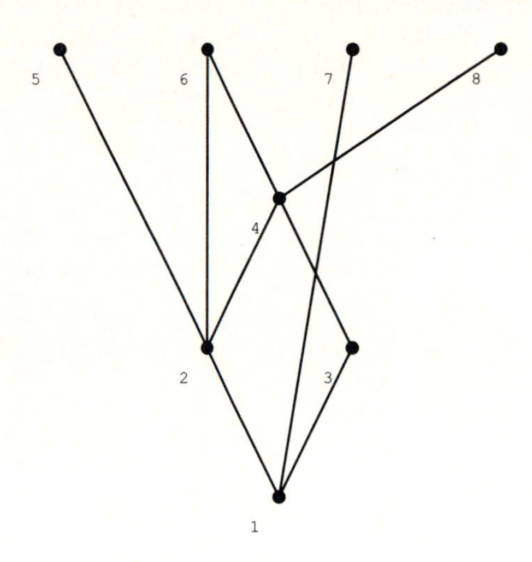

Because there are four maximal elements in the partial order, any chain partition requires at least four chains to partition the elements. This partition demonstrates that four chains suffice.

```
In[27]:= MinimumChainPartition[d]
Out[27]= {{1, 2, 4, 8}, {3, 6}, {5}, {7}}
```

The prime numbers form an antichain in the divisibility lattice and in fact are a maximum antichain for this partial order. By Dilworth's theorem, the length of the maximum antichain equals the size of the minimum chain partition for any partial order.

```
In[28]:= MaximumAntichain[d]
Out[28]= {2, 3, 5, 7}
```

■ 6.4.3 Maximal Matchings

Matching can be generalized to arbitrary graphs, and while the basic idea is the same the algorithms become much more complicated. Berge's theorem [Ber57] states that a matching is maximal if and only if it contains no *augmenting path.*

Instead of using augmenting paths to find the *maximum* matching in an arbitrary graph, we use a greedy algorithm to find a *maximal* matching, meaning one which cannot be enlarged by simply adding an edge.

```
MaximalMatching[g_Graph] :=
        Block[{match={}},
                Scan[
                        (If [Intersection[#,match]=={}, match=Join[match,#]])&,
                        ToUnorderedPairs[g]
                ];
                Partition[match,2]
        ]
```

Finding a Maximal Matching in a Graph

Petersen proved that every 3-regular graph without bridges has a perfect matching. This graph illustrates that introducing bridges can prevent perfect matchings.

```
In[29]:= ShowLabeledGraph[
g=ReadGraph["graphs/nomatching"] ];
```

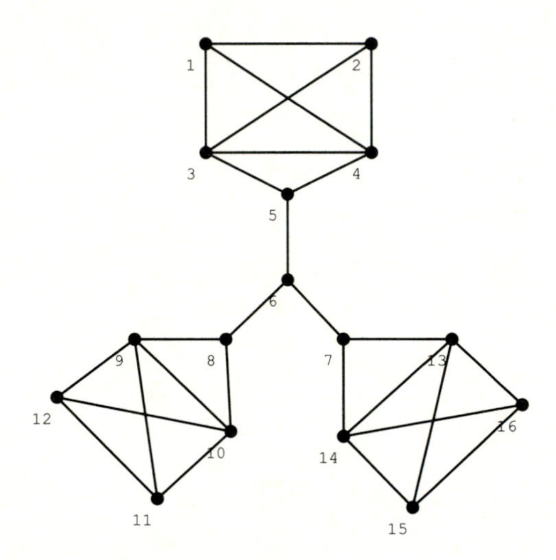

Since a perfect matching on the previous graph would have eight edges, and the maximal matching has seven, in this case the maximal matching is a maximum matching.

```
In[30]:= MaximalMatching[g]

Out[30]= {{1, 2}, {3, 4}, {5, 6}, {7, 13}, {8, 9},
  {10, 11}, {14, 15}}
```

Any maximal matching of a K_n is a maximum matching, and perfect if n is even.

```
In[31]:= MaximalMatching[K[8]]

Out[31]= {{1, 2}, {3, 4}, {5, 6}, {7, 8}}
```

A perfect matching on a graph is a 1-regular subgraph of order n. In general, a k-*factor* of a graph is a k-regular subgraph of order n. k-factors are a generalization of complete matchings since complete matchings are 1-factors. A graph is k-*factorable* if it is the union of disjoint k-factors.

6.4.4 Stable Marriages

Not all matching problems are most naturally described in terms of graphs. Perhaps the most amusing example is the *stable marriage problem.*

Given a set of n men and n women, it is desired to marry them off, one man to one woman. As in the real world, each man has an opinion of each woman, and ranks them in terms of desirability from 1 to n. The women do the same to the men. Now suppose they are all married off and there exists couples $\{m_1, w_1\}$ and $\{m_2, w_2\}$. If m_1 prefers w_2 to w_1 and w_2 prefers m_1 to her current spouse m_2, domestic bliss is doomed. Such a marriage is *unstable* because m_1 and w_2 would run off to be with each other. The goal of the stable marriage problem is to find a way to match men and women subject to their preference functions, such that the matching is stable. Obviously, stability is a desirable property, but can it always be achieved?

Gale and Shapely [GS62] proved that for *any* set of preference functions, a stable marriage exists. Even if one person is so unattractive that everyone ranks that person last, he or she can be assigned a spouse s undesirable enough to all others that no one would be willing to give up his or her spouse to rescue s. The proof is algorithmic. Starting from their favorite, unmarried men take turns proposing to the high rated woman they have not proposed to yet. If a woman gets more than one proposal, she takes the one that is best, leaving the loser unmarried. Eventually, everyone gets married, since a woman can't turn down a proposal unless she has a better one, and further, this marriage is stable, since each man always proposes to the highest ranked woman who hasn't rejected him yet. Thus no man can better his lot with further proposals.

The Gale-Shapely algorithm is used in matching hospitals to interns, and has lead to a well-developed theory of stable marriage [GI89].

```
StableMarriage[mpref_List,fpref_List] :=
        Block[{n=Length[mpref],freemen,cur,i,w,husband},
                freemen = Range[n];
                cur = Table[1,{n}];
                husband = Table[n+1,{n}];
                While[ freemen != {},
                        {i,freemen}={First[freemen],Rest[freemen]};
                        w = mpref[[ i,cur[[i]] ]];
                        If[BeforeQ[ fpref[[w]], i, husband[[w]] ],
                                If[husband[[w]] != n+1,
                                        AppendTo[freemen,husband[[w]] ]
                                ];
                                husband[[w]] = i,
                                cur[[i]]++;
```

```
                                AppendTo[freemen,i]
                        ];
                ];
                InversePermutation[ husband ]
        ] /; Length[mpref] == Length[fpref]

BeforeQ[l_List,a_,b_] :=
        If [First[l]==a, True, If [First[l]==b, False, BeforeQ[Rest[l],a,b] ] ]
```

The Gale-Shapely Algorithm for Stable Marriage

The Gale-Shapely algorithm finds a stable marriage for any set of preference functions. The ith element of the returned permutation is the wife of man i.

```
In[32]:= TableForm[{ men=Table[RandomPermutation[9],{9}],
women=Table[RandomPermutation[9],{9}]}]

Out[32]//TableForm=

 3  4  7  2  9  1  8  5  6     6  1  5  4  3  8  2  7  9
 5  6  8  2  9  3  7  1  4     3  8  1  6  5  9  2  4  7
 9  7  3  2  8  4  1  5  6     9  4  6  7  5  8  3  1  2
 3  6  9  7  5  8  2  1  4     7  6  4  8  9  2  1  3  5
 2  6  5  8  1  4  3  7  9     8  7  2  3  4  9  6  1  5
 4  6  7  5  9  8  3  2  1     1  7  8  9  6  3  4  2  5
 3  2  1  8  5  9  4  7  6     1  7  5  3  8  9  2  4  6
 3  4  2  5  8  9  1  7  6     5  9  3  2  1  7  8  4  6
 6  9  1  5  3  8  7  4  2     6  3  2  5  1  4  8  9  7
```

Because of the proposal sequence, the Gale-Shapely algorithm yields the *male-optimal* marriage, under which each man gets his best possible match in any stable marriage.

```
In[33]:= StableMarriage[men,women]

Out[33]= {7, 5, 9, 3, 8, 4, 1, 2, 6}
```

The sexual bias can be reversed by simply exchanging the roles of men and women. The inverse permutation returns who the men are married to, and it can be verified that where the couples differ from the previous matching the men are less well off and the women happier.

```
In[34]:= InversePermutation[ StableMarriage[women,men] ]

Out[34]= {7, 1, 2, 3, 8, 9, 4, 5, 6}
```

6.5 Planar Graphs

Planar graphs are graphs which can be embedded in the plane with no pair of edges crossing. There is an interesting connection between convex polyhedra in E^3 and planar embeddings. For a given polyhedron, replace one of the vertices with a lightbulb, make the faces of the polyhedron out of glass, and make the edges where the faces meet out of lead. These leaded edges cast lines as shadows, which meet at the shadows of the vertices of the polyhedron or, if incident on the light, extend on to infinity. Since the polyhedron is convex, any ray originating from the light passes through exactly one other point on the polyhedron, so no two lines can cross.

The graphs of the five *Platonic* solids, the convex polyhedra whose faces are all regular polygons, are all represented somewhere in this book. These are the *tetrahedron* (four vertices, four triangular faces), the *cube* (eight vertices, six square faces), the *octahedron* (six vertices, eight triangular faces), the *icosahedron* (twelve vertices, twenty triangular faces), and the *dodecahedron* (twenty vertices, twelve pentagonal faces).

Although interesting, this correspondence does not give us an algorithm for testing planarity. Two graphs are *homeomorphic* if they can be obtained from the same graph by replacing edges with paths. Kuratowski's theorem [Kur30] states that a graph is planar if and only if it contains no subgraph homeomorphic to K_5 or $K_{3,3}$.

6.5.1 Testing Planarity

Although there are several efficient algorithms [CHT90, Eve79, HT74] for testing planarity, all are difficult to implement. Most of these algorithms are based on ideas from an old $O(n^3)$ algorithm by Auslander and Parter [AP61]. Their algorithm is based on the observation that if a graph is planar, for any cycle there is a way to embed it in the plane so that the vertices of this cycle lie on a circle. This follows from the fact that an embedding on a sphere is equivalent to an embedding in the plane, as discussed above. If this cycle is then deleted from the graph, the connected components which are left are called *bridges*. Obviously, if any bridge is non-planar, the graph is non-planar. Further, the graph is non-planar if these bridges interlock in the wrong way.

Some simplification can be done by breaking the graph into connected components, and using Euler's formula to reject graphs with too many edges.

```
PlanarQ[g_Graph] :=
        Apply[
                And,
                Map[(PlanarQ[InduceSubgraph[g,#]])&, ConnectedComponents[g]]
```

```
        ] /; !ConnectedQ[g]

PlanarQ[g_Graph] := False /;  (M[g] > 3 V[g]-6)
PlanarQ[g_Graph] := True /;   (M[g] < V[g] + 3)
PlanarQ[g_Graph] := PlanarGivenCycle[ g, Rest[FindCycle[g]] ]

PlanarGivenCycle[g_Graph, cycle_List] :=
        Block[{b, j, i},
                {b, j} = FindBridge[g, cycle];
                If[ InterlockQ[j, cycle],
                        False,
                        Apply[And, Table[SingleBridgeQ[b[[i]],j[[i]]], {i,Length[b]}]]
                ]
        ]

SingleBridgeQ[b_Graph, {_}] := PlanarQ[b]

SingleBridgeQ[b_Graph, j_List] :=
        PlanarGivenCycle[ JoinCycle[b,j],
                Join[ ShortestPath[b,j[[1]],j[[2]]], Drop[j,2]] ]

JoinCycle[g1_Graph, cycle_List] :=
        Block[{g=g1},
                Scan[(g = AddEdge[g,#])&, Partition[cycle,2,1] ];
                AddEdge[g,{First[cycle],Last[cycle]}]
        ]
```

Testing the Planarity of a Graph

Each bridge consists of a set of zero or more vertices forming a connected component, as well as a set of *junction points* joining the bridge to the cycle. There will be zero vertices in the bridge when it is just an edge spanning the cycle. Unless the connected component forming a bridge is itself not planar, the junction points for each bridge determines whether they interlock, since the connected component can be shrunk to an arbitrarily small point.

Complications emerge in extracting bridges since different bridges can share junction points. Because of the need to keep the vertex labelings consistent, `InduceSubgraph` cannot be used to extract bridges, and `IsolateSubgraph` takes its place.

```
FindBridge[g_Graph, cycle_List] :=
    Block[{rg = RemoveCycleEdges[g, cycle], b, bridge, j},
        b = Map[
                (IsolateSubgraph[rg,g,cycle,#])&,
                Select[ConnectedComponents[rg], (Intersection[#,cycle]=={})&]
        ];
```

```
        b = Select[b, (!EmptyQ[#])&];
        j = Join[
                Map[Function[bridge,Select[cycle, MemberQ[Edges[bridge][[#]],1]&] ], b],
                Complement[
                        Select[ToUnorderedPairs[g],
                                (Length[Intersection[#,cycle]] == 2)&],
                        Map[Sort, Partition[Append[cycle,First[cycle]],2,1]]
                ]
        ];
        {b, j}
    ]

RemoveCycleEdges[g_Graph, c_List] :=
        FromOrderedPairs[
                Select[ ToOrderedPairs[g], (Intersection[c,#] === {})&],
                Vertices[g]
        ]

IsolateSubgraph[g_Graph,orig_Graph,cycle_List,cc_List] :=
        Block[{eg=ToOrderedPairs[g], og=ToOrderedPairs[orig]},
                FromOrderedPairs[
                        Join[
                                Select[eg, (Length[Intersection[cc,#]] == 2)&],
                                Select[og, (Intersection[#,cycle]!={} &&
                                        Intersection[#,cc]!={})&]
                        ],
                        Vertices[g]
                ]
        ]
```

Extracting Bridges for Planarity Testing

If there are two bridges across this cycle which interlock, meaning that they cannot both be drawn inside the cycle without crossing, one can be drawn outside and one can be drawn inside, so the graph is planar. A graph is non-planar if and only if *three* bridges mutually interlock.

The interlock testing is done by partitioning the bridges into those which get embedded inside the cycle and those that go outside. For each bridge, it goes on the inside unless it locks with a bridge already on the inside list. If so, it will be placed on the outside, unless it locks with a bridge already on the outside list. Such a conflict means the graph is non-planar.

```
InterlockQ[ bl_List, c_List ] :=
        Block[{in = out = {}, code, jp, bridgelist = bl },
                While [ bridgelist != {},
                        {jp, bridgelist} = {First[bridgelist],Rest[bridgelist]};
```

```
                    code = Sort[ Map[(Position[c, #][[1,1]])&, jp] ];
                    If[ Apply[ Or, Map[(LockQ[#,code])&, in] ],
                            If [ Apply[Or, Map[(LockQ[#,code])&, out] ],
                                    Return[True],
                                    AppendTo[out,code]
                            ],
                            AppendTo[in,code]
                    ]
            ];
            False
        ]

LockQ[a_List,b_List] := Lock1Q[a,b] || Lock1Q[b,a]

Lock1Q[a_List,b_List] :=
        Block[{bk, aj},
                bk = Min[ Select[Drop[b,-1], (#>First[a])&] ];
                aj = Min[ Select[a, (# > bk)&] ];
                (aj < Max[b])
        ]
```

Testing Whether Three or More Bridges Interlock

By definition, neither $K_{3,3}$ nor K_5 is planar.

```
In[35]:= PlanarQ[K[5]] || PlanarQ[K[3,3]]
Out[35]= False
```

Every planar graph on nine vertices has a non-planar complement [BHK62].

```
In[36]:= PlanarQ[ GraphComplement[GridGraph[3,3]] ]
Out[36]= False
```

The current embedding has no effect on whether `PlanarQ` rules planar or non-planar.

```
In[37]:= PlanarQ[ RandomTree[10] ]
Out[37]= True
```

The graph associated with the tetrahedron is K_4, the cube is the product of K_2 and C_4, and the octahedron is $K_{2,2,2}$. A proof [Mes83] that there can exist no other platonic solids follows from Euler's formula that for any planar graph with V vertices, E edges, and F faces, $V - E + F = 2$.

```
In[38]:= PlanarQ[K[4]] && PlanarQ[K[2,2,2]] &&
PlanarQ[GraphProduct[K[2],Cycle[4]]]
Out[38]= True
```

■ 6.5.2 Planar Embeddings

The previous section gave an algorithm for testing whether a graph is planar. Since the bridges were partitioned into inside and outside relative to the cycle, this algorithm can be adapted to return a planar embedding.

However, we will not give such an algorithm. Our graph-drawing procedures are limited to representing edges by straight lines. Although it has been shown [Far48] that every planar graph has a straight-line embedding, the resultant embeddings are distorted enough that they are usually not very attractive. Vertex positions which admit planar embeddings do not necessarily admit straight-line embeddings. However, [dFPP88] provide an algorithm for constructing a straight line embedding of an order n graph, positioning each vertex as a point on a $2n - 4 \times n - 2$ grid.

Other invariants generalize the notation of planarity. An *outerplanar graph* is a graph which can be embedded in the plane such that all vertices lie on the outer face. A graph is outerplanar if and only if it contains no subgraph homeomorphic to K_5 or $K_{2,3}$. The *genus* of a graph is the minimum number of handles which must be added to the plane to embed the graph without any crossings. Planar graphs have genus zero. The *thickness* of a graph G is the minimum number of planar subgraphs of G whose union is G. A simple but effective lower bound on the thickness of a graph $t(G) \geq \lceil \frac{m}{3n-6} \rceil$ follows from Euler's formula and the pigeonhole principle.

The *crossing number* of a graph is the minimum number of overlapping edges which must appear in any plane drawing of the graph. The standard embedding of complete bipartite graphs makes no attempt to minimize the number of crossings. In fact, the crossing number of $K_{3,4}$ is 2, instead of 17.

```
In[39]:= ShowGraph[ K[3,4] ];
```

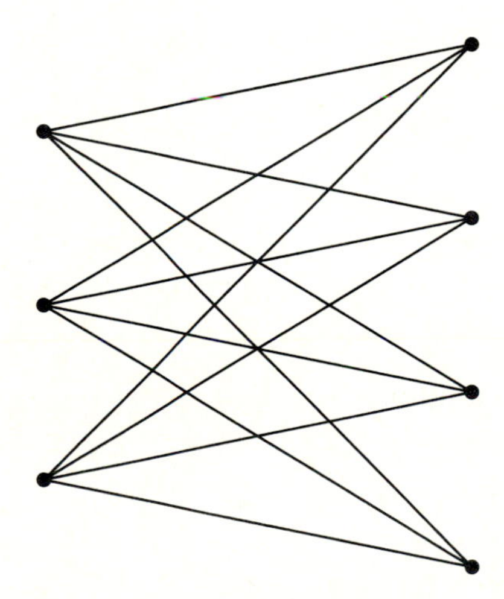

6.6 Exercises and Research Problems

Exercises

1. Modify the implementation of Floyd's algorithm so that it computes the actual shortest path between each pair instead of just the cost. Represent this information in a concise format, such as using a parent relation.

2. Use the *Mathematica* function `LinearProgramming` to find the maximum flow through a network by setting up a system of inequalities such that the flow through each edge is bounded by its capacity and the flow into each vertex equals the flow out, maximizing the flow into the sink. How does the efficiency of this routine compare to `NetworkFlow`?

3. Write a function to construct a random Euclidean graph as follows. Use `RandomVertices` to generate a list of n points in the unit square. Now construct a weighted complete graph where the cost of each edge is the distance between the embedding of the vertices.

   Experiment with computing shortest-path spanning trees, minimum spanning trees, and traveling salesman tours on random Euclidean graphs. Are these subgraphs always planar embeddings?

4. Write a function to partition the edges of K_{2n} into n spanning trees.

5. Modify `PlanarQ` so that it positions the vertices of the graph in a way which permits a planar embedding. These positions do not have to permit a straight-line embedding.

6. Develop and implement a simple algorithm for finding the maximum matching in a tree, which is special case of bipartite matching. Find a tree for which `MaximalMatching` does not find the maximum matching.

7. Implement Prim's algorithm for finding minimum spanning trees and compare it with `MinimumSpanningTree` for efficiency on a variety of graphs.

8. Write a function to test whether a complete matching is stable, given the male and female preference functions.

9. The *divorce digraph* [GI89] is a binary relation associated with an instance of the stable marriage problem. The vertices correspond to the $n!$ matchings. There is a directed edge $\{a, b\}$ between matchings a and b if b results from an unstable pair in a leaving their respective spouses and marrying each other, with the rejects getting paired off. Stable marriages are vertices with out-degree zero in the divorce digraph.

Write a function to construct the divorce digraph of a stable marriage instance. Verify that the following male $\{\{1,2,3\},\{1,2,3\},\{2,1,3\}\}$ and female $\{\{3,2,1\},\{2,3,1\},\{1,2,3\}\}$ preference functions result in a divorce digraph which contains a directed cycle. Further, show that whatever permutations represent male 1's and female 3's preference functions, the divorce digraph contains a cycle.

Research Problems

1. Does is there always exist a directed path from any unstable marriage to a stable one in the divorce digraph of the previous problem [GI89]?
2. What is the expected size of the maximum matching in a tree on n vertices?
3. How hard is it to find the "square root" [Muk67] of a graph G, meaning a graph whose square is G?
4. Find a simple and thus easily implemented linear time planarity testing algorithm [ET89]. The algorithm in [CHT90] represents a step in this direction.

Appendix

This appendix consists of four parts: a section on preliminary material, a glossary of the main functions defined in this book, a bibliography, and an index.

Getting the most out of this book requires knowing some discrete mathematics and some *Mathematica*, and it reasonable to expect that most readers will be more familiar with one than the other. The next section discusses what is expected with respect to both, and how to pick up what you are missing. It includes basic references in combinatorics, graph theory, and the analysis of algorithms. The bulk of this section will be devoted to programming in *Mathematica* which, although it might look scary, will come easily if you have had some experience programming.

Also included is a reference guide to all major functions defined in the text and available in machine-readable format. Functions which were intended for internal use are not included in the glossary of functions.

About the illustration overleaf:

Circulants are an interesting class of graphs which includes complete graphs and cycles as special cases. Since the ith row of the adjacency matrix is a circular shift of $i-1$ places from the first row, a circulant graph possesses a high degree of symmetry. This random circulant graph was constructed using `ShowGraph[ CirculantGraph[20, RandomSubset[Range[10]]] ];`.

Mathematical Preliminaries

The most important prerequisite that this book assumes is a curiousity about discrete structures. Given sufficient interest anyone, even with little or no background in combinatorics and graph theory, can gain some appreciation for the subject just by experimenting with the functions in this book. More background is necessary to truly understand why all the functions work, for many are quite subtle and not fully discussed in the text. For this reason, extensive references are provided.

The equivalent of one course in combinatorics and graph theory should be sufficient to understand how the various structures relate to each other, and some experience with the analysis of algorithms will be useful to understand the computational techniques. If you do not have all of this background, the rest of this section will tell you how to get it.

Combinatorics

Our coverage of combinatorics is somewhat idiosyncratic. We expect the reader to know what a recurrence relation is and how to formulate one, but don't get into issues of how to solve them. See [GKP89, Lue80] for detailed treatments on solving recurrence relations. Recurrences are used for counting structures as well as for formulating algorithms to construct them.

There are three texts on elementary combinatorics which we heartily recommend. Tucker [Tuc84], Roberts [Rob84], and Liu [Liu85] all treat both combinatorics and graph theory on a level sufficient for our purposes. More advanced texts include Stanley [Sta86] and Goulden/Jackson [GJ83].

We know of two books which, like this one, emphasize the actual implementations of combinatorial algorithms. Nijenhuis and Wilf [NW78, Wil89] discuss in greater detail the theory associated with about two dozen of the algorithms which appear within this book, and provide implementations in Fortran. Stanton and White [SW86] present Pascal programs for operations on permutations, partitions, and Young tableaux, to motivate the underlying theory.

Graph Theory

All the elementary combinatorics texts mentioned above discuss some graph theory, sufficient for appreciating most of the material in this book. However, there are several excellent graph theory books available. The grandfather of them all is Harary [Har69], and the recent book by Buckley and Harary [BH90] can be interpreted as a new edition of this classic. Gould [Gou88] should appeal to computer scientists much as Bondy and Murty [BM76] appeals to mathematicians. Chartrand [Cha85] is more appropriate for those seeking an elementary treatment of some introductory topics.

Algorithms

Progress in the design and analysis of combinatorial algorithms is one of the true success stories in computer science. Since the mid-1960's, paradigms for the development of efficient algorithms have been identified, as well as techniques for establishing that no decent algorithm likely exists for a particular problem. The algorithms we implement use such techniques as dynamic programming, backtracking, and divide and conquer, so to understand them requires understanding these paradigms.

There are two generations of texts on the design and analysis of algorithms. Representatives of the first generation such as Aho, Hopcroft, and Ullman [AHU74], Reingold, Nievergelt, and Deo [RND77], and Horowitz and Sahni [HS78] certainly suffice for the purposes of this book, although they miss out on such recent developments as computational geometry, amortized analysis, and parallel and randomized algorithms. The second generation is characterized by Manber [Man89], Brassard and Bratley [BB88], Sedgewick [Sed88], and Baase [Baa88].

The time complexity of an algorithm is described using the big-oh notation, which is discussed in any algorithm text and [GKP89]. Basically, it expresses the time an algorithm will take in the worst case as a function of the *size* of the input. For example, in a sorting problem the size of the input is the number of keys n which are to be sorted. Thus if we say that a sorting algorithm is $O(n^2)$ on a RAM machine, we mean that the time which the algorithm takes on a conventional computer will be upper bounded by some second-degree polynomial.

Many of the properties of graphs which we are interested in are *NP-complete*, meaning that not only is no algorithm for solving the problem in polynomial-time known, it can be established with a high degree of certainty that no efficient algorithm for solving the problem exists. Although discussed in all of the above texts, the definitive treatment of NP-completeness is Garey and Johnson [GJ79].

The techniques for analyzing algorithms have been applied to popular music [Knu84], with amusing results.

Mathematica Preliminaries

The most complete reference on *Mathematica* is [Wol88], which includes a full language description as well as a cogent explanation of the philosophy of *Mathematica* and how it compares to other programming languages. Our style of including numerous examples using the function being discussed is shamelessly stolen from this manual. For any serious work with *Mathematica*, it is necessary to have a copy of the language description in front of you.

With a programming language as rich as *Mathematica*, comprising about 800 different functions, the novice is likely to be overwhelmed with the number of different ways to do things. In fact, it takes a good deal of experience to find the *correct* way to write a particular program. The functions in this book provide good examples of real *Mathematica* programs, since many of them have been rewritten several times for clarity and efficiency. Maeder's *Programming in Mathematica* [Mae89] is an excellent tutorial, and is written by one of the original developers of the system. Particularly useful is the treatment of such system-specific features as packages and defaults, as well as the *Mathematica* functional style of programming.

To the reader who has never programmed in any language before, we recommend studying Abelson and Sussman [ASS85]. They present an introduction to many of the more sophisticated ideas associated with programming via Scheme, a dialect of the language Lisp. To a greater extent than imperative programming languages, such as Fortran, Pascal and C, the Lisp way of programming is the *Mathematica* way.

The self-help facility in *Mathematica* provides a brief description of each function available on-line. Typing `?Append` gives you a description of function `Append`, while `??Append` provides additional information which may or may not be useful. Included with the distribution of our combinatorial and graph theoretic functions are documentation strings for all the major functions, so these can be accessed in an identical manner.

Packages such as the one defined in this book must be loaded into *Mathematica*, after which the new functions are indistinguishable from built-in *Mathematica* functions.

```
In[1]:= <<Combinatorica
Out[1]= Global`
```

The help string for each of the functions we develop is identical to the description in the glossary of procedures in the back of the book.

```
In[2]:= ?Permute
Permute[l,p] permutes list l according to permutation p.
```

For these new functions, the implementation is given when ?? is used. These implementations are not formatted as nicely as in the text.

```
In[2]:= ??Permute

Permute[l,p] permutes list l according to permutation p.
Permute/: Permute[l_List, (p_)?PermutationQ] := l[[p]]
```

The * character is a wildcard, matching everything. With wildcards, the names of all matched functions are returned.

```
In[2]:= ?Add*

AddEdge                    AddToEncroachingLists
AddTo                      AddVertex
```

It is beyond the scope of this book to provide a thorough introduction to the *Mathematica* language. However, in the interests of making our treatment self-contained, what follows is a terse guide to the major constructs of the programming language. Since the discussion is limited to the facilities we use frequently, graphics and symbolic computation will not be included.

The Structure of Functions

Mathematica is perhaps best thought of as a *functional* programming language. All functions in *Mathematica* are call-by-value, meaning that it is impossible to write a function which changes the value of its arguments. Although it is possible to change the value of a global variable within a *Mathematica* function, it is usually not a great idea. The correct way to get a result from a function is to return it as a value.

Since the text of this book contains well over two hundred syntactically correct *Mathematica* functions, we limit the discussion here to a single generic function:

```
FunctionName[arg1_Integer?OddQ, Graph, argn_List] :=
        Block[{v1, vk},
                v1 = vk = arg1;
                vk
        ] /; MemberQ[argn,arg1]
```

Example *Mathematica* Function

This function could be invoked using `FunctionName[3,Graph,{1,3}]`, with the arguments bound to names `arg1` and `argn`. The underscore _ signifies that a pattern is to be matched, with `_Integer?OddQ` to match an object with *head* `Integer` such that the predicate `OddQ` returns `True`. Further, the entire function matches only if the membership condition is satisfied, where such conditions are defined with `/;`. `Block` is an optional structure used to create local variables, in this case names `v1` and `vk`.

The value which a function returns is either the value of the last statement executed within the function, or the value of x if Return[x] is invoked. When invoked as above, this function should return 3. Statements within the body of a function are separated by ;, which returns a Null value.

Mathematica uses dynamic scoping to associate names with variables, meaning that the name is bound to the most recently created, active instance of the variable. Thus if a variable is not declared in the Block of the current function, the Block of the function which called it is checked, and so on up the calling chain until the global name space is reached. Variables become dereferenced when a Block is exited, or when cleared with Clear[x].

A nice property of *Mathematica* which made this entire project feasible, is that it is an *extensible* language, meaning that new functions are invoked in the same way as built-in functions. Thus the procedures we create can be thought of as an extension to *Mathematica* for discrete mathematics. When using the package, there is no reason why you need be aware these functions are *not* part of *Mathematica* once they are loaded.

We shall observe the standard *Mathematica* conventions for naming functions. Every name is completely spelled out, with no abbreviations used. The first letter of each word is capitalized and multiword titles are concatenated together. These rules serve to insure logical consistency, at the expense of a little typing. For example, the name of the function to find the inverse of a permutation is InversePermutation, not Inversepermutation, not InversePerm, and definitely not InPerm. Predicates, functions which return True or False, all have names which end with Q. Thus, the name of the function which tests whether a list is a permutation is PermutationQ.

■ Mathematical Operations

Mathematica has a large number of mathematical operations available, most of which are self-explanatory, since they follow the naming convention discussed above.

The basic binary arithmetic operations are +, -, *, /, and ^ for exponentiation. The functional equivalences of these are Plus, Subtract, Times, Divide, and Power. The assignment operator is = or Set.

There is a full range of trigonometric functions, including the basic Sin, Cos, and Tan. Ceiling and Floor round real numbers up or down, while Abs finds the absolute value of its argument. Pseudo-random integers and reals over a specified interval can be generated with Random. Most of the 800 functions built in to *Mathematica* are of this flavor, and we avoid giving a complete list here.

There are assignment operators for the basic arithmetic operations: +=, -=, *=, and /=.

```
In[2]:= (x=10; x*=10)
Out[2]= 100
```

As in the C language, there are pre- and post-increment and decrement operators, which return the value after or before it changes.

```
In[3]:= {++x,x++}
Out[3]= {101, 101}
```

This is the first middle binomial coefficient $\binom{2n}{n}$ with residue $(-1)^n$ when $2n+1$ is composite, thus answering an open problem in [GKP89].

```
In[4]:= Mod[ Binomial[5906,2953], 5907]
Out[4]= 5906
```

Mathematica has two main data types for representing numbers. By default, exact rational arithmetic is used for all computation. This means that arbitrary precision integers are used, with fractions reduced to lowest terms. If this level of precision is unnecessary, `Real` numbers are much faster and more convenient, with `N[x]` being the way to convert `x` to `Real` with the default precision.

Observe that `Pi` remains unexpanded without `N`, since any expansion results in a loss of precision. The number of terms in the expansion of π can be specified with `N`.

```
In[5]:= {N[Pi,100], Pi}
Out[5]= {3.14159265358979323846264338327950288419716939937\
    510582097494459230781640628620899862803482534211706 8,
   Pi}
```

Working with arbitrary precision arithmetic takes some getting used to, for irrational quantities are not evaluated as numbers until treated with `N`. Fortunately, this is a book about *discrete* mathematics, and so these problems do not arise often.

```
In[6]:= Sqrt[28]
Out[6]= 2 Sqrt[7]
```

Zeno's paradox states that someone repeatedly moving halfway to the goal line never scores a touchdown. The `Sum` command implements the $\sum$ notation of mathematics. Observe that the result is given as a rational number, since it was not specified otherwise.

```
In[7]:= Sum[ 1/(2^i), {i,1,100}]
```

$$Out[7]= \frac{1267650600228229401496703205375}{1267650600228229401496703205376}$$

`N` is used to convert a rational number to a real. Computations on reals are much faster since there is no reduction to lowest terms.

```
In[8]:= Sum[ N[ 1/(2^i) ], {i,1,100}]
Out[8]= 1.
```

The Product command is analogous to the $\prod$ notation, here computing 100!. Observe that the iterator notation starts with $i = 1$ by default, incrementing by 1 each time.

```
In[9]:= Product[i, {i,100}]
Out[9]= 93326215443944152681699238856266700490715968264 38\
  16214685929638952175999932299156089414639761565182862 53\
  697920827223758251185210916864000000000000000000000000
```

All major matrix operations are available in *Mathematica*.

```
In[10]:= {Eigenvalues[IdentityMatrix[4]],
Det[IdentityMatrix[4]]}
Out[10]= {{1, 1, 1, 1}, 1}
```

Dot products are used for matrix multiplication.

```
In[11]:= TableForm[{{1},{2},{3}} . {{1,2,3}}]
                     1  2  3
                     2  4  6
Out[11]//TableForm=  3  6  9
```

A wide variety of number theoretic functions also prove useful.

```
In[12]:= Divisors[16] &&
FactorInteger[4332749032784732342]
Out[12]= {1, 2, 4, 8, 16} &&
  {{2, 1}, {1459, 1}, {1484835172304569, 1}}
```

■ List Manipulation

The basic data structure in *Mathematica* is the *list*, which has a tremendous amount of versatility. Permutations are represented as lists, as are partitions and subsets. Matrices and Young tableaux are lists of lists, while Graph structures are a collection of two lists of lists.

Mathematica provides several ways to access a particular element or set of elements from a list, by presenting the illusion that a list is an array. The ith element of a list l is given by `l[[i]]` or `Part[l,i]`, and so by letting a counter i run from 1 to `Length[l]` each element of the list can be accessed in a sequential fashion. A list of the first n elements of l is given by `Take[l,n]`, while the last n elements are `Take[l,-n]`. A list *without* the first n elements of l is obtained by `Drop[l,n]`. Various special cases of these operations are available and convenient, such as `First`, `Last`, and `Rest`.

Lists can be enlarged with `Join`, `Append`, and `Prepend` as will be shown below. Nested lists can be restructured by `Flatten`, which removes the innermost levels of parentheses from a list. The nesting associated with regular, two-dimensional lists such as matrices can be rearranged with `Transpose`. A circular shift of the elements of a list is achieved by `RotateLeft` or `RotateRight`. The elements of a list can be placed in the appropriate canonical order with `Sort` and then reversed with Reverse. Replace and ReplaceAll can be used to edit arbitrary structures according to transformation rules.

It is important to realize that all these operations leave the original list unchanged. Besides =, very few operations in *Mathematica* have side effects. AppendTo and PrependTo are exceptions; they have the same syntax as Append and Prepend but change the value of the first argument. Using the array notation, the ith element of a list can be changed, as in l[[i]] = 5;, but for technical reasons *Mathematica* rewrites the entire list l when doing this, so modifying an element of a list is in fact a linear-time operation.

First and Rest are equivalent to CAR and CDR in old versions of Lisp. Range[n] is a very useful function which returns a list of the first n integers.

```
In[13]:= {First[Range[10]], Rest[Range[10]]}
Out[13]= {1, {2, 3, 4, 5, 6, 7, 8, 9, 10}}
```

These two restructuring operations give the same results, since the first-level list contains only two elements, and most operations by default only work on first-level structures. The % is shorthand for the result of the previous input.

```
In[14]:= {Reverse[%], RotateLeft[%,1]}
Out[14]= {{{2, 3, 4, 5, 6, 7, 8, 9, 10}, 1},
  {{2, 3, 4, 5, 6, 7, 8, 9, 10}, 1}}
```

Join merges two lists on equal terms, while append makes the first argument the last element of the second argument, which must be a list.

```
In[15]:= TableForm[ {Join[{1,2,3},{4,5,6}],
Append[{1,2,3},{4,5,6}], Prepend[{1,2,3},{4,5,6}]}]
                    {1, 2, 3, 4, 5, 6}
                    {1, 2, 3, {4, 5, 6}}
Out[15]//TableForm= {{4, 5, 6}, 1, 2, 3}
```

Regular two-dimensional list structures, such as matrices, can be transposed.

```
In[16]:= Transpose[{{1,2},{3,4},{5,6}}]
Out[16]= {{1, 3, 5}, {2, 4, 6}}
```

Partition is a very useful function for partitioning the elements of a list into regular-sized structures. Here the list is partitioned into two element lists, where successive lists have an overlap of one element.

```
In[17]:= Partition[{a,b,c,d,e,f},2,1]
Out[17]= {{a, b}, {b, c}, {c, d}, {d, e}, {e, f}}
```

All major set operations are supported as operations on lists. Lists which represent sets are sorted, with the multiplicities removed.

```
In[18]:= Union[ Intersection[{5,2,3,2},{1,2,3}],
Complement[{1,2,3,4,5},{1,2}] ]
Out[18]= {2, 3, 4, 5}
```

Mathematica has a notion of Infinity which can be used to simplify computations with Min and Max operations.

```
In[19]:= Max[{3, 5, 5.6, Infinity, 2}]
Out[19]= Infinity
```

■ Iteration

To exploit the fact that computers execute programs faster than people can write them, a language must have some facility for looping, or executing a block of code more than once. In a structured language, this is usually done through *iteration*. *Mathematica* has statements similar to `While` and `For` from C and `Do` as in Fortran. Unique to *Mathematica* is `Table`, which constructs a list where the ith element is some prescribed function of i.

Since data structures in *Mathematica* are lists, it is often necessary to perform some function on each element of a list, or to compute some function of the entire list. The right way to do this is with `Map` and `Apply`. `Map[f,l]`, where `f` is some function and `l` as a list, returns a list where the ith element is `f[ l[[i]] ]`. `Apply[f,l]` invokes `f` exactly once, where the ith argument of `f` is taken as `l[[i]]`. `Map` can be emulated in a clumsy way using `Table`, but `Apply[f,l]` is *not* the same as `f[l]`, since there is an extra set of brackets around `l` we don't want.

`EvenQ` is a function which returns `True` if the argument is even, so mapping it over the integers alternates `False`, `True`.

```
In[20]:= Map[ EvenQ, Range[10]]
Out[20]= {False, True, False, True, False, True, False,
  True, False, True}
```

Here `Map` and `Apply` work together to test whether `Range[10]` consists of all even numbers.

```
In[21]:= Apply[ And, Map[EvenQ, Range[10]] ]
Out[21]= False
```

Without the `Apply`, `And` doesn't do its thing.

```
In[22]:= And[ Map[EvenQ,Range[10]] ]
Out[22]= {False, True, False, True, False, True, False,
  True, False, True}
```

Both `Map-Apply` and `Apply-Map` can be useful paradigms for working with lists. This computes the sums of each sublist.

```
In[23]:= Map[(Apply[Plus,#])&,{{1,2},{3,4},{5,6}}]
Out[23]= {3, 7, 11}
```

`Scan` is identical to `Map` except that a list is not constructed. Any useful computation with `Scan` is done via side-effects.

Because `Map`, `Scan`, and `Apply` take functions as arguments, making full use of them requires a convenient mechanism for declaring short functions. Such short functions are called *pure* functions because no name is associated with them. An example of a pure function in *Mathematica* is `(Apply[Plus,#])&`, as declared above. The argument to this nameless function is `#`, which is treated in the same way as any other parameter. An alternative way to define this pure function is `Function[n,Apply[Plus,n]]`, which can be used to prevent `#` from being confused with nested pure functions.

Here is the previous example with an alternative representation of the pure function.

```
In[24]:= Map[ Function[n,Apply[Plus,n]],
{{1,2},{3,4},{5,6},{7,8}}]

Out[24]= {3, 7, 11, 15}
```

The most natural way to implement many of the algorithms in this book is through *recursion*, breaking the problem into smaller problems of the same type. If the subproblems are truly of the same type, they can be solved by calling the same function which solved the bigger one. For this reason, recursive implementations of programs follow almost directly from an inductive definition of a problem [Man89]. *Mathematica* supports recursion, and we use it often as a fast and elegant alternative to iteration. As a consequence of how *Mathematica* is implemented, a recursive program is often faster than an equivalent iterative one.

■ Ten Little n-Sums

To illustrate the range of different control structures available in *Mathematica*, and the effect that the right choice can have on efficiency, below are given ten different ways to sum up the first thousand positive integers. You might be surprised at the timing results, although of course this is one particular version of *Mathematica* running on one particular machine. The timings are given as ratios with respect to the best time. Your mileage may vary.

The fastest way to compute this sum is to construct the list of integers and use `Apply`. The reason is that no explicit iteration is being performed, with the moral being that using built-in functions like `Range` coupled with functional operators like `Map` and `Apply` is usually the right way to go.

```
In[25]:= best = Timing[Apply[Plus,Range[1000]]]

Out[25]= {0.1 Second, 500500}
```

The `Do` loop is the most straightforward iteration structure. There is no exit from the loop until the prescribed number of iterations takes place, even with a `Return`.

```
In[26]:= Timing[sum=0; Do[sum+=i,{i,1000}]; sum] / best

Out[26]= {6.2, 1}
```

The `For` loop has initialization, test, increment, and body sections, all of which are separated by commas. The price for this extra flexibility over the `Do` is efficiency.

```
In[27]:= Timing[ For[sum=0; i=1, i<=1000, i++, sum += i];
sum] / best

Out[27]= {8.4, 1}
```

The `While` loop is not as natural for this problem, since the counter must be maintained explicitly, but in fact `For` is simply a disguised `While`.

```
In[28]:= Timing[ sum=0; i=1; While[i<=1000, sum+=(i++)];
sum] / best

Out[28]= {8.5, 1}
```

`Scan` is even less natural for this problem. A list of the numbers from 1 to 1000 is constructed and then traversed, maintaining a running sum.

```
In[29]:= Timing[ sum=0; Scan[(sum+=#)&, Range[1000]]; sum]
/ best
Out[29]= {6.7, 1}
```

`Sum` is the cleanest way to compute this sum, since it is a directed implementation of the $\sum$ operator. It is faster than most other implementations, but is *not* the fastest way to solve the problem. `Clear[i]` is necessary to remove the value of `i`. The iteration operations are often quite fussy about this.

```
In[30]:= Timing[ Clear[i]; Sum[i,{i,1,1000}] ] / best
Out[30]= {2.1, 1}
```

Our second class of implementations is based on list construction operations. With respect to summations, this means building a list of all partial sums, before extracting the one which is desired. Although this is a roundabout way to perform the computation, it is instructive since list construction operations will be used throughout the book more often than iteration.

`Table` is the simplest way to construct a list.

```
In[31]:= Timing[sum=0; Last[ Table[sum+=i,{i,1000}] ] ] /
best
Out[31]= {6.6, 1}
```

`Accumulate` applies a function to each of the prefixes of the list, returning a list of the results. It is surprising how competitive this function is with `Apply`, considering you get 999 sums for free.

```
In[32]:= Timing[ Last[ Accumulate[Plus, Range[1000]] ] ] /
best
Out[32]= {3.2, 1}
```

Although being used inappropriately in this example, `Map` is the best way to apply a function to each element of the list. Realize that the increment function is being applied to 1000 elements in this list, making a new list of which the first 999 elements are ignored, and *still* `Map` is competitive with the iteration structures. `Map` and `Apply` usually work together well to produce a single quantity.

```
In[33]:= Timing[ sum=0; Last[ Map[(sum+=#)&, Range[1000]]
] ] / best
Out[33]= {6.7, 1}
```

As Gauss noticed while still a child, the sum of the first n integers is $n(n+1)/2$. This illustrates that improving the algorithm is always more important than improving the implementation.

```
In[34]:= Timing[ Binomial[1001,2] ]
Out[34]= {0., 500500}
```

Conditionals

Conditional statements change the flow of control during the execution of a program, and are the basis of any interesting computation. In the imperative style of *Mathematica* programming, the `If` statement is used to select between alternatives. No explicit "then" or "else" is specified, as the appropriate clause is identified by position in `If[condition, then, else]`. The else clause is optional. `Switch` generalizes `If` to more than two-way branches.

A variety of unary predicates (`OddQ`, `EvenQ`, `MatrixQ`) and binary predicates (`MemberQ`, `OrderedQ`) are included with *Mathematica*, most but not all of which obey the convention that the names of predicates end with `Q`. Exceptions include `Positive` and `Negative`. Particularly important are the relational operations `>`, `<`, `<=`, `>=`, and `==`, and the Boolean connectives `And`, `Or`, `Not`, with shorthand forms of `&&`, `||`, and `!`, respectively.

One subtlety is the distinction between `SameQ` and `Equal`. When two variables `x` and `y` have not been bound to specific quantities, in a strict sense the question of whether they are "equal" has no answer. `SameQ` or `===` tests whether its two arguments are identical, and always returns `True` or `False`.

Since `X` is not bound, no boolean value results.

```
In[35]:= X == 5
Out[35]= X == 5
```

Leaving out one equal sign makes this an assignment instead of a comparison, and can lead to hard-to-find bugs.

```
In[36]:= X = 5
Out[36]= 5
```

Now either `SameQ` or `Equal` will return `True`.

```
In[37]:= X === 5 && X == 5
Out[37]= True
```

Several higher-order constructs permit us to get conditional behavior without the explicit use of `If`. `Position` identifies which elements in a list satisfy a given condition, in a format which can be used by the conditional mapping function `MapAt`. `Select` returns a list of which elements in a given list satisfy a particular criterion. `Count` computes how often a particular element occurs in a list.

The odd positions in this 11-element list contain the even values. The positions are given within brackets because `Position` traverses nested structures. The pattern match is necessary when using `Position`.

```
In[38]:= Position[Range[10,20], _?EvenQ]
Out[38]= {{1}, {3}, {5}, {7}, {9}, {11}}
```

`Select` returns the possibly empty set of elements satisfying the condition.

```
In[39]:= Select[Range[10,20],EvenQ]
Out[39]= {10, 12, 14, 16, 18, 20}
```

Glossary of Functions

AcyclicQ

AcyclicQ[g] returns True if graph g is acyclic. AcyclicQ[g,Directed] returns True if g is a directed acyclic graph.

AcyclicQ[g,Undirected] is a synonym for AcyclicQ[g]. See Page 189 ▪ See also: FindCycle, TreeQ.

AddEdge

AddEdge[g,{x,y}] returns graph g with a new undirected edge $\{x,y\}$, while AddEdge[g,{x,y},Directed] returns graph g with a new directed edge $\{x,y\}$.

AddEdge[g,{x,y},Undirected] will also add an undirected edge to g. See Page 83 ▪ See also: AddVertex, DeleteEdge.

AddVertex

AddVertex[g] adds a disconnected vertex to graph g.

▪ See Page 84 ▪ See also: AddEdge, DeleteVertex.

AllPairsShortestPath

AllPairsShortestPath[g] returns a matrix, where the (i,j)th entry is the length of the shortest path in g between vertices i and j.

Floyd's algorithm is used for graphs with negative cost edges, so the result is correct unless there are negative cost cycles. See Page 228 ▪ See also: ShortestPath, ShortestPathSpanningTree.

ArticulationVertices

ArticulationVertices[g] returns a list of all articulation vertices in graph g, vertices whose removal will disconnect the graph.

▪ See Page 177 ▪ See also: BiconnectedComponents, Bridges.

Automorphisms

Automorphisms[g] finds the automorphism group of a graph g, the set of isomorphisms of g with itself.

Automorphisms returns the automorphism group in a format suitable for use with Polya. See Page 184 ▪ See also: Isomorphisms, PermutationGroupQ.

Backtrack

`Backtrack[s,partialQ,solutionQ]` performs a backtrack search of the state space s, expanding a partial solution so long as *partialQ* is `True` and returning the first complete solution, as identified by *solutionQ*.

Each solution is assumed to consist of a permutation of `Length[s]` elements, where the possible values for the ith element is given by `s[[i]]`. `Backtrack[s,partialQ,solutionQ,All]` continues the search to find all solutions. See Page 12 ▪ See also: `DistinctPermutations`, `Isomorphism`.

BiconnectedComponents

`BiconnectedComponents[g]` returns a list of the biconnected components of graph g.

▪ See Page 177 ▪ See also: `ArticulationVertices`, `BiconnectedQ`, `Bridges`.

BiconnectedQ

`BiconnectedQ[g]` returns True if graph g is biconnected.

▪ See Page 177 ▪ See also: `ArticulationVertices`, `BiconnectedComponents`, `Bridges`.

BinarySearch

`BinarySearch[l,k,f]` searches sorted list l for key k and returns the the position of l containing k, with f a function which extracts the key from an element of l.

A fractional index is returned if k is not found. The default extraction function f is `Identity`. See Page 15 ▪ See also: `SelectionSort`.

BinarySubsets

`BinarySubsets[l]` returns all subsets of l ordered according to the binary string defining each subset.

▪ See Page 41 ▪ See also: `Subsets`.

BipartiteMatching

`BipartiteMatching[g]` returns the list of edges associated with a maximum matching in bipartite graph g.

▪ See Page 240 ▪ See also: `MaximumAntichain`, `MinimumChainPartition`, `StableMarriage`.

BipartiteQ

`BipartiteQ[g]` returns `True` if graph g is bipartite.

▪ See Page 213 ▪ See also: `TwoColoring`, `K`.

BreadthFirstTraversal

`BreadthFirstTraversal[g,v]` performs a breadth-first traversal of graph g starting from vertex v, and returns a list of vertices in the order in which they were encountered.

`BreadthFirstTraversal[g,v,Edge]` lists edges in the order encountered on the breadth-first search, until all vertices have been visited. See Page 95 ▪ See also: `DepthFirstTraversal`.

Bridges

`Bridges[g]` returns a list of the bridges of graph g, the edges whose removal disconnects the graph.

▪ See Page 177 ▪ See also: `ArticulationVertices`, `BiconnectedComponents`, `BiconnectedQ`.

CartesianProduct

`CartesianProduct[l1,l2]` returns the Cartesian product of lists *l1* and *l2*.

▪ See Page 132 ▪ See also: `GraphJoin`.

CatalanNumber

`CatalanNumber[n]` computes the nth Catalan number, for a positive integer n.

The Catalan numbers are defined by the recurrence $C_n = \sum_{k=0}^{n-1} C_k C_{n-1-k}$. See Page 71 ▪ See also: `NumberOfTableaux`.

ChangeEdges

`ChangeEdges[g,e]` constructs a graph with the adjacency matrix e and the embedding of graph g.

▪ See Page 81 ▪ See also: `ChangeVertices`.

ChangeVertices

`ChangeVertices[g,v]` constructs a graph with the adjacency matrix of graph g and the list v as its embedding.

▪ See Page 81 ▪ See also: `ChangeEdges`.

ChromaticNumber

`ChromaticNumber[g]` computes the chromatic number of the graph, the fewest number of colors necessary to color the graph.

Since the chromatic number problem is NP-complete, this function is doomed to be slow even on small graphs. See Page 211 ▪ See also: `ChromaticPolynomial`, `VertexColoring`.

`ChromaticPolynomial`

`ChromaticPolynomial[g,z]` returns the chromatic polynomial $P(z)$ of graph g, which counts the number of ways to color g with exactly z colors.

Since the chromatic number problem is NP-complete, this function is doomed to be slow even on small graphs. See Page 210 ▪ See also: `ChromaticNumber`, `VertexColoring`.

`CirculantGraph`

`CirculantGraph[n,l]` constructs a circulant graph on n vertices, meaning the ith vertex is adjacent to the $(i+j)$th and $(i-j)$th vertex, for each j in list l.

▪ See Page 140 ▪ See also: `Cycle`, `K`.

`CircularVertices`

`CircularVertices[n]` constructs a list of n points equally spaced on a circle.

`CircularVertices[g]` performs a circular embedding of graph g. See Page 104 ▪ See also: `Cycle`, `K`, `ChangeVertices`.

`CliqueQ`

`CliqueQ[g,c]` returns `True` if the list of vertices c defines a clique in graph g.

▪ See Page 217 ▪ See also: `MaximumClique`, `Turan`.

`CodeToLabeledTree`

`CodeToLabeledTree[l]` constructs the unique labeled tree on n vertices from the Prufer code l, which consists of a list of $n-2$ integers from 1 to n.

▪ See Page 151 ▪ See also: `LabeledTreeToCode`, `RandomTree`.

`Cofactor`

`Cofactor[m,{i,j}]` calculates the (i,j)th cofactor of matrix m.

▪ See Page 235 ▪ See also: `NumberOfSpanningTrees`.

`CompleteQ`

`CompleteQ[g]` returns `True` if graph g is complete.

▪ See Page 89 ▪ See also: `K`, `EmptyQ`.

`Compositions`

`Compositions[n,k]` returns a list of all compositions of integer n into k parts.

The compositions are returned in lexicographic order. See Page 61 ▪ See also: `NextComposition`, `RandomComposition`.

ConnectedComponents

`ConnectedComponents[g]` returns the vertices of graph g partitioned into connected components.

`InduceSubgraph` can be used to isolate each individual subgraph. See Page 171 ▪ See also: `BiconnectedComponents`, `ConnectedQ`, `StronglyConnectedComponents`, `WeaklyConnectedComponents`.

ConnectedQ

`ConnectedQ[g]` returns `True` if undirected graph g is connected. `ConnectedQ[g,Directed]` and `ConnectedQ[g,Undirected]` returns `True` if g is strongly or weakly connected, respectively.

▪ See Page 171 ▪ See also: `ConnectedComponents`, `StronglyConnectedComponents`, `WeaklyConnectedComponents`.

ConstructTableau

`ConstructTableau[p]` performs the bumping algorithm repeatedly on each element of permutation p, resulting in a distinct Young tableau.

▪ See Page 65 ▪ See also: `DeleteFromTableau`, `InsertIntoTableau`.

Contract

`Contract[g,{x,y}]` gives the graph resulting from contracting edge $\{x, y\}$ of graph g.

▪ See Page 91 ▪ See also: `ChromaticPolynomial`, `InduceSubgraph`.

Cycle

`Cycle[n]` constructs the cycle on n vertices, a 2-regular connected graph.

▪ See Page 144 ▪ See also: `AcyclicQ`, `RegularGraph`.

DeBruijnSequence

`DeBruijnSequence[a,n]` constructs a de Bruijn sequence on the alphabet described by list a, the shortest sequence such that every string of length n on a occurs as a contiguous subrange of the sequence.

▪ See Page 195 ▪ See also: `EulerianCycle`, `Strings`.

DegreeSequence

`DegreeSequence[g]` returns the sorted degree sequence of graph g.

▪ See Page 157 ▪ See also: `GraphicQ`, `RealizeDegreeSequence`.

DeleteCycle

DeleteCycle[g,c] deletes undirected cycle c from graph g. DeleteCycle[g,c,Directed] deletes directed cycle c from graph g.

A cycle of n vertices is specified as a list of length $n+1$, with the first element the same as the last element. DeleteCycle[g,c,Undirected] is a synonym for DeleteCycle[g,c]. See Page 189 ▪ See also: ExtractCycles, FindCycle.

DeleteEdge

DeleteEdge[g,{x,y}] returns graph g minus undirected edge $\{x, y\}$, while DeleteEdge[g,{x,y},Directed] returns graph g minus directed edge {x,y}.

DeleteEdge[g,{x,y},Undirected] will also delete an undirected edge from g. See Page 83 ▪ See also: AddEdge, DeleteVertex.

DeleteFromTableau

DeleteFromTableau[t,r] deletes the last element of row r from Young tableaux t.

▪ See Page 65 ▪ See also: ConstructTableau, InsertIntoTableau.

DeleteVertex

DeleteVertex[g,v] deletes vertex v from graph g.

▪ See Page 84 ▪ See also: AddVertex, DeleteEdge.

DepthFirstTraversal

DepthFirstTraversal[g,v] performs a depth-first traversal of graph g starting from vertex v, and returns a list of vertices in the order in which they were encountered.

DepthFirstTraversal[g,v,Edge] lists edges in the order encountered on the depth-first search, until all vertices have been visited. See Page 96 ▪ See also: BreadthFirstTraversal.

DerangementQ

DerangementQ[p] tests whether permutation p is a derangement, a permutation without a fixed point.

▪ See Page 33 ▪ See also: Derangements, NumberOfDerangements.

Derangements

Derangements[p] constructs all derangements of permutation p.

Derangements[n], where n is an integer, constructs all derangements on n elements. See Page 33 ▪ See also: DerangementQ, NumberOfDerangements.

Diameter

`Diameter[g]` computes the diameter of graph g, the length of the longest shortest path between two vertices of g.

▪ See Page 107 ▪ See also: `Eccentricity`, `Radius`.

DilateVertices

`DilateVertices[v,d]` multiplies each coordinate of each vertex position in list l by d, thus dilating the embedding.

`DilateVertices[g,d]` dilates the vertices of graph g. See Page 113 ▪ See also: `NormalizeVertices`, `RotateVertices`, `TranslateVertices`.

DistinctPermutations

`DistinctPermutations[l]` returns all permutations of the multiset described by list l.

The other available permutation generating functions will produce duplicates when confronted with a multiset. See Page 13 ▪ See also: `LexicographicPermutations`, `MinimumChangePermutations`.

Distribution

`Distribution[l,set]` lists the frequency of occurrence of each element of *set* in list l.

`Distribution[l]` takes `Union[l]` as the set. See Page 106 ▪ See also: `RankedEmbedding`.

DurfeeSquare

`DurfeeSquare[p]` computes the number of rows involved in the Durfee square of partition p, the side of the largest sized square contained within the Ferrers diagram of p.

▪ See Page 56 ▪ See also: `FerrersDiagram`, `TransposePartition`.

Eccentricity

`Eccentricity[g]` computes the eccentricity of each vertex v of graph g, the length of the longest shortest path from v.

`Eccentricity[g,v]` computes the eccentricity of vertex v. See Page 107 ▪ See also: `Diameter`, `GraphCenter`, `AllPairsShortestPath`.

EdgeColoring

`EdgeColoring[g]` uses Brelaz's heuristic to find a good, but not necessarily minimal, edge coloring of graph g.

▪ See Page 216 ▪ See also: `VertexColoring`.

EdgeConnectivity

`EdgeConnectivity[g]` computes the minimum number of edges whose deletion from graph g disconnects it.

▪ See Page 178 ▪ See also: `NetworkFlow`, `VertexConnectivity`.

Edges

`Edges[g]` returns the adjacency matrix of graph g.

- See Page 81 ▪ See also: `M`, `Vertices`.

EmptyGraph

`EmptyGraph[n]` generates an empty graph on n vertices.

- See Page 140 ▪ See also: `K`, `EmptyQ`.

EmptyQ

`EmptyQ[g]` returns `True` if graph g contains no edges.

- See Page 89 ▪ See also: `EmptyGraph`, `CompleteQ`.

EncroachingListSet

`EncroachingListSet[p]` constructs the encroaching list set associated with permutation p.

Other supported measures of presortedness include `Inversions`, `Runs`, and `LongestIncreasingSubsequence`. See Page 76 ▪ See also: `Tableaux`.

EquivalenceClasses

`EquivalenceClasses[r]` identifies the equivalence classes among the elements of matrix r.

`EquivalenceClasses[g]` returns the equivalence classes defined by the adjacency matrix of graph g. See Page 18 ▪ See also: `EquivalenceRelationQ`.

EquivalenceRelationQ

`EquivalenceRelationQ[r]` returns `True` if the matrix r defines an equivalence relation. `EquivalenceRelationQ[g]` tests whether the adjacency matrix of graph g defines an equivalence relation.

- See Page 18 ▪ See also: `EquivalenceClasses`.

EulerianCycle

`EulerianCycle[g]` finds an Eulerian circuit of undirected graph g if one exists. `EulerianCycle[g,Directed]` finds an Eulerian circuit of directed graph g if one exists.

A cycle of n vertices is specified as a list of length $n+1$, with the first element the same as the last element. `EulerianCycle[g,Undirected]` is a synonym for `EulerianCycle[g]`. See Page 194 ▪ See also: `DeBruijnSequence`, `EulerianQ`, `HamiltonianCycle`.

■ EulerianQ

EulerianQ[g] returns True if graph g is Eulerian, meaning there exists a tour which includes each edge exactly once. EulerianQ[g,Directed] returns True if directed graph g is Eulerian.

EulerianQ[g,Undirected] is a synonym for EulerianQ[g]. See Page 193 ▪ See also: EulerianCycle, HamiltonianQ.

■ Eulerian

Eulerian[n,k] computes the number of permutations of length n with k runs.

The Eulerian numbers are defined by the recurrence $\left\langle {n \atop k} \right\rangle = k \left\langle {n-1 \atop k} \right\rangle + (n-k+1) \left\langle {n-1 \atop k-1} \right\rangle$. See Page 30 ▪ See also: Runs.

■ ExactRandomGraph

ExactRandomGraph[n,e] constructs a random labeled graph of exactly e edges and n vertices.

▪ See Page 155 ▪ See also: RandomGraph, NthPair, RealizeDegreeSequence.

■ ExpandGraph

ExpandGraph[g,n] expands graph g to n vertices by adding disconnected vertices.

▪ See Page 129 ▪ See also: AddVertex, InduceSubgraph.

■ FerrersDiagram

FerrersDiagram[p] draws a Ferrers diagram of integer partition p.

The size of the dots automatically scales to fill the window. See Page 54 ▪ See also: Partitions, TransposePartition.

■ FindCycle

FindCycle[g] finds a list of vertices which define an undirected cycle in graph g. FindCycle[g,Directed] finds a directed cycle in graph g.

A cycle of n vertices is specified as a list of length $n+1$, with the first element the same as the last element. FindCycle[g,Undirected] is a synonym for FindCycle[g]. See Page 188 ▪ See also: AcyclicQ, DeleteCycle, ExtractCycles.

■ FromAdjacencyLists

FromAdjacencyLists[l] constructs an adjacency matrix representation for a graph with adjacency lists l, using a circular embedding. FromAdjacencyLists[l,v] uses v as the embedding for the resulting graph.

▪ See Page 86 ▪ See also: ToAdjacencyLists, ToOrderedPairs.

FromCycles

`FromCycles[c]` restores a cycle structure c to the original permutation.

The cycle structure is described by a list of cycles. See Page 21 ▪ See also: `ToCycles`, `HideCycles`, `RevealCycles`.

FromInversionVector

`FromInversionVector[v]` reconstructs the unique permutation with inversion vector v.

▪ See Page 27 ▪ See also: `ToInversionVector`, `Inversions`.

FromOrderedPairs

`FromOrderedPairs[l]` constructs an adjacency matrix representation from a list of ordered pairs l, using a circular embedding. `FromOrderedPairs[l,v]` uses v as the embedding for the resulting graph.

Specifying the embedding ensures that the reconstructed graph has the same number of vertices as the original. See Page 87 ▪ See also: `ToOrderedPairs`, `ToUnorderedPairs`, `FromUnorderedPairs`.

FromUnorderedPairs

`FromUnorderedPairs[l]` constructs an adjacency matrix representation from a list of unordered pairs l, using a circular embedding. `FromUnorderedPairs[l,v]` uses v as the embedding for the resulting graph.

Specifying the embedding ensures that the reconstructed graph has the same number of vertices as the original. See Page 88 ▪ See also: `ToOrderedPairs`, `FromOrderedPairs`, `ToUnorderedPairs`.

FunctionalGraph

`FunctionalGraph[f,n]` constructs the functional digraph on n vertices defined by integer function f.

▪ See Page 164 ▪ See also: `IntervalGraph`, `MakeGraph`.

Girth

`Girth[g]` computes the length of the shortest cycle in unweighted graph g.

`Girth` works only on simple graphs. See Page 190 ▪ See also: `FindCycle`, `ShortestPath`.

GraphCenter

`GraphCenter[g]` returns a list of the vertices of graph g with minimum eccentricity.

▪ See Page 108 ▪ See also: `AllPairsShortestPath`, `Eccentricity`.

GraphComplement

`GraphComplement[g]` returns the complement of graph g.

▪ See Page 93 ▪ See also: `SelfComplementaryQ`.

GraphDifference

`GraphDifference[g,h]` constructs the graph resulting from subtracting the adjacency matrix of graph g from that of graph h.

The difference is only defined when `V[g] = V[h]`. See Page 131 ▪ See also: `GraphSum`, `GraphProduct`.

GraphIntersection

`GraphIntersection[g,h]` constructs the graph defined by the edges which are in both graph g and graph h.

Graphs g and h must have the same number of vertices to be successfully intersected. See Page 131 ▪ See also: `GraphJoin`, `GraphUnion`.

GraphJoin

`GraphJoin[g,h]` constructs the join of graphs g and h.

The join of two graphs is their union with the addition of all possible edges between the two graphs. See Page 132 ▪ See also: `GraphProduct`, `GraphUnion`.

GraphPower

`GraphPower[g,k]` computes the kth power of graph g, meaning there is an edge between any pair of vertices of g with a path between them of length at most k.

▪ See Page 230 ▪ See also: `ShortestPath`.

GraphProduct

`GraphProduct[g,h]` constructs the product of graphs g and h.

The embedding of the product graph attempts to reflect its structure. See Page 133 ▪ See also: `GraphDifference`, `GraphSum`.

GraphSum

`GraphSum[g,h]` constructs the graph resulting from adding the adjacency matrices of graphs g and h.

The sum is defined only when `V[g] = V[h]`. See Page 131 ▪ See also: `GraphDifference`, `GraphProduct`.

GraphUnion

`GraphUnion[g,h]` constructs the union of graphs g and h. `GraphUnion[n,g]` constructs n copies of graph g, where n is an integer.

▪ See Page 129 ▪ See also: `GraphIntersection`, `GraphJoin`.

GraphicQ

GraphicQ[s] returns True if the list of integers s is graphic, and thus represents a degree sequence of some graph.

▪ See Page 157 ▪ See also: DegreeSequence, RealizeDegreeSequence.

GrayCode

GrayCode[l] constructs a binary reflected Gray code on set l.

Subsets also constructs a Gray code. See Page 43 ▪ See also: BinarySubsets, LexicographicSubsets.

GridGraph

GridGraph[n,m] constructs an $n \times m$ grid graph, the product of paths on n and m vertices.

▪ See Page 147 ▪ See also: GraphProduct, Path.

HamiltonianCycle

HamiltonianCycle[g] finds a Hamiltonian cycle in graph g if one exists.
HamiltonianCycle[g,All] returns all Hamiltonian cycles of graph g.

A cycle of n vertices is specified as a list of length $n+1$, with the first element the same as the last element. Since the Hamiltonian cycle problem is NP-complete, this function can be inefficient even on small graphs. See Page 196 ▪ See also: EulerianCycle, HamiltonianQ.

HamiltonianQ

HamiltonianQ[g] returns True if there exists a Hamiltonian cycle in graph g, in other words, if there exists a cycle which visits each vertex exactly once.

▪ See Page 196 ▪ See also: EulerianQ, HamiltonianCycle.

Harary

Harary[k,n] constructs the minimal k-connected graph on n vertices.

▪ See Page 179 ▪ See also: EdgeConnectivity, VertexConnectivity.

HasseDiagram

HasseDiagram[g] constructs a Hasse diagram of the relation defined by directed acyclic graph g.

Hasse diagrams are most appropriate for representing partial orders, but any acyclic graph can be displayed. See Page 206 ▪ See also: PartialOrderQ, TransitiveReduction.

HeapSort

HeapSort[l] performs a heap sort on the items of list l.

The built-in *Mathematica* function Sort will be much faster. See Page 38 ▪ See also: Heapify, SelectionSort.

Heapify

`Heapify[p]` builds a heap from permutation p.

`Heapify[p,k]` merges position k into two heaps rooted at $2k$ and $2k+1$. See Page 35 ▪ See also: `HeapSort`, `RandomHeap`.

HideCycles

`HideCycles[c]` canonically encodes the cycle structure c into a unique permutation.

Note that this is not does not produce the same permutation as `FromCycles`. See Page 22 ▪ See also: `ToCycles`, `FromCycles`, `RevealCycles`.

Hypercube

`Hypercube[n]` constructs an n-dimensional hypercube.

▪ See Page 148 ▪ See also: `GrayCode`.

IdenticalQ

`IdenticalQ[g,h]` returns `True` if graphs g and h have identical adjacency matrices.

Graphs which are identical are isomorphic but not necessarily vise versa. See Page 182 ▪ See also: `IsomorphicQ`, `Isomorphism`.

IncidenceMatrix

`IncidenceMatrix[g]` returns the (0,1)-incidence matrix of graph g, which has a row for each vertex and column for each edge and $(v,e)=1$ if and only if vertex v is incident upon edge e.

▪ See Page 135 ▪ See also: `LineGraph`.

IndependentSetQ

`IndependentSetQ[g,i]` returns `True` if the vertices in list i define an independent set in graph g.

▪ See Page 219 ▪ See also: `CliqueQ`, `VertexCoverQ`, `MaximumIndependentSet`.

Index

`Index[p]` returns the index of permutation p, the sum of all subscripts j such that $p[j]$ is greater than $p[j+1]$.

▪ See Page 29 ▪ See also: `Inversions`.

InduceSubgraph

`InduceSubgraph[g,s]` constructs the subgraph of graph g induced by the list of vertices s.

▪ See Page 90 ▪ See also: `Contract`.

InsertIntoTableau

`InsertIntoTableau[e,t]` inserts integer e into Young tableau t using the bumping algorithm.

- See Page 64 ▪ See also: `ConstructTableau`, `DeleteFromTableau`.

IntervalGraph

`IntervalGraph[l]` constructs the interval graph defined by the list of intervals l.

- See Page 163 ▪ See also: `FunctionalGraph`, `MakeGraph`.

InversePermutation

`InversePermutation[p]` yields the multiplicative inverse of permutation `p`.

- See Page 18 ▪ See also: `Permute`.

Inversions

`Inversions[p]` counts the number of inversions in permutation p.

Other supported measures of presortedness include `Runs`, `EncroachingListSet`, and `LongestIncreasingSubsequence`. See Page 29 ▪ See also: `FromInversionVector`, `ToInversionVector`.

InvolutionQ

`InvolutionQ[p]` returns `True` if permutation p is its own inverse.

- See Page 32 ▪ See also: `InversePermutation`, `NumberOfInvolutions`.

IsomorphicQ

`IsomorphicQ[g,h]` returns `True` if graphs g and h are isomorphic.

- See Page 183 ▪ See also: `IdenticalQ`, `Isomorphism`.

Isomorphism

`Isomorphism[g,h]` returns an isomorphism between graphs g and h if one exists.

`Isomorphism[g,h,All]` finds all isomorphisms between the two graphs. See Page 182 ▪ See also: `Automorphisms`, `IdenticalQ`, `IsomorphicQ`.

Josephus

`Josephus[n,m]` generates the inverse of the permutation defined by executing every mth member in a circle of n men.

- See Page 34 ▪ See also: `InversePermutation`.

KSubsets

KSubsets[l,k] returns all subsets of set l containing exactly k elements, ordered lexicographically.

- See Page 44 ▪ See also: LexicographicSubsets, NextKSubset, RandomKSubset.

K

K[n] creates a complete graph on n vertices. K[a,b,c,...,k] creates a complete k-partite graph of the prescribed shape.

- See Page 140 ▪ See also: CompleteQ, EmptyGraph.

LabeledTreeToCode

LabeledTreeToCode[g] reduces the tree g to its Prufer code.

- See Page 151 ▪ See also: CodeToLabeledTree, Strings.

LexicographicPermutations

LexicographicPermutations[l] constructs all permutations of list l in lexicographic order.

The list of elements l is assumed to be lexicographically ordered. See Page 4 ▪ See also: RankPermutation, NthPermutation, NextPermutation.

LexicographicSubsets

LexicographicSubsets[l] returns all subsets of set l in lexicographic order.

- See Page 43 ▪ See also: NthSubset, Subsets.

LineGraph

LineGraph[g] constructs the line graph of graph g.

- See Page 136 ▪ See also: IncidenceMatrix.

LongestIncreasingSubsequence

LongestIncreasingSubsequence[p] find the longest increasing scattered subsequence of permutation p.

Runs can be used to identify the longest increasing contiguous subsequence of permutation p. See Page 74 ▪ See also: Inversions, TableauClasses.

M

M[g] gives the number of edges in undirected graph g.

M[g,Directed] gives the number of directed edges in graph g. See Page 81 ▪ See also: Edges, V.

MakeGraph

MakeGraph[v,f] constructs the binary relation defined by function f on all pairs of elements of list v.

- See Page 161 ▪ See also: FunctionalGraph, IntervalGraph.

MakeSimple

MakeSimple[g] returns an undirected, unweighted graph derived from directed graph g.

- See Page 94 ▪ See also: MakeUndirected, SimpleQ.

MakeUndirected

MakeUndirected[g] returns a graph with an undirected edge for each directed edge of graph g.

- See Page 94 ▪ See also: MakeSimple, UndirectedQ.

MaximalMatching

MaximalMatching[g] returns the list of edges associated with a maximal matching of graph g.

Note that the returned matching is not guaranteed to be the maximum matching in the graph, just one which cannot be expanded by simply adding an edge. See Page 244 ▪ See also: BipartiteMatching.

MaximumAntichain

MaximumAntichain[g] returns a largest set of unrelated vertices in partial order g.

- See Page 242 ▪ See also: BipartiteMatching, MinimumChainPartition, PartialOrderQ.

MaximumClique

MaximumClique[g] finds the largest clique in graph g.

Since the maximum clique problem is NP-complete, this function is doomed to be inefficient even with very small graphs. See Page 217 ▪ See also: CliqueQ, MinimumVertexCover, MaximumIndependentSet.

MaximumIndependentSet

MaximumIndependentSet[g] finds the largest independent set of graph g.

- See Page 219 ▪ See also: IndependentSetQ, MaximumClique, MinimumVertexCover.

MaximumSpanningTree

MaximumSpanningTree[g] uses Kruskal's algorithm to find a maximum spanning tree of graph g.

- See Page 233 ▪ See also: MinimumSpanningTree, NumberOfSpanningTrees.

MinimumChainPartition

MinimumChainPartition[g] partitions partial order g into a minimum number of chains.

■ See Page 242 ■ See also: BipartiteMatching, MaximumAntichain, PartialOrderQ.

MinimumChangePermutations

MinimumChangePermutations[l] constructs all permutations of list l such that adjacent permutations differ by only one transposition.

The transposition order is not cyclic, in that the last permutation is not necessarily a single transposition removed from the first. See Page 11 ■ See also: DistinctPermutations, LexicographicPermutations.

MinimumSpanningTree

MinimumSpanningTree[g] uses Kruskal's algorithm to find a minimum spanning tree of graph g.

■ See Page 233 ■ See also: MaximumSpanningTree, NumberOfSpanningTrees, ShortestPathSpanningTree.

MinimumVertexCover

MinimumVertexCover[g] finds the minimum vertex cover of graph g.

Since the vertex cover problem is NP-complete, this function is doomed to be slow even for very small graphs. See Page 218 ■ See also: VertexCoverQ, MaximumClique, MaximumIndependentSet.

MultiplicationTable

MultiplicationTable[l,f] constructs the complete transition table defined by the binary relation function f on the elements of list l.

The multiplication table contains the index of the appropriate elements of l, with a 0 whenever f isn't closed on l. See Page 17 ■ See also: PermutationGroupQ.

NetworkFlowEdges

NetworkFlowEdges[g,source,sink] returns the adjacency matrix showing the distribution of the maximum flow from *source* to *sink* in graph g.

■ See Page 237 ■ See also: NetworkFlow.

NetworkFlow

NetworkFlow[g,source,sink] finds the maximum flow through directed graph g from *source* to *sink*.

■ See Page 237 ■ See also: EdgeConnectivity, NetworkFlowEdges, VertexConnectivity.

NextComposition

NextComposition[l] constructs the integer composition which follows l in a canonical order.

This order is *not* lexicographic as used in Compositions. See Page 61 ▪ See also: Compositions, RandomComposition.

NextKSubset

NextKSubset[l,s] computes the k-subset of list l which appears after k-subsets s in lexicographic order.

▪ See Page 45 ▪ See also: KSubsets, RandomKSubset.

NextPartition

NextPartition[p] returns the integer partition following p in reverse lexicographic order.

Reverse lexicographic order is also used by Partitions. See Page 53 ▪ See also: Partitions, RandomPartition.

NextSubset

NextSubset[l,s] constructs the subset of l following subset s in canonical order.

The canonical order is defined by BinarySubsets. See Page 41 ▪ See also: NthSubset, RankSubset.

NextTableau

NextTableau[t] returns the tableau of shape t which follows t in lexicographic order.

▪ See Page 68 ▪ See also: RandomTableau, Tableaux.

NormalizeVertices

NormalizeVertices[v] returns a list of vertices with the same structure as v but with all coordinates of all points between 0 and 1.

NormalizeVertices[g] returns an embedding of graph g with its embedding normalized. See Page 113 ▪ See also: DilateVertices, RotateVertices, TranslateVertices.

NthPair

NthPair[n] returns the nth unordered pair of positive integers, when sequenced to minimize the size of the larger integer.

NthPair can be used to select a random, non-identical pair of integers from 1 to m with NthPair[Random[Integer,1,Binomial[m,2]]]. See Page 155 ▪ See also: Contract, ExactRandomGraph.

NthPermutation

NthPermutation[n,l] returns the nth lexicographic permutation of list l.

The index n must be an integer, although not necessarily positive. See Page 5 ▪ See also: LexicographicPermutations, RankPermutation.

NthSubset

NthSubset[n,l] returns the nth subset of list l in canonical order.

NthSubset[n,m] returns the nth subset of $\{1, 2, ..., m\}$. The canonical order is defined by BinarySubsets. See Page 41 ▪ See also: NextSubset, RankSubset.

NumberOfCompositions

NumberOfCompositions[n,k] counts the number of distinct compositions of integer n into k parts.

▪ See Page 60 ▪ See also: Compositions, RandomComposition.

NumberOfDerangements

NumberOfDerangements[n] counts the derangements on n elements, the permutations without any fixed points.

The number of derangements is given by the recurrence $d_n = (n-1)(d_{n-1} + d_{n-2})$. See Page 33 ▪ See also: DerangementQ, Derangements.

NumberOfInvolutions

NumberOfInvolutions[n] counts the number of involutions on n elements.

The number of involutions is given by the recurrence $t_n = t_{n-1} + (n-1)t_{n-2}$. See Page 32 ▪ See also: InvolutionQ.

NumberOfPartitions

NumberOfPartitions[n] counts the number of distinct integer partitions of n.

The built-in *Mathematica* function PartitionsP is equivalent but much faster. See Page 57 ▪ See also: Partitions, RandomPartition.

NumberOfPermutationsByCycles

NumberOfPermutationsByCycles[n,m] returns the number of permutations of length n with exactly m cycles.

▪ See Page 23 ▪ See also: Polya.

NumberOfSpanningTrees

NumberOfSpanningTrees[g] computes the number of distinct labeled spanning trees of graph g.

▪ See Page 236 ▪ See also: MinimumSpanningTree, Cofactor.

NumberOfTableaux

`NumberOfTableaux[p]` uses the hook length formula to count the number of Young tableaux with shape defined by partition p.

`NumberOfTableaux[n]` counts the total number of Young tableaux which can be formed with n elements. See Page 70 ▪ See also: `CatalanNumber`, `Tableaux`.

OrientGraph

`OrientGraph[g]` assigns a direction to each edge of a bridgeless, undirected graph g, so that the graph is strongly connected.

▪ See Page 174 ▪ See also: `ConnectedQ`, `StronglyConnectedComponents`.

PartialOrderQ

`PartialOrderQ[g]` returns `True` if the binary relation defined by the adjacency matrix of graph g is a partial order, meaning it is transitive, reflexive, and anti-symmetric.

▪ See Page 203 ▪ See also: `HasseDiagram`, `TransitiveQ`.

PartitionQ

`PartitionQ[p]` returns `True` if p is an integer partition.

The order in which the positive integers appear is not considered. See Page 51 ▪ See also: `Partitions`.

Partitions

`Partitions[n]` constructs all partitions of integer n in reverse lexicographic order.

`Partitions[n,k]` constructs all integer partitions of n with largest part at most k. See Page 51 ▪ See also: `NextPartition`, `RandomPartition`.

Path

`Path[n]` constructs a tree consisting only of a path on n vertices.

▪ See Page 147 ▪ See also: `GridGraph`, `ShortestPath`.

PerfectQ

`PerfectQ[g]` returns true is g is a perfect graph, meaning that for every induced subgraph of g the size of the largest clique equals the chromatic number.

Since no polynomial time algorithm for this problem is known, this function is doomed to be slow even on small graphs. See Page 219 ▪ See also: `ChromaticNumber`, `MaximumClique`.

PermutationGroupQ

`PermutationGroupQ[l]` returns `True` if the list of permutations l forms a permutation group.

▪ See Page 19 ▪ See also: `Automorphisms`, `MultiplicationTable`.

PermutationQ

`PermutationQ[p]` returns `True` if p represents a permutation and `False` otherwise.

Permutations are defined as arrangements of the integers from 1 to n. See Page 3 ▪ See also: `Permute`.

Permute

`Permute[l,p]` permutes list l according to permutation p.

`Permute` is the multiplication operator associated with permutation groups. See Page 3 ▪ See also: `InversePermutation`, `PermutationQ`.

PlanarQ

`PlanarQ[g]` returns `True` if graph g is planar, meaning it can be drawn in the plane so no two edges cross.

Whether the embedding of g is planar is not considered by `PlanarQ`. See Page 247 ▪ See also: `ShowGraph`.

Polya

`Polya[g,m]` returns the polynomial giving the number of colorings, with m colors, of a structure defined by the permutation group g.

If c is an integer, `Polya[g,c]` returns the polynomial evaluated at c. See Page 25 ▪ See also: `Automorphisms`, `PermutationGroupQ`.

PseudographQ

`PseudographQ[g]` returns `True` if graph g is a pseudograph, meaning it contains self-loops.

▪ See Page 89 ▪ See also: `RemoveSelfLoops`.

RadialEmbedding

`RadialEmbedding[g]` constructs a radial embedding of graph g, radiating from the center of the graph.

`RadialEmbedding[g,v]` constructs a radial embedding with v as the center. This embedding is always planar for trees. See Page 108 ▪ See also: `RandomTree`, `RootedEmbedding`.

Radius

`Radius[g]` computes the radius of graph g, the minimum eccentricity of any vertex of g.

▪ See Page 108 ▪ See also: `Diameter`, `Eccentricity`.

RandomComposition

`RandomComposition[n,k]` constructs a random composition of integer n into k parts.

▪ See Page 60 ▪ See also: `Compositions`, `NumberOfCompositions`.

RandomGraph

RandomGraph[n,p,{l,h}] constructs a random labeled graph on n vertices with an edge probability of p and edge weights of integers drawn uniformly at random from the range (l, h). RandomGraph[n,p,{l,h},Directed] similarly constructs a random directed graph.

RandomGraph[n,p] constructs an unweighted random labeled graph on n vertices with edge probability p. RandomGraph[n,p,Directed] constructs a random directed graph on n vertices with directed edge probability p. See Page 154 ▪ See also: ExactRandomGraph, RealizeDegreeSequence.

RandomHeap

RandomHeap[n] constructs a random heap on n elements.

▪ See Page 36 ▪ See also: Heapify, HeapSort.

RandomKSubset

RandomKSubset[l,k] returns a random subset of set l with exactly k elements.

RandomKSubset[n,k] constructs of random subset of Range[n] for integer n. See Page 45 ▪ See also: KSubsets, NextKSubset.

RandomPartition

RandomPartition[n] constructs a random partition of integer n.

▪ See Page 58 ▪ See also: NumberOfPartitions, Partitions.

RandomPermutation

RandomPermutation[n] returns a random permutation of length n.

▪ See Page 7 ▪ See also: NthPermutation.

RandomSubset

RandomSubset[l] creates a random subset of set l.

RandomKSubset can be used to construct a random subset of a desired cardinality. See Page 42 ▪ See also: NthSubset, Subsets.

RandomTableau

RandomTableau[p] constructs a random Young tableau of shape p.

▪ See Page 72 ▪ See also: NextTableau, Tableaux.

RandomTree

RandomTree[n] constructs a random labeled tree on n vertices.

Each labeled tree occurs with equal probability, but random labeled trees are not the same as random unlabeled trees. See Page 152 ▪ See also: CodeToLabeledTree, TreeQ.

RandomVertices

`RandomVertices[g]` assigns a random embedding to graph g.

`RandomVertices[n]` constructs a list of n points randomly selected from the unit square. See Page 155 ▪ See also: `RandomGraph`.

RankGraph

`RankGraph[g,l]` partitions the vertices into classes based on the shortest geodesic distance to a member of list l.

▪ See Page 105 ▪ See also: `RankedEmbedding`.

RankPermutation

`RankPermutation[p]` computes the rank of permutation p in lexicographic order.

This is the inverse operation of `NthPermutation`. See Page 5 ▪ See also: `LexicographicPermutations`, `NthPermutation`.

RankSubset

`RankSubset[l,s]` computes the rank, in canonical order, of subset s of set l.

The canonical order is defined by `BinarySubsets`. See Page 41 ▪ See also: `NextSubset`, `NthSubset`.

RankedEmbedding

`RankedEmbedding[g,l]` performs a ranked embedding of graph g, with the vertices ranked in terms of geodesic distance from a member of list l.

▪ See Page 106 ▪ See also: `RankGraph`.

ReadGraph

`ReadGraph[f]` reads a graph represented as edge lists from file f, and returns the graph as a graph object.

The file format is equivallent to what is used by SPREMB [EFK88]. See Page 120 ▪ See also: `WriteGraph`.

RealizeDegreeSequence

`RealizeDegreeSequence[s]` constructs a semirandom graph with degree sequence s.

Since `RealizeDegreeSequence` produces semirandom graphs, the result is non-deterministic unless the random number generator is reseeded. The resulting graphs are not randomly selected in any formal sense. `RealizeDegreeSequence[s,n]` reseeds the random number generator with integer n as the seed. See Page 158 ▪ See also: `GraphicQ`, `DegreeSequence`.

RegularGraph

`RegularGraph[k,n]` constructs a semirandom k-regular graph on n vertices, if such a graph exists.

Since `RealizeDegreeSequence` is used to produce semirandom graphs, the result is non-deterministic unless the random number generator is reseeded. The resulting graphs are not randomly selected in any formal sense. See Page 159 ▪ See also: `RealizeDegreeSequence`, `RegularQ`.

RegularQ

`RegularQ[g]` returns `True` if g is a regular graph.

▪ See Page 159 ▪ See also: `DegreeSequence`, `RegularGraph`.

RemoveSelfLoops

`RemoveSelfLoops[g]` constructs a graph g with the same edges except for any self-loops.

▪ See Page 89 ▪ See also: `PseudographQ`, `SimpleQ`.

RevealCycles

`RevealCycles[p]` unveils the canonical hidden cycle structure of permutation p.

Note that this is not the cycle structure of permutation p, but under the assumption that p is the result of a `HideCycles` operation. See Page 22 ▪ See also: `ToCycles`, `FromCycles`, `RevealCycles`.

RootedEmbedding

`RootedEmbedding[g,v]` constructs a rooted embedding of graph g with vertex v as the root.

This embedding is planar for trees and wheels. See Page 111 ▪ See also: `RadialEmbedding`.

RotateVertices

`RotateVertices[v,theta]` rotates each vertex position in list v by *theta* radians around the origin (0,0).

`RotateVertices[g,theta]` rotates the embedding of graph g by `theta` radians about the origin (0,0). See Page 113 ▪ See also: `RotateVertices`, `TranslateVertices`.

Runs

`Runs[p]` partitions p into contiguous increasing subsequences.

Other supported measures of presortedness include `Inversions`, `EncroachingListSet`, and `LongestIncreasingSubsequence`. See Page 30 ▪ See also: `Eulerian`.

SamenessRelation

`SamenessRelation[l]` constructs a binary relation from a list of permutations l which is an equivalence relation if l is a permutation group.

▪ See Page 19 ▪ See also: `EquivalenceRelationQ`, `PermutationGroupQ`.

SelectionSort

`SelectionSort[l,f]` sorts list l using ordering function f.

The built-in *Mathematica* function `Sort` is much faster than `SelectionSort`. See Page 15 ▪ See also: `BinarySearch`, `HeapSort`.

SelfComplementaryQ

`SelfComplementaryQ[g]` returns `True` if graph g is self-complementary, meaning it is isomorphic to its complement.

▪ See Page 187 ▪ See also: `GraphComplement`, `Isomorphism`.

ShakeGraph

`ShakeGraph[g,d]` performs a random perturbation of the vertices of graph g, with each vertex moving at most a distance d from its original position.

`ShakeGraph` can be used to place the vertices of a graph in general position, so all edges are unambiguously displayed. Fractional distances usually perturb the graph sufficiently. If `ShakeGraph` is invoked with one argument, the maximum perturbation is $d = 0.1$. See Page 114 ▪ See also: `ShowGraph`, `SpringEmbedding`.

ShortestPathSpanningTree

`ShortestPathSpanningTree[g,v]` constructs the shortest-path spanning tree originating from v, so that the shortest path in graph g from v to any other vertex is the path in the tree.

The tree is returned as a graph with the same embedding as g. See Page 226 ▪ See also: `AllPairsShortestPath`, `MinimumSpanningTree`, `ShortestPath`.

ShortestPath

`ShortestPath[g,start,end]` finds the shortest path between vertices *start* and *end* in graph g.

The path is defined by a list of the vertices from *start* to *end*. See Page 226 ▪ See also: `AllPairsShortestPath`, `ShortestPathSpanningTree`.

ShowGraph

`ShowGraph[g]` displays graph g according to its embedding. `ShowGraph[g,Directed]` displays directed graph g according to its embedding, with arrows illustrating the orientation of each edge.

The display is automatically scaled to fill the window. See Page 100 ▪ See also: `ShowLabeledGraph`.

ShowLabeledGraph

`ShowLabeledGraph[g]` displays graph g according to its embedding, with each vertex labeled with its vertex number. `ShowLabeledGraph[g,l]` uses the ith element of list l as the label for vertex i.

The display is automatically scaled to fill the window. See Page 103 ▪ See also: `ShowGraph`.

■ SignaturePermutation

`SignaturePermutation[p]` gives the signature of permutation p.

The built-in *Mathematica* function `Signature` computes the equivalent function more quickly. See Page 24 ■ See also: `MinimumChangePermutations`, `ToCycles`.

■ SimpleQ

`SimpleQ[g]` returns `True` if g is a simple graph, meaning it is unweighted and contains no self-loops.

■ See Page 89 ■ See also: `PseudographQ`, `UnweightedQ`.

■ Spectrum

`Spectrum[g]` gives the eigenvalues of graph g.

■ See Page 85 ■ See also: `Edges`.

■ SpringEmbedding

`SpringEmbedding[g]` beautifies the embedding of graph g by modeling the graph as a system of springs.

`SpringEmbedding[g,i,increment]` runs the heuristic for i iterations with the maximum change per iteration governed by *increment*. See Page 116 ■ See also: `ShakeGraph`, `ShowGraph`.

■ StableMarriage

`StableMarriage[mpref,fpref]` finds the male optimal stable marriage defined by lists of permutations describing male and female preferences.

■ See Page 245 ■ See also: `BipartiteMatching`, `MaximalMatching`.

■ Star

`Star[n]` constructs a star on n vertices, which is a tree with one vertex of degree $n-1$.

■ See Page 146 ■ See also: `TreeQ`, `Wheel`.

■ StirlingFirst

`StirlingFirst[n,k]` computes the Stirling numbers of the first kind.

This function is defined by the recurrence $\left[{n \atop k}\right] = (n-1)\left[{n-1 \atop k}\right] + \left[{n-1 \atop k-1}\right]$. This definition is different than the built-in function `StirlingS1`. See Page 23 ■ See also: `NumberOfPermutationsByCycles`.

StirlingSecond

StirlingSecond[n,k] computes the Stirling numbers of the second kind.

This function is defined by the recurrence $\left\{ {n \atop k} \right\} = k\left\{ {n-1 \atop k} \right\} + \left\{ {n-1 \atop k} \right\}$. This function is identical to the built-in function StirlingS2. See Page 23 ▪ See also: NumberOfPermutationsByCycles.

Strings

Strings[l,n] constructs all possible strings of length n from the elements of list l.

The strings are returned in lexicographic order as defined by the order of list l. See Page 40 ▪ See also: CodeToLabeledTree, DistinctPermutations.

StronglyConnectedComponents

StronglyConnectedComponents[g] returns the strongly connected components of directed graph g.

▪ See Page 172 ▪ See also: ConnectedQ, WeaklyConnectedComponents.

Subsets

Subsets[l] returns all subsets of set l.

Subsets[n] constructs all subsets of Range[n] for an integer n. These subsets are sequenced to form a Gray code. See Page 43 ▪ See also: BinarySubsets, GrayCode, LexicographicSubsets.

TableauClasses

TableauClasses[p] partitions the elements of permutation p into classes according to their initial columns during Young tableaux construction.

▪ See Page 74 ▪ See also: InsertIntoTableau, LongestIncreasingSubsequence.

TableauQ

TableauQ[t] returns True if and only if t represents a Young tableau.

▪ See Page 63 ▪ See also: RandomTableau, Tableaux.

Tableaux

Tableaux[p] constructs all tableaux whose shape is given by integer partition p.

Tableaux[n] constructs all tableaux on n elements. The tableaux are listed in lexicographic order by shape. See Page 69 ▪ See also: NextTableau, RandomTableau.

TableauxToPermutation

TableauxToPermutation[t1,t2] constructs the unique permutation associated with Young tableaux *t1* and *t2*, where both tableaux have the same shape.

▪ See Page 66 ▪ See also: InsertIntoTableau, DeleteFromTableau.

`ToAdjacencyLists`

`ToAdjacencyLists[g]` constructs an adjacency list representation for graph g.

The ith element of the returned structure is a list of the vertices adjacent to vertex i of g. See Page 86 ▪ See also: `FromAdjacencyLists`, `ToOrderedPairs`.

`ToCycles`

`ToCycles[p]` returns the cycle structure of permutation p.

The cycle structure is described by a list of cycles. See Page 20 ▪ See also: `FromCycles`, `HideCycles`, `RevealCycles`.

`ToInversionVector`

`ToInversionVector[p]` computes the inversion vector associated with permutation p.

▪ See Page 27 ▪ See also: `FromInversionVector`, `Inversions`.

`ToOrderedPairs`

`ToOrderedPairs[g]` constructs a list of ordered pairs representing the edges of undirected graph g.

▪ See Page 87 ▪ See also: `ToUnorderedPairs`, `FromOrderedPairs`, `FromUnorderedPairs`.

`ToUnorderedPairs`

`ToUnorderedPairs[g]` constructs a list of vertex pairs representing graph g, with one pair per undirected edge.

▪ See Page 87 ▪ See also: `ToOrderedPairs`, `FromOrderedPairs`, `FromUnorderedPairs`.

`TopologicalSort`

`TopologicalSort[g]` returns a permutation of the vertices of directed acyclic graph g such that an edge $\{i, j\}$ implies vertex i appears before vertex j.

▪ See Page 208 ▪ See also: `AcyclicQ`, `PartialOrderQ`.

`TransitiveClosure`

`TransitiveClosure[g]` finds the transitive closure of graph g, the superset of g which contains edge $\{x, y\}$ iff there is a path from x to y.

▪ See Page 203 ▪ See also: `TransitiveQ`, `TransitiveReduction`.

`TransitiveQ`

`TransitiveQ[g]` returns `True` if graph g defines a transitive relation.

`TransitiveQ[r]` returns True if the matrix r defines a transitive relation. See Page 18 ▪ See also: `PartialOrderQ`, `TransitiveClosure`, `TransitiveReduction`.

TransitiveReduction

`TransitiveReduction[g]` finds the smallest graph which has the same transitive closure as g.

`TransitiveReduction` is only guaranteed to find the smallest reduction when g is a directed acyclic graph. See Page 204 ▪ See also: `HasseDiagram`, `TransitiveClosure`.

TranslateVertices

`TranslateVertices[v,{x,y}]` adds the vector $\{x, y\}$ to each vertex in list v.

`TranslateVertices[g,{x,y}]` translates the embedding of graph g by adding the vector $\{x, y\}$ to the position of each vertex. See Page 113 ▪ See also: `NormalizeVertices`.

TransposePartition

`TransposePartition[p]` reflects a partition p of k parts along the main diagonal, creating a partition with maximum part k.

▪ See Page 55 ▪ See also: `DurfeeSquare`, `FerrersDiagram`.

TransposeTableau

`TransposeTableau[t]` reflects a Young tableau t along the main diagonal, creating a different tableau.

▪ See Page 63 ▪ See also: `Tableaux`.

TravelingSalesmanBounds

`TravelingSalesmanBounds[g]` computes upper and lower bounds on the minimum cost traveling salesman tour of graph g.

These functions provide better bounds on graphs which satisfy the triangle inequality. See Page 201 ▪ See also: `TravelingSalesman`, `TriangleInequalityQ`.

TravelingSalesman

`TravelingSalesman[g]` finds the optimal traveling salesman tour in graph g.

Since the traveling salesman problem is NP-complete, this function can be inefficient even on very small graphs. See Page 199 ▪ See also: `HamiltonianCycle`, `TravelingSalesmanBounds`.

TreeQ

`TreeQ[g]` returns `True` if graph g is a tree.

▪ See Page 189 ▪ See also: `AcyclicQ`, `RandomTree`.

TriangleInequalityQ

TriangleInequalityQ[g] returns True if the weight function defined by the adjacency matrix of graph g satisfies the triangle inequality.

TriangleInequalityQ[m] tests if square matrix m observes the triangle inequality. See Page 200 ▪ See also: AllPairsShortestPath, TravelingSalesmanBounds.

Turan

Turan[n,p] constructs the Turan graph, the extremal graph on n vertices which does not contain K_p.

▪ See Page 144 ▪ See also: K, MaximumClique.

TwoColoring

TwoColoring[g] finds a two-coloring of graph g if g is bipartite.

▪ See Page 213 ▪ See also: BipartiteQ, K.

UndirectedQ

UndirectedQ[g] returns True if graph g is undirected.

▪ See Page 94 ▪ See also: MakeUndirected.

UnweightedQ

UnweightedQ[g] returns True if all entries in the adjacency matrix of graph g are zero or one.

▪ See Page 89 ▪ See also: SimpleQ.

V

V[g] gives the order or number of vertices of graph g.

▪ See Page 81 ▪ See also: M, Vertices.

VertexColoring

VertexColoring[g] uses Brelaz's heuristic to find a good, but not necessarily minimal, vertex coloring of graph g.

▪ See Page 214 ▪ See also: ChromaticNumber, ChromaticPolynomial, EdgeColoring.

VertexConnectivity

VertexConnectivity[g] computes the minimum number of vertices whose deletion from graph g disconnects it.

▪ See Page 178 ▪ See also: EdgeConnectivity, Harary, NetworkFlow.

■ VertexCoverQ

`VertexCoverQ[g,c]` returns `True` if the vertices in list c define a vertex cover of graph g.

▪ See Page 218 ▪ See also: `CliqueQ`, `IndependentSetQ`, `MinimumVertexCover`.

■ Vertices

`Vertices[g]` returns the embedding of graph g.

▪ See Page 81 ▪ See also: `Edges`, `V`.

■ WeaklyConnectedComponents

`WeaklyConnectedComponents[g]` returns the weakly connected components of directed graph g.

▪ See Page 172 ▪ See also: `ConnectedQ`, `StronglyConnectedComponents`.

■ Wheel

`Wheel[n]` constructs a wheel on n vertices, which is the join of `K[1]` and `Cycle[n-1]`.

▪ See Page 146 ▪ See also: `Cycle`, `Star`.

■ WriteGraph

`WriteGraph[g,f]` writes graph g to file f using an edge list representation.

The file format is equivallent to what is used by SPREMB [EFK88]. See Page 121 ▪ See also: `ReadGraph`.

References

[AGU72] A. Aho, M. R. Garey, and J. D. Ullman. The transitive reduction of a directed graph. *SIAM J. Computing*, 1:131–137, 1972.

[AH77] K. I. Appel and W. Haken. Every planar map is four colorable. II: Reducibility. *Illinois J. Math.*, 21:491–567, 1977.

[AH86] K. I. Appel and W. Haken. The four color proof suffices. *Mathematical Intelligencer*, 8-1:10–20,58, 1986.

[AHK77] K. I. Appel, W. Haken, and J. Koch. Every planar map is four colorable. I: Discharging. *Illinois J. Math.*, 21:429–490, 1977.

[AHU74] A. Aho, J. Hopcroft, and J. Ullman. *The Design and Analysis of Computer Algorithms.* Addison-Wesley, Reading, Mass., 1974.

[AMU88] M. J. Atallah, G. K. Manacher, and J. Urrutia. Finding a minimum independent dominating set in a permutation graph. *Discrete Applied Mathematics*, 21:177–183, 1988.

[And76] G. Andrews. *The Theory of Partitions.* Addison-Wesley, Reading, Mass., 1976.

[AP61] L. Auslander and S. Parter. On imbedding graphs in the sphere. *J. Mathematics and Mechanics*, 10:517–523, 1961.

[ASS85] H. Abelson, G. J. Sussman, and J. Sussman. *Structure and Interpretation of Computer Programs.* MIT Press, Cambridge, Mass., 1985.

[Baa88] S. Baase. *Computer Algorithms.* Addison-Wesley, Reading, Mass., 1988.

[BB88] G. Brassard and P. Bratley. *Algorithmics.* Prentice-Hall, Englewood Cliffs, N.J., 1988.

[BC67] M. Behzad and G. Chartrand. No graph is perfect. *Amer. Math. Monthly*, 74:962–963, 1967.

[BC76] J. A. Bondy and V. Chvátal. A method in graph theory. *Discrete Math.*, 15:111–136, 1976.

[BC87] W. W. R. Ball and H. S. M. Coxeter. *Mathematical Recreations and Essays.* Dover, New York, 1987.

[BD85] R. Brigham and R. Dutton. A compilation of relations between graph invariants. *Networks*, 15:73–105, 1985.

[Bei68] L. W. Beineke. Derived graphs and digraphs. In H. Sachs, H. Voss, and H. Walther, editors, *Beiträge zur Graphentheorie*, pages 17–33, Teubner, Leipzig, 1968.

[Ber57] C. Berge. Two theorems in graph theory. *Proc. Nat. Acad. Sci. USA*, 43:842–844, 1957.

[BH90] F. Buckley and F. Harary. *Distances in Graphs.* Addison-Wesley, Redwood City, Calif., 1990.

[BHK62] J. Battle, F. Harary, and Y. Kodama. Every planar graph with nine points has a nonplanar complement. *Bull. Amer. Math. Soc.*, 68:569–571, 1962.

[Big74] N. L. Biggs. *Algebraic Graph Theory.* Cambridge University Press, London, 1974.

[Bir12] G. D. Birkhoff. A determinant formula for the number of ways of coloring a map. *Ann. of Math.*, 14:42–46, 1912.

[BK87] A. Brandstadt and D. Kratsch. On domination problems for permutation and other graphs. *Theoretical Computer Science*, 54:181–198, 1987.

[BL46] G. D. Birkhoff and D. C. Lewis. Chromatic polynomials. *Trans. Amer. Math. Soc.*, 60:355–451, 1946.

[BL62] J. Bolland and C. Lekkerkerer. Representation of a finite graph by a set of intervals on the real line. *Fund. Math.*, 51:45–64, 1962.

[BL76] K. Booth and G. Leuker. Testing the consecutive ones property, interval graphs, and graph planarity using PQ-trees. *J. Comp. System Sciences*, 13:335–379, 1976.

[BLW76] N. L. Biggs, E. K. Lloyd, and R. J. Wilson. *Graph Theory 1736-1936.* Clarendon Press, Oxford, 1976.

[BM76] J. A. Bondy and U. S. R. Murty. *Graph Theory with Applications.* North-Holland, New York, 1976.

[Bol78] B. Bollobás. *Extremal Graph Theory.* Academic Press, London, 1978.

[Bol79] B. Bollobás. *Graph Theory.* Springer-Verlag, New York, 1979.

[Bre79] D. Brelaz. New methods to color the vertices of a graph. *Comm. ACM*, 22:251–256, 1979.

[Bro41] R. L. Brooks. On coloring the nodes of a network. *Proc. Cambridge Philos. Soc.*, 37:194–197, 1941.

[BS89] B. Birgisson and G. Shannon. *GraphView: An Extensible Interactive Platform for Manipulating and Displaying Graphs.* Technical Report 295, Computer Science Department, Indiana University, Bloomington, Ind., December 1989.

[BT80] F. Boesch and R. Tindell. Robbins' theorem for mixed graphs. *Amer. Math. Monthly*, 87:716–719, 1980.

[BW82] A. Björner and M. Wachs. Bruhat order of Coxeter groups and shellability. *Advances in Math.*, 43:87–100, 1982.

[Can69] J. J. Cannon. Computers in group theory. *Comm. ACM*, 12:3–12, 1969.

[Can90] J. J. Cannon. The group theory system Cayley. *J. Symbolic Computation*, 1990.

[Cay89] A. Cayley. A theorem on trees. *Quart. J. Math.*, 23:376–378, 1889.

[CDN89] C. J. Colbourn, R. P. J. Day, and L. D. Nel. Unranking and ranking spanning trees of a graph. *J. Algorithms*, 10:271–286, 1989.

[CDS80] D. Cvetković, M. Doob, and H. Sachs. *Spectra of Graphs.* Academic Press, New York, 1980.

[CG70] D. G. Corneil and C. C. Gottlieb. An efficient algorithm for graph isomorphism. *J. ACM*, 17:51–64, 1970.

[Cha68] G. Chartrand. On Hamiltonian line graphs. *Trans. Amer. Math. Soc.*, 134:559–566, 1968.

[Cha73] P. J. Chase. Transposition graphs. *SIAM J. Computing*, 2:128–133, 1973.

[Cha82] S. Chaiken. A combinatorial proof of the all-minors matrix tree theorem. *SIAM J. Alg. Disc. Methods*, 3:319–329, 1982.

[Cha85] G. Chartrand. *Introductory Graph Theory.* Dover, New York, 1985.

[Chr71] N. Christofides. An algorithm for the chromatic number of a graph. *Computer J.*, 14:38–39, 1971.

[CHT90] J. Cai, X. Han, and R. Tarjan. *New Solutions to Four Planar Graph Problems.* Technical Report, New York University, 1990.

[Chv70] V. Chvátal. A note on coefficients of chromatic polynomials. *J. Combinatorial Theory*, 9:95–96, 1970.

[CLR87] F. Chung, T. Leighton, and A. Rosenberg. Embedding graphs in books: a layout problem with applications to VLSI design. *SIAM J. Algebraic and Discrete Methods*, 8:33–58, 1987.

[CM70] S. Crespi-Reghizzi and R. Morpuro. A language for treating graphs. *Comm. ACM*, 13:319–323, 1970.

[CM78] M. Capobianco and J. Molluzzo. *Examples and Counterexamples in Graph Theory.* North-Holland, New York, 1978.

[CT78] V. Chvátal and C. Thomassen. Distances in orientations of graphs. *J. Combinatorial Theory B*, 24:61–75, 1978.

[Cve81] D. Cvetković. A project for using computers in further development of graph theory. In G. Chartrand, Y. Alavi, D. L. Goldsmith, L. Lesniak-Foster, and D. R. Lick, editors, *The Theory and Applications of Graphs*, pages 285–296, Wiley, New York, 1981.

[Cve84] D. Cvetković. *Graph: An Expert System for the Classification and Extension of Knowledge in the Field of Graph Theory.* Technical Report, Faculty of Electrical Engineering, University of Belgrade, Belgrade, Yugoslavia, 1984.

[dB46] N. G. de Bruijn. A combinatorial problem. *Koninklijke Nederlandse Akademie v. Wetenschappen*, 49:758–764, 1946.

[dFPP88] H. de Fraysseix, J. Pach, and R. Pollack. Small sets supporting Fary embeddings of planar graphs. In *Proc. of the 20th Symposium on the Theory of Computing*, pages 426–433, ACM, 1988.

[DHRT87] M. Dao, M. Habib, J. P. Richard, and D. Tallot. Cabri, an interactive system for graph manipulation. In *Graph Theoretic Concepts in Computer Science*, pages 58–67, Lecture Notes in Computer Science, Springer-Verlag, New York, 1987.

[Dij59] E. W. Dijkstra. A note on two problems in connection with graphs. *Numerische Math.*, 1:269–271, 1959.

[Dil50] R. P. Dilworth. A decomposition theorem for partially ordered sets. *Ann. of Math.*, 51:161–166, 1950.

[DK67] L. Danzer and V. Klee. Lengths of snakes in boxes. *J. Combinatorial Theory*, 2:258–265, 1967.

[Ead84] P. Eades. A heuristic for graph drawing. *Congressus Numerantium*, 42:149–160, 1984.

[Ech88] M. Echeandia. *New Functions for LILA Interpreter and Kernal for Graphpack.* Technical Report 88-25, Dept. of Computer Science, Rensselaer Polytechnic Institute, Troy, N.Y., October 1988.

[Eco89] U. Eco. *Foucault's Pendulum.* Harcourt Brace Jovanovich, San Diego, 1989.

[EFK88] P. Eades, I. Fogg, and D. Kelly. *SPREMB: A System for Developing Graph Algorithms.* Technical Report, Department of Computer Science, University of Queensland, St. Lucia, Queensland, Australia, 1988.

[EG60] P. Erdös and T. Gallai. Graphs with prescribed degrees of vertices. *Mat. Lapok (Hungarian)*, 11:264–274, 1960.

[Egg75] R. B. Eggleton. Graphic sequences and graphic polynomials. In A. Hajnal, editor, *Infinite and Finite Sets*, pages 385–392, North-Holland, Amsterdam, 1975.

[EJ73] J. Edmonds and E. L. Johnson. Matching, Euler tours, and the Chinese postman. *Math. Programming*, 5:88–124, 1973.

[EK72] J. Edmonds and R. M. Karp. Theoretical improvements in algorithmic efficiency for network flow problems. *J. ACM*, 19:248–264, 1972.

[Erd61] P. Erdös. Graph theory and probability II. *Canad. J. Math.*, 13:346–352, 1961.

[ES35] P. Erdös and G. Szekeres. A combinatorial problem in geometry. *Compositio Math.*, 2:464–470, 1935.

[ES74] P. Erdös and J. Spencer. *Probabilistic Methods in Combinatorics.* Academic Press, New York, 1974.

[ET75] S. Even and R. E. Tarjan. Network flow and testing graph connectivity. *SIAM J. Computing*, 4:507–518, 1975.

[ET89] P. Eades and R. Tamassia. *Algorithms for Drawing Graphs: An Annotated Bibliography.* Technical Report CS-89-09, Department of Computer Science, Brown University, Providence, R.I., February 1989.

[Eve79] S. Even. *Graph Algorithms.* Computer Science Press, Rockville, Md., 1979.

[Faj87] S. Fajtlowicz. On conjectures of Graffiti II. *Congressus Numerantium*, 60:187–197, 1987.

[Faj88] S. Fajtlowicz. On conjectures of Graffiti. *Discrete Mathematics*, 72:113–118, 1988.

[Faj90] S. Fajtlowicz. *Written on the Wall.* Technical Report, Dept. of Mathematics, University of Houston, January 11, 1990.

[Far48] I. Fáry. On straight line representation of planar graphs. *Acta. Sci. Math. Szeged*, 11:229–233, 1948.

[FF62] L. R. Ford and D. R. Fulkerson. *Flows in Networks.* Princeton University Press, Princeton, N.J., 1962.

[FHL80] M. Furst, J. Hopcroft, and E. Luks. Polynomial time algorithms for permutation groups. In *Proc. Symp. Foundations of Computer Science*, pages 36–41, IEEE, 1980.

[FHM65] D. R. Fulkerson, A. J. Hoffman, and M. H. McAndrew. Some properties of graphs with multiple edges. *Canad. J. Math.*, 17:166–177, 1965.

[Fle74] H. Fleischner. The square of every two-connected graph is Hamiltonian. *J. Combinatorial Theory B*, 16:29–34, 1974.

[Flo62] R. W. Floyd. Algorithm 97: shortest path. *Comm. ACM*, 5:345, 1962.

[Flo64] R. W. Floyd. Algorithm 245: treesort 3. *Comm. ACM*, 7:701, 1964.

[Fol67] J. Folkman. Regular line-symmetric graphs. *J. Combinatorial Theory*, 3:215–232, 1967.

[Fou60] J. D. Foulkes. Directed graphs and assembly schedules. In *Proc. Symp. Appl. Math.*, pages 281–289, Amer. Math. Soc., 1960.

[Fre83] P. Freeman. The secretary problem and extensions: a review. *International Statistical Review*, 51:188–208, 1983.

[Fru39] R. Frucht. Herstellung von Graphen mit vorgegebener abstrakter Gruppe. *Compositio Math.*, 6:239–250, 1939.

[FT87] M. L. Fredman and R. E. Tarjan. Fibonacci heaps on their uses in network optimization. *J. ACM*, 34:596–615, 1987.

[Ful65] D. R. Fulkerson. Upsets in round robin tournaments. *Canad. J. Math.*, 17:957–969, 1965.

[Ful71] D. R. Fulkerson. Blocking and anti-blocking pairs of polyhedra. *Math. Programming*, 1:168–194, 1971.

[FZ82] D. Franzblau and D. Zeilberger. A bijective proof of the hook-length formula. *J. Algorithms*, 3:317–342, 1982.

[Gal59] T. Gallai. Über extreme Punkt- und Kantenmengen. *Ann. Univ. Sci. Budapest, Eötvös Sect. Math.*, 2:133–138, 1959.

[Gas67] B. J. Gassner. Sorting by replacement selection. *Comm. ACM*, 10:89–93, 1967.

[GGM58] H. Gupta, A. E. Gwyther, and J. C. P. Miller. *Tables of Partitions.* Volume 4, Royal Society Mathematical Tables, London, 1958.

[GH61] R. E. Gomery and T. C. Hu. Multiterminal network flows. *J. SIAM*, 9:551–570, 1961.

[GH64] P. Gilmore and A.J. Hoffman. A characterization of comparability graphs and interval graphs. *Canad. J. Math.*, 16:539–548, 1964.

[GH85] R. L. Graham and P. Hell. On the history of the minimum spanning tree problem. *Ann. History of Computing*, 7:43–57, 1985.

[GI89] D. Gusfield and R. Irving. *The Stable Marriage Problem.* MIT Press, Cambridge, Mass., 1989.

[Gil58] E. N. Gilbert. Gray codes and paths on the n-cube. *Bell System Tech. J.*, 37:815–826, 1958.

[GJ79] M. Garey and D. Johnson. *Computers and Intractability: a Guide to the Theory of NP-Completeness.* W. H. Freeman, San Francisco, 1979.

[GJ83] I. P. Goulden and D. M. Jackson. *Combinatorial Enumeration*. John Wiley and Sons, New York, 1983.

[GKP89] R. Graham, D. Knuth, and O. Patashnik. *Concrete Mathematics*. Addison-Wesley, Reading, Mass., 1989.

[Gol66] S. W. Golomb. *Shift Register Sequences*. Holden-Day, San Francisco, 1966.

[Gol80] M. C. Golumbic. *Algorithmic Graph Theory and Perfect Graphs*. Academic Press, New York, 1980.

[Goo46] I. J. Good. Normal recurring decimals. *J. London Math. Soc.*, 21:167–172, 1946.

[Gou88] R. Gould. *Graph Theory*. Benjamin Cummings, Menlo Park, Calif., 1988.

[GP79] W. Gates and C. Papadimitriou. Bounds for sorting by prefix reversal. *Discrete Math.*, 27:47–57, 1979.

[Gra53] F. Gray. Pulse code communication. United States Patent Number 2,632,058, March 17, 1953.

[GS62] D. Gale and L. S. Shapley. College admissions and the stability of marriage. *Amer. Math. Monthly*, 69:9–14, 1962.

[Gup66] R. P. Gupta. The chromatic index and the degree of a graph. *Notices of the Amer. Math. Soc.*, 13:66T–429, 1966.

[Hak62] S. Hakimi. On the realizability of a set of integers as degrees of the vertices of a graph. *SIAM J. Appl. Math.*, 10:496–506, 1962.

[Hal35] P. Hall. On representatives of subsets. *J. London Math. Soc.*, 10:26–30, 1935.

[Har62] F. Harary. The maximum connectivity of a graph. *Proc. Nat. Acad. Sci. USA*, 48:1142–1146, 1962.

[Har69] F. Harary. *Graph Theory*. Addison-Wesley, Reading, Mass., 1969.

[Hav55] V. Havel. A remark on the existence of finite graphs (Czech.). *Časopis Pest. Mat.*, 80:477–480, 1955.

[Hea63] B. R. Heap. Permutations by interchanges. *Computer J.*, 6:293–294, 1963.

[Hie73] C. Hierholzer. Ueber die Möglichkeit, einen Linienzug ohne Wiederholung und ohne Unterbrechnung zu umfahren. *Math. Ann.*, 6:30–42, 1873.

[HK75] J. Hopcroft and R. Karp. An $n^{5/2}$ algorithm for maximum matching in bipartite graphs. *SIAM J. Computing*, 225–231, 1975.

[HM66] F. Harary and L. Moser. The theory of round robin tournaments. *Amer. Math. Monthly*, 73:231–246, 1966.

[HN65] F. Harary and C. J. A. Nash-Williams. On Eulerian and Hamiltonian graphs and line graphs. *Canad. Math. Bull.*, 8:701–709, 1965.

[Hol81] I. Holyer. The NP-completeness of edge colorings. *SIAM J. Computing*, 10:718–720, 1981.

[Hol88] D. F. Holt. The Cayley group theory system. *Notices of the American Mathematical Society*, 35:1135–1140, 1988.

[HP66] F. Harary and E. M. Palmer. The smallest graph whose group is cyclic. *Czech. Math. J.*, 16:70–71, 1966.

[HP73] F. Harary and E. M. Palmer. *Graphical Enumeration.* Academic Press, New York, 1973.

[HR73] F. Harary and R. Read. Is the null graph a pointless concept? In *Graphs and Combinatorics Conference*, George Washington University, Springer-Verlag, 1973.

[HS78] E. Hororwitz and S. Sahni. *Fundamentals of Computer Algorithms.* Computer Science Press, Rockville, Md., 1978.

[HT73] J. Hopcroft and R. Tarjan. Algorithm 447: efficient algorithms for graph manipulation. *Comm. ACM*, 16:372–378, 1973.

[HT74] J. Hopcroft and R. Tarjan. Efficient planarity testing. *J. ACM*, 21:549–568, 1974.

[Joh63] S. M. Johnson. Generation of permutations by adjacent transpositions. *Math. Computation*, 17:282–285, 1963.

[Kin72] C. A. King. A graph-theoretic programming language. In R. Read, editor, *Graph Theory and Computing*, pages 63–75, Academic Press, New York, 1972.

[Kir47] G. Kirchhoff. Über die Auflösung der Gleichungen, auf welche man bei der untersuchung der linearen verteilung galvanischer Ströme geführt wird. *Ann. Phys. Chem.*, 72:497–508, 1847.

[KK89] T. Kamada and S. Kawai. An algorithm for drawing general undirected graphs. *Information Processing Letters*, 31:7–15, 1989.

[Kli82] P. Klingsberg. A Gray code for compositions. *J. Algorithms*, 3:41–44, 1982.

[Knu67] D. E. Knuth. Oriented subtrees of an arc digraph. *J. Combinatorial Theory*, 3:309–314, 1967.

[Knu73a] D. E. Knuth. *Fundamental Algorithms.* Volume 1 of *The Art of Computer Programming*, Addison-Wesley, Reading, Mass., second edition, 1973.

[Knu73b] D. E. Knuth. *Sorting and Searching.* Volume 3 of *The Art of Computer Programming*, Addison-Wesley, Reading, Mass., 1973.

[Knu84] D. E. Knuth. The complexity of songs. *Comm. ACM*, 27:344–346, 1984.

[Knu88] D. E. Knuth. The four volume problem: how should algorithms be compared? In *Fourth SIAM Conference on Discrete Mathematics*, Cathedral Hill Hotel, San Francisco, 1988.

[Kru56] J. B. Kruskal. On the shortest spanning subtree of a graph and the traveling salesman problem. *Proc. Amer. Math. Soc.*, 7:48–50, 1956.

[Kur30] K. Kuratowski. Sur le problème des courbes gauches en topologie. *Fund. Math.*, 15:217–283, 1930.

[Kwa62] M. K. Kwan. Graphic programming using odd or even points. *Chinese Math.*, 1:273–277, 1962.

[LBB81] G. N. Lewis, N. J. Boynton, and F. W. Burton. Expected complexity of fast search with uniformly distributed data. *Information Processing Letters*, 13:4–7, 1981.

[Lev73] L. A. Levin. Universal searching problems. *Problems of Information Transmission*, 9:265–266, 1973.

[Lin65] S. Lin. Computer solutions of the traveling salesman problem. *Bell System Tech. J.*, 44:2245–2269, 1965.

[Liu68] C. L. Liu. *Introduction to Combinatorial Mathematics.* McGraw-Hill, New York, 1968.

[Liu85] C. L. Liu. *Elements of Discrete Mathematics.* McGraw-Hill, New York, second edition, 1985.

[LLKS85] E. Lawler, J. K. Lenstra, A. H. G. Rinnooy Kan, and D. B. Shmoys. *The Traveling Salesman Problem.* John Wiley and Sons, New York, 1985.

[LNS85] R. Lipton, S. North, and J. Sandberg. A method for drawing graphs. In *Proc. First ACM Symposium on Computational Geometry*, pages 153–160, 1985.

[Lov68] L. Lovász. On chromatic number of finite set-systems. *Acta. Math. Acad. Sci. Hungar.*, 19:59–67, 1968.

[Lov72] L. Lovász. Normal hypergraphs and the perfect graph conjecture. *Discrete Math.*, 2:253–267, 1972.

[Lov75] L. Lovász. Three short proofs in graph theory. *J. Combinatorial Theory B*, 19:111–113, 1975.

[LP86] L. Lovász and M. D. Plummer. *Matching Theory.* North-Holland, Amsterdam, 1986.

[Luc91] E. Lucas. *Récréations Mathématiques.* Cauthier-Villares, Paris, 1891.

[Lue80] G. S. Lueker. Some techniques for solving recurrences. *Computing Surveys*, 12:419–436, 1980.

[Luk80] E. M. Luks. Isomorphism of bounded valence can be tested in polynomial time. In *Proc. of the 21st Annual Symposium on Foundations of Computing*, pages 42–49, IEEE, 1980.

[Mac60] P. MacMahon. *Combinatory Analysis.* Volume I and II reprinted in one volume, Chelsea Publishing, New York, 1960.

[Mae89] R. Maeder. *Programming in Mathematica.* Addison-Wesley, Redwood City, Calif., 1989.

[Mal88] S. M. Malitz. Genus g graphs have pagenumber $O(\sqrt{g})$. In *Proc. 29th Symp. Foundations of Computer Science*, pages 458–468, IEEE, 1988.

[Man85a] H. Mannila. Measures of presortedness and optimal sorting algorithms. *IEEE Transactions on Computers*, 34:318–325, 1985.

[Man85b] B. Manvel. Extremely greedy coloring algorithms. In F. Harary and J. Maybee, editors, *Graphs and Applications*, pages 257–270, Wiley, New York, 1985.

[Man89] U. Manber. *Introduction to Algorithms.* Addison-Wesley, Reading, Mass., 1989.

[McK90] B. McKay. *Nauty User's Guide.* Technical Report TR-CS-90-02, Department of Computer Science, Australian National University, 1990.

[Men27] K. Menger. Zur allgemeinen Kurventheorie. *Fund. Math.*, 10:95–115, 1927.

[Mes83] B. E. Meserve. *Fundamental Concepts of Geometry.* Dover, New York, 1983.

[MMI72] D. W. Matula, G. Marble, and J. D. Isaacson. Graph coloring algorithms. In R. Read, editor, *Graph Theory and Computing*, pages 109–122, Academic Press, New York, 1972.

[MO63] L. E. Moses and R. V. Oakford. *Tables of Random Permutations.* Stanford University Press, Stanford, Calif., 1963.

[Moo59] E. F. Moore. The shortest path through a maze. In *Proc. Internat. Symp. Switching Th., Part II*, pages 285–292, Harvard University Press, Cambridge MA, 1959.

[Moo68] J. W. Moon. *Topics on Tournaments.* Holt, Rinehart, and Winston, New York, 1968.

[MS80] J. I. Munro and H. Suwanda. Implicit data structures for fast search and update. *J. Computer and System Sciences*, 21:236–250, 1980.

[Muk67] A. Mukhopadhyay. The square root of a graph. *J. Combinatorial Theory*, 2:290–295, 1967.

[NW78] A. Nijenhuis and H. Wilf. *Combinatorial Algorithms.* Academic Press, New York, second edition, 1978.

[OR81] R. J. Opsut and F. S. Roberts. On the fleet maintenance, mobile radio frequency, task assignment, and traffic phasing problems. In G. Chartrand, Y. Alavi, D. L. Goldsmith, L. Lesniak-Foster, and D. R. Lick, editors, *The Theory and Applications of Graphs*, pages 479–492, Wiley, New York, 1981.

[Ore60] O. Ore. A note on Hamiltonian circuits. *Amer. Math. Monthly*, 67:55, 1960.

[Orl88] J. B. Orlin. A faster strongly polynomial minimum cost flow algorithm. In *Proc. 20th ACM Symposium Theory of Computing*, pages 377–387, 1988.

[Pal85] E. M. Palmer. *Graphical Evolution: An Introduction to the Theory of Random Graphs.* Wiley-Interscience, New York, 1985.

[PMN88] C. G. Ponder, P. C. McGeer, and A. P. Ng. Are applicative languages inefficient? *SIGPLAN Notices*, 23(6):135–139, 1988.

[Pol37] G. Polya. Kombinatorische anzahlbestimmungen für gruppen, graphen und chemische verbindugen. *Acta Math.*, 68:145–254, 1937.

[Pri57] R. C. Prim. Shortest connection networks and some generalizations. *Bell System Tech. J.*, 36:1389–1401, 1957.

[Pro89] J. Propp. Some variants of Ferrers diagrams. *J. Combinatorial Theory A*, 52:98–128, 1989.

[Pru18] H. Prüfer. Neuer beweis eines Satzes über Permutationen. *Arch. Math. Phys.*, 27:742–744, 1918.

[PS82] C. H. Papadimitriou and K. Steiglitz. *Combinatorial Optimization: Algorithms and Complexity.* Prentice-Hall, Englewood Cliffs, N.J., 1982.

[Rea68] R. C. Read. An introduction to chromatic polynomials. *J. Combinatorial Theory*, 4:52–71, 1968.

[Rea69] R. Read. Teaching graph theory to a computer. In W. Tutte, editor, *Recent Progress in Combinatorics*, pages 161–173, Academic Press, New York, 1969.

[Red34] L. Rédei. Ein Kombinatorischer Satz. *Acta Litt. Szeged*, 7:39–43, 1934.

[RND77] E. M. Reingold, J. Nievergelt, and N. Deo. *Combinatorial Algorithms.* Prentice-Hall, Englewood Cliffs, N.J., 1977.

[Rob39] H. E. Robbins. A theorem on graphs with an application to a problem of traffic control. *Amer. Math. Monthly*, 46:281–283, 1939.

[Rob78] F. S. Roberts. *Graph Theory and Its Applications to Problems of Society.* NSF-CBMS Monograph No. 29, SIAM, Philadephia, 1978.

[Rob84] F. S. Roberts. *Applied Combinatorics.* Prentice-Hall, Englewood Cliffs, N.J., 1984.

[Ron84] C. Ronse. *Feedback Shift Registers.* Volume 146 of *Lecture Notes in Computer Science*, Springer-Verlag, Berlin, 1984.

[RS89] V. Reddy and S. Skiena. *Frequencies of Large Distances in Integer Lattices.* Technical Report, Department of Computer Science, State University of New York, Stony Brook, June 1989.

[RSL77] D. J. Rosenkrantz, R. E. Stearns, and P. M. Lewis. An analysis of several heuristics for the traveling salesman problem. *SIAM J. Computing*, 6:563–581, 1977.

[RT81] E. Reingold and J. Tilford. Tidier drawings of trees. *IEEE Trans. Software Engineering*, 7:223–228, 1981.

[Rys57] H. J. Ryser. Combinatorial properties of matrices of zeros and ones. *Canad. J. Math.*, 9:371–377, 1957.

[Sab60] G. Sabidussi. Graph multiplication. *Math. Z.*, 72:446–457, 1960.

[Sac62] H. Sachs. Über selbstkomplementäre Graphen. *Publ. Math. Debrecen*, 9:270–288, 1962.

[Sav89] C. Savage. Gray code sequences of partitions. *J. Algorithms*, 10:577–595, 1989.

[Sch61] C. Schensted. Longest increasing and decreasing subsequences. *Canadian J. Math.*, 13:179–191, 1961.

[SD76] D. C. Schmidt and L. E. Druffel. A fast backtracking algorithm to test directed graphs for isomorphism using distance matrices. *J. ACM*, 23:433–445, 1976.

[Sed77] R. Sedgewick. Permutation generation methods. *Computing Surveys*, 9:137–164, 1977.

[Sed88] R. Sedgewick. *Algorithms*. Addison-Wesley, Reading, Mass., second edition, 1988.

[SK86] T. L. Saaty and P. C. Kainen. *The Four-Color Problem*. Dover, New York, 1986.

[Ski88] S. Skiena. Encroaching lists as a measure of presortedness. *BIT*, 28:775–784, 1988.

[Ski89] S. Skiena. Reconstructing graphs from cut-set sizes. *Info. Processing Letters*, 32:123–127, 1989.

[SL87] H. Shultz and B. Leonard. Strategies for hiring a secretary. *UMAP Journal*, 8:301–305, 1987.

[SR83] K. Supowit and E. Reingold. The complexity of drawing trees nicely. *Acta Informatica*, 18:377–392, 1983.

[Sta71] R. Stanley. Theory and application of plane partitions I, II. *Studies in Applied Math.*, 50:167–188, 259–279, 1971.

[Sta86] R. P. Stanley. *Enumerative Combinatorics*. Volume 1, Wadsworth & Brooks/Cole, Monterey, Calif., 1986.

[SW86] D. Stanton and D. White. *Constructive Combinatorics*. Springer-Verlag, New York, 1986.

[Sze43] T. Szele. Kombinatorische Untersuchungen über den gerichteten vollständigen Graphen. *Mat. Fiz. Lapok*, 50:223–256, 1943.

[Sze83] G. Szekeres. Distribution of labeled trees by diameter. *Lecture Notes in Mathematics*, 1036:392–397, 1983.

[Tai80] P. G. Tait. Remarks on the colouring of maps. *Proc. Royal Soc. Edinburgh*, 10:729, 1880.

[Tar72] R. E. Tarjan. Depth-first search and linear graph algorithms. *SIAM J. Computing*, 1:146–160, 1972.

[Tar75] R. E. Tarjan. Efficiency of a good but not linear set union algorithm. *J. ACM*, 22:215–225, 1975.

[Tar83] R. E. Tarjan. *Data Structures and Network Algorithms.* Society for Industrial and Applied Mathematics, Philadelphia, 1983.

[Tho56] C. B. Thompkins. Machine attacks on problems whose variables are permutations. In *Proc. Symposium Applied Mathematics*, page 203, American Mathematical Society, 1956.

[Tro62] H. F. Trotter. Perm (algorithm 115). *Comm. ACM*, 5:434–435, 1962.

[Tuc84] A. Tucker. *Applied Combinatorics.* John Wiley and Sons, New York, second edition, 1984.

[Tur41] P. Turán. On an extremal problem in graph theory. *Mat. Fiz. Lapok*, 48:436–452, 1941.

[Tut46] W. T. Tutte. On Hamilton circuits. *J. London Math. Soc.*, 21:98–101, 1946.

[Tut61] W. T. Tutte. A theory of 3-connected graphs. *Indag. Math.*, 23:441–455, 1961.

[Tut70] W. T. Tutte. On chromatic polynomials and the golden ratio. *J. Combinatorial Theory*, 9:289–296, 1970.

[Tut72] W. T. Tutte. Non-Hamiltonian planar maps. In R. Read, editor, *Graph Theory and Computing*, pages 295–301, Academic Press, New York, 1972.

[Vau80] J. Vaucher. Pretty printing of trees. *Software Practice and Experience*, 10:553–561, 1980.

[Viz64] V. G. Vizing. On an estimate of the chromatic class of a p-graph (in Russian). *Diskret. Analiz*, 3:23–30, 1964.

[vRW65] A. van Rooij and H. Wilf. The interchange graph of a finite graph. *Acta Math. Acad. Sci. Hungar.*, 16:263–269, 1965.

[WH60] P. D. Whiting and J. A. Hillier. A method for finding the shortest route through a road network. *Operational Res. Quart.*, 11:37–40, 1960.

[Whi32] H. Whitney. Congruent graphs and the connectivity of graphs. *American J. Mathematics*, 54:150–168, 1932.

[Wil84] H. Wilf. Backtrack: an $o(1)$ expected time algorithm for the graph coloring problem. *Information Processing Letters*, 18:119–121, 1984.

[Wil85a] H. Wilf. Graphs and their spectra: old and new results. *Congressus Numerantium*, 50:37–43, 1985.

[Wil85b] R. J. Wilson. *Introduction to Graph Theory*. Longman, Essex, England, third edition, 1985.

[Wil86] R. J. Wilson. An Eulerian trail through Königsberg. *J. Graph Theory*, 10:265–275, 1986.

[Wil89] H. Wilf. *Combinatorial Algorithms: An Update*. Society for Industrial and Applied Mathematics, Philadelphia, 1989.

[Wol88] S. Wolfram. *Mathematica*. Addison-Wesley, Redwood City, Calif., 1988.

[Wor84] N. Wormald. Generating random regular graphs. *J. Algorithms*, 5:247–280, 1984.

[WS79] C. Wetherell and A. Shannon. Tidy drawing of trees. *IEEE Trans. Software Engineering*, 5:514–520, 1979.

Index

Abelian group, 17
Abelson, H., 259
Abs, 261
Accumulate, 267
Acyclic graphs, 188
AcyclicQ, 189, 270
AddEdge, 83, 270
AddToEncroachingLists, 75
AddVertex, 84, 270
Adjacency lists, 81, 86
Adjacency matrices, 81
    circulant graphs, 140
    incidence matrices, 136
    modifications, 83
    powers, 230
    spanning trees, 235
    square, 158
    sum and difference, 131
Aho, A., 176, 203, 258
Algol, 123
Algorithmic graph theory, 224
Algorithms, 267
    biconnected components, 175
    bipartite matching, 240
    coloring, 214
    connectivity, 178
    Eulerian cycles, 194
    girth, 191
    Hamiltonian cycles, 196
    isomorphism, 182
    network flow, 237
    orienting graphs, 174
    permutation graphs, 28
    planarity, 247
    shortest paths, 225
    sorting, 6
    stable matchings, 245
    strong connectivity, 172
    transitive closure, 203
AllPairsShortestPath, **228**, 252, **270**
All-pairs shortest paths, algorithms, 228
    isomorphism testing, 182
    radial embedding, 109
Alphabetical order, 4
And, 268
Andrews, G., 53
Antichain, 241
Appel, K., 210
Append, 259, 263
AppendTo, 264
Apple Macintosh, 122
Applicative machines, 9, 48
Apply, 265, 266, 267
Arc, 80
Arc reversals, 48
Arguments, 260
Around the world game, 198
Array notation, 3, 8, 264
Articulation vertex, 175
ArticulationVertices, 177, 270
Ascending sequences, 30
Ascents, 31
ASCII, 120
Atallah, M., 28
Augmenting paths, 237, 243
Auslander, L., 247
Automorphism, 184
Automorphism group, 19, 25, 122
Automorphisms, 26, **184**, **270**
Average distance, 125

Baase, S., 258
Back edges, 96, 188
Backtrack, **12**, 196, 220, **271**
Backtracking, 12
    traveling salesman, 199
Balanced parentheses, 63
Ball, W., 198
Battle, J., 250
Behzad, M., 158
Beineke, L., 138
Berge, C., 243
Best-first search, 225
BiconnectedComponents, 177, 271
Biconnected components, 175
BiconnectedQ, 177, 271
Biconnectivity, 180
Biggs, N., 85, 162, 193
Bijection, compositions, 61
    cycle representation, 22
    index, 30
    inversion vectors, 27
    involutions, 67
    partitions, 53, 55
    permutations, 5
    subsets, 41
    trees, 151, 236

Young tableaux, 63, 64
Binary heaps, 35
Binary relations, equivalence relations, 18
partial orders, 203
Binary representation, 41
`BinarySearch`, **15**, 75, **271**
Binary search, 15
`BinarySubsets`, **41**, 47, **271**
Binary trees, 35
Bindings, 268
`Binomial`, 262, 267
Binomial coefficients, Catalan numbers, 71
compositions, 60
counting heaps, 36
counting k-subsets, 44
Gauss, 267
lattice paths, 230
middle, 262
multinomial coefficients, 14
Binomial distribution, 42
Biparential heaps, 71
Bipartite graphs, 142
cliques, 218
coloring, 213, 220
contraction, 92
embeddings, 106
grid graphs, 148
hypercubes, 150
isomorphisms, 183
perfect, 219
scheduling, 216
spectrum, 85
transpositions, 10
`BipartiteMatching`, 170, **240**, **271**
Bipartite matching, 237, 240, 252
`BipartiteQ`, 10, **213**, **271**
Birgisson, B., 122
Birkhoff, G., 210
Björner, A., 145
`Block`, 97, 260
Boesch, F., 175
Bolland, J., 163
Bollobás, B., 136, 143, 156
Bondy, J., 180, 197, 221, 257
Book embeddings, 124
Boolean algebra, 170, 207
Boolean operators, 268
Booth, K., 164
Boynton, N., 16
Branch, 80
Brandstadt, A., 28
Brassard, G., 258
Bratley, P., 258
Breadth-first search, **95**, 124
girth, 191
ranked embeddings, 106
shortest paths, 225
`BreadthFirstTraversal`, 95, 272
Brelaz, D., 214
Brelaz's algorithm, 214, 220
Bridgeless graphs, 174
`Bridges`, 177, 272
Bridges, 171, **177**
Fleury's algorithm, 193
Königsberg, 192
matchings, 244
orienting graphs, 174
planarity testing, 247
Brigham, R., 123
Brooks, R., 215
Buckley, F., 140, 257
Bugs, 268
Bumping procedure, 64, 73
Burton, F., 16
Business world, 108

CABRI, 122
Cages, 191, 221
Cai, J., 247, 253
Calendar, 198
Candelabra, 227
Canine, 180
Cannon, J., 17
Capobianco, M., 129, 166
CAR, 264
`CartesianProduct`, 132, 272
Cartesian products, 132, 161
Casinos, 221
`CatalanNumber`, 71, 272
Catalan numbers, 71
CAYLEY, 17
Cayley, A., 151
CDR, 264
`Ceiling`, 16, 261
Center, 107, 124, 227
Chaiken, S., 235
Chain, 241
`ChangeEdges`, 81, 272
`ChangeVertices`, 81, 272
Character strings, 40

Chartrand, G., 138, 158, 198, 257
Chase, P., 10
Chessboard, 148
Chinese postman problem, 194
Christofides, N., 214
`ChromaticNumber`, 211, 272
Chromatic number, 170, 210
    conjecture, 125
    girth, 215
    perfect graphs, 219
    wheel, 212
`ChromaticPolynomial`, 210, 273
Chromatic polynomial, **210**, 221
Chung, F., 124
Chvatal, V., 175, 197, 221
`CirculantGraph`, 140, 273
Circulant graphs, 140
    bipartite, 99
    cycle, 144
    Harary graph, 179
    k-partite, 166
Circular embedding, 99, 104, 124
`CircularVertices`, 104, 273
Circumference, 192, 220
C language, 262, 265
Class, 73
Claw, 138
`Clear`, 261, 267
Clique, 90, 217
    chromatic number, 215
    perfect graphs, 219
    Turán graph, 143
`CliqueQ`, 217, 273
Codes, 42
`CodeToLabeledTree`, 151, 273
Coefficients, 212, 221
`Cofactor`, 235, 273
Cofactor, 235
Colbourn, C., 236
Color degree, 214
Coloring, graphs, 210
Combination locks, 195
Combinations, 2, 40
Combinatorial explosion, 40
Combinatorics, 2
Commutative operation, 134
Compiler optimization, 210
`Complement`, 264
Complement, 93
    complete graph, 141
    connectivity, 171
    perfect graph, 219
    sum, 131
    vertex cover, 218
Complete bipartite graphs, 166
    construction, 142
    cutsets, 99
    edge chromatic number, 216
    edge connectivity, 178
    embeddings, 98
    girth, 191
    Hamiltonian, 197
    join, 132
    network flow, 238
    regular, 159
    star, 146
    vertex connectivity, 179
Complete directed acyclic graph, 209
Complete graphs, chromatic polynomial, 211
    coloring, 215
    connectivity, 180
    construction, 140
    degree sequence, 158
    embeddings, 104
    Eulerian, 193
    examples, 82, 86, 88, 89, 122, 161
    intersection, 131
    notation, 128
    powers, 231
    properties, 143, 166
Complete k-partite graphs, 142
    coloring, 215
    complement, 93
    rotation, 114
    vertex cover, 218
`CompleteQ`, **89**, 131, **273**
Complete subgraph, 217
`Compositions`, 61, 273
Compositions, 50
Computational geometry objects, 98
Computer networks, 145, 149, 175
Conditional statements, 268
Conjectures, 123
`ConnectedComponents`, 171, 274
Connected components, 171
    chromatic polynomial, 212
    embedding, 118
    traversal, 96
`ConnectedQ`, 171, 172, 173, 274
Connectivity, 170

directed graphs, 172
random graphs, 220
spectrum, 85
strong, 172
undirected graphs, 154
weak, 172
`ConstructTableau`, **65**, 78, **274**
Containment, 170
Contiguous subsequence, 75
`Contract`, 91, 274
Contraction, 91, 124, 210
Convex embeddings, 104
Convex polyhedra, 247
Corneil, D., 182
`Cos`, 261
`CostOfPath`, 199
`Count`, 268
Counting, compositions, 60
de Bruijn sequences, 196
derangements, 33
heaps, 36
inversions, 28
involutions, 32
necklaces, 25
partitions, 77
permutations by cycle, 23
runs, 30
spanning trees, 235
subsets, 40
Young tableaux, 70
Cover number, 219
Coxeter, H., 198
Crespi-Reghizzi, S., 123
Cross edges, 96
Crossing number, 251
Cube, geometric, 149, 247, 250
power, 230
Cubic graphs, 177, 185, 194
Cut-sets, 99
Cvetković, D., 85, 122
`Cycle`, 144, 274
Cycle, 144
chromatic polynomial, 212
connectivity, 180
example, 132
induced, 149
line graph, 137, 216
non-interval graphs, 164, 166
self-complementary graphs, 187
spanning trees, 236
traversal, 95
Cycles in permutations, 20, 33
Cyclic graphs, 188, 206
Cyclic groups, 185

DAGs, 190, 208
Danzer, L., 149
Dao, M., 122
Data structures, 38, 81, 232
Day, R., 236
de Bruijn, N., 195
`DeBruijnSequence`, 195, 274
de Bruijn sequence, 195
Decrement operators, 262
de Fraysseix, H., 251
Degree matrix, 235
`DegreeSequence`, 157, 274
Degree sequence, 157
Eulerian graphs, 192
isomorphisms, 182, 184
Degree set, 167
`DeleteCycle`, 189, 275
`DeleteEdge`, 83, 275
`DeleteFromTableau`, 65, 275
`DeleteVertex`, 84, 275
Deo, N., 258
Depth-first search, **96**, 124, 171, 172, 188
`DepthFirstTraversal`, 96, 275
`DerangementQ`, 33, 275
`Derangements`, 33, 275
Derangements, 33
Dereferencing, 261
`Det`, 263
Determinant, 235
Diagonal matrix, 235
`Diameter`, 107, 276
Diameter, 107, 187
cycle, 230
spanning tree, 227
Difference, graphs, 131
`Digits`, 41
`Dijkstra`, 225
Dijkstra, E., 225
Dijkstra's algorithm, 225
`DilateVertices`, 113, 276
Dilworth, R., 241
Dilworth's theorem, 241
Directed acyclic graphs, 190, 208
Directed cycles, 188
Directed graphs, 93, 101

Disconnected graphs, 159, 212
Disoriented graphs, 94
Distinct parts, 56, 70
Distinct parts, partition, 77
`DistinctPermutations`, 13, 276
Distinct permutations, 12
`Distribution`, 60, **106**, **276**
Distribution, uniform, 9, 37, 42, 46, 73
`Divide`, 261
Divisibility relation, 209, 243
`Divisors`, 263
`DivisorSigma`, 77
Divorce digraph, 252
`Do`, 265, 266
Documentation, 259
Dodecahedron, 198, 247
Dominance relation, 52
Doob, M., 85
`Dot`, 263
Dots, 53, 60
Drawing graphs, 82, 99, 184
Drawing polygons, 101
Drawing trees, 107
`Drop`, 263
Druffel, L., 182
Dual graph, 147
`DurfeeSquare`, 56, 276
Durfee square, 56, 78
Dutton, R., 123
Dynamic programming, Eulerian numbers, 31
 partitions, 57
 shortest paths, 228
 Stirling numbers, 23
Dynamic scoping, 261

e, 34
Eades, P., 98, 99, 104, 116, 122, 253
`Eccentricity`, 107, 276
Eccentricity, 107
Echeandia, M., 122
Eco, U., 143
Edge, 80
 chromatic number, 216
 coloring, 216
 connectivity, 177
 contraction, 91
 deletion, 83
 insertion, 83
 interchange operation, 166
`EdgeColoring`, 216, 276
`EdgeConnectivity`, **178**, 220, **276**
`Edges`, 81, 277
Edges, 9
Edmonds, J., 194, 237
Efficiency, 5, 7, 84
Eggleton, R., 166
`Eigenvalues`, 263
Eigenvalues, 85
 conjectures, 125
`Element`, 199, 203, 233
Embeddings, 81
 circular, 104
 complete k-partite graphs, 143
 convex, 247
 from other embeddings, 83
 hypercube, 166
 misleading, 139
 product, 134
 ranked, 105
 torus, 192
`EmptyGraph`, 140, 277
Empty graph, 141
 chromatic polynomial, 211
 join, 132
`EmptyQ`, **89**, 131, **277**
`EncroachingListSet`, 76, 277
Encroaching list sets, 75, 78
Envelopes, 34
`Equal`, 268
`EquivalenceClasses`, 18, 277
Equivalence classes, 18, 182
`EquivalenceRelationQ`, 18, 277
Equivalence relations, 18
Erdos, P., 75, 156, 157, 215
Euclidean graphs, 201, 252
Euclidean traveling salesman problem, 200
`Eulerian`, 30, 278
`EulerianCycle`, 194, 277
Eulerian cycle, interesting, 221
Eulerian graphs, 138, **192**
Eulerian numbers, 30
`EulerianQ`, 193, 278
Euler, L., 51, 57, 192
Euler's formula, 247, 250
Even cycles, 213, 241
Even partitions, 77
Even permutations, 24
`EvenQ`, 265, 268
Even, S., 239, 247
`ExactRandomGraph`, 155, 278

Execution, 34
Exercises, 47, 77, 124, 166, 220, 252
`Expand`, 212
`ExpandGraph`, 129, 278
Extensible languages, 261
`ExtractCycles`, 189
Extremal graph theory, 143, 179

Fáry, I., 100, 251
Factorial, 263
`FactorInteger`, 263
Factor of a graph, 244
Fajtlowicz, S., 123, 125
Falling factorial function, 212
`False`, 265
Feedback shift registers, 196, 221
Ferrars diagrams, 53
`FerrersDiagram`, 50, **54**, **278**
Ferrers diagrams, 53, 78
Files, 119
`FindCycle`, 188, 278
`FindSet`, 232
`First`, 263, 264
Fixed-points, 18, 21, 32, 33
`Flatten`, 263
Fleischner, H., 231
Fleury's algorithm, 193
`Floor`, 16, 261
Floyd, R., 35, 228
Floyd's algorithm, 228, 252
Fogg, I., 98, 122
Folkman, J., 186
`For`, 265, 266
Ford, L., 178, 237
Fortran, 123, 265
Forward edges, 96
Foucault's Pendulum, 143
Foulkes, J., 175
Four color problem, 198, 210, 220
Franzblau, D., 70
Fredman, M., 235
Freeman, P., 48
Free trees, 107, 151
French plurals, 64
`FromAdjacencyLists`, 86, 278
`FromCycles`, 21, 279
`FromInversionVector`, 27, 279
`FromOrderedPairs`, 87, 279
`FromParent`, 188
`FromUnorderedPairs`, 88, 279
Frucht graph, 185
Frucht, R., 185
Fulkerson, D., 157, 166, 178, 219, 237
`Function`, 265
`FunctionalGraph`, 164, 279
Functional graphs, 164
Furst, M., 20

Gale, D., 245
Gallai, T., 157, 219
Garey, M., 181, 203, 232, 258
Gassner, B., 30
Gates, W., 48
Gauss, C., 267
GED, 119, 122
Generating, complete graphs, 140
  complete k-partite graphs, 142
  cycles, 144
  distinct permutations, 12
  grid graphs, 147
  heaps, 47
  hypercubes, 148
  k-subsets, 44
  permutations, 3
  random graphs, 155
  random k-subsets, 45
  random permutations, 6
  random subsets, 42
  regular graphs, 159
  stars, 145
  strings, 40
  subsets, 41
  trees, 151
  wheels, 146
  Young tableaux, 64, 68
Genus, 251
Geodesic, 225, 237
Geometric transformations, 113
Gilbert, E., 43
Gilmore, P., 163
`Girth`, **190**, 220, **279**
Girth, 190
  cages, 191, 221
  chromatic numbers, 215
Goddyn, L., 167
Golomb, S., 196
Golumbic, M., 221
Gomery, R., 239
Good, I., 195
Goodman, E., 48

Gottlieb, C., 182
Goulden, I., 257
Gould, R., 215, 221, 257
GRAFFITI, 123
  conjectures, 125
Graham, R., 24, 31, 34, 232, 257, 258, 262
`Graph`, **81**, 263
Graph, 80
GRAPH, 122
Graphbase, 123
`GraphCenter`, 108, 279
`GraphComplement`, 93, 279
`GraphDifference`, 131, 280
Graphical enumeration, 153
Graphic degree sequences, 166
`GraphicQ`, 157, 281
Graphics, 100, 103
Graphic sequences, 157
`GraphIntersection`, **131**, 166, **280**
`GraphJoin`, **132**, 166, **280**
`GraphPower`, 230, 280
`GraphProduct`, **133**, 166, **280**
Graphs, 9
  center, 124
  coloring, 210
  complement, 93, 129, 218
  display, 100
  embeddings, 134
  matching, 240
  product, 148
  spectrum, 85
  traversal, 95
`GraphSum`, 131, 280
`GraphUnion`, 84, 118, **129**, 166, 171, **280**
GraphView, 122
GRAPPLE, 122
`GrayCode`, **43**, 47, **281**
Gray code, 47
  compositions, 62
  partitions, 53, 78
  subsets, 42, 149
Gray, F., 42
Greedy algorithm, 213
`GridGraph`, 94, 107, **147**, **281**
Grid graphs, 147
  embedding, 166
  independent sets, 219
  powers, 226, 230
Groups, 17, 184
Gupta, H., 57
Gupta, R., 216
Gusfield, D., 252
Gwyther, A., 57

Habib, M., 122
Hakan, W., 210
Hakimi, S., 157
Hall, M., 27
Hall, P., 240
Hall's marriage theorem, 240
`HamiltonianCycle`, 148, **196**, **281**
Hamiltonian cycles, 196
  backtracking, 220
  biconnectivity, 231
  circumference, 192
  dodecahedron, 198
  grid graphs, 148
  hypercube, 43, 149
  knight's tour, 166
  NP-completeness, 194
  tournaments, 175
  transposition graph, 10, 11
Hamiltonian graphs, biconnectivity, 177
  bipartite, 197
  grid graphs, 148
  hypercube, 149
  line graphs, 138
  squares, 231
  transpositions, 10, 11
Hamiltonian paths, 175
`HamiltonianQ`, 196, 281
Hamilton, Sir William, 198
Han, X., 247, 253
`Harary`, 179, 281
Harary, F., 80, 137, 138, 140, 141, 153, 156, 175, 179, 185, 250, 257
Harary graph, 179
`HasseDiagram`, 52, **206**, **281**
Hasse diagram, 170, 206
Havel, V., 157
Heap, B., 10
`Heapify`, 35, 282
Heaps, 35, 47
  biparential, 71
`HeapSort`, 38, 281
Heapsort, 38
Heawood graph, 192
Heirarchy, 110
Hell, P., 232
Help, 259

Heuristics, coloring, 214
isomorphism, 182
traveling salesman, 201
Hidden cycle representation, 22, 47
HideCycles, 22, 282
Hierholzer, C., 194
Hillier, J., 225
Hoffman, A., 163, 166
Holt, D., 17
Holyer, I., 216
Homeomorphism, 247
Hooke's law, 116
Hook length formula, 70
Hopcroft, J., 20, 96, 176, 240, 247, 258
Horowitz, E., 258
Hospitals, 245
Hu, T., 239
Hypercube, 148, 282
Hypercubes, 148
embedding, 166
Gray code, 43
matching, 241

IBM Personal Computer, 122
Icosahedron, 121, 247
Identical graphs, 181
IdenticalQ, 182, 282
IdentityMatrix, 263
Identity operation, 22, 28, 131
Identity permutation, 17, 21, 32
If, 268
Illusion, 80, 90
Imperative languages, 268
Improving embeddings, 112
IncidenceMatrix, 135, 282
Incidence matrix, 136
Inclusion, 207
Increasing subsequence, 73
Increment operators, 262
In-degree, 192, 206
Independence number, 219
Independent set, 217, 219, 232
IndependentSetQ, 219, 282
Index, 29, 282
Index, permutation, 29
Induced subgraphs, 90, 166
InduceSubgraph, 84, **90**, 171, **282**
Infinity, 264
InitializeUnionFind, 232
InsertIntoTableau, 64, 283
Insertion, Young tableaux, 64, 73
Integer partitions, 51
Interlocking bridges, 247, 249
InterlockQ, 250
Interns, 245
Intersection, 264
Intersection, graphs, 130
Intersection graphs, 135
IntervalGraph, **163**, 166, **283**
Interval graphs, 163
Invariants, 154, 170
InversePermutation, **18**, 67, **283**
Inverse permutation, 17, 67
inversions, 29
involution, 32
Josephus, 35
Inversion poset, 162
Inversions, 29, 283
Inversions, 27, 37, 47
Inversion vectors, 27, 47
InvolutionQ, 32, 283
Involutions, 20, 21, **32**, 67
Irrational networks, 237
Irrational numbers, 262
Irving, R., 252
Isaacson, J., 215
IsolateSubgraph, 248, 249
IsomorphicQ, 134, **183**, **283**
Isomorphism, 182, 283
Isomorphism, 137, 138, 181

Jackson, D., 257
Johnson, D., 181, 232, 258
Johnson, E., 194
Johnson, S., 11
Join, 263, 264
Join, graphs, 118, 131, 146
Josephus, 34, 283
Josephus permutations, 34, 48
Journal of Algorithms, 181
Junction, 80
Junction points, 248

K, 80, 82, 124, **140**, **142**, **284**
Kainen, P., 210
Kaliningrad, 193
Kamada, T., 116
Karp, R., 237, 240
Kawai, S., 116
K-connectivity, 177, 178, 179, 237

K-edge-connectivity, 177
Kelly, D., 98, 122
K-factor, 244
King, C., 123
Kirchhoff, G., 136, 235
Klee, V., 149
Klingsberg, P., 62
Knight's tour graph, 166
Knuth, D., 14, 24, 28, 30, 31, 32, 34, 37, 64, 70, 123, 196, 257, 258, 262
Königsberg, 136, 192
Koch, J., 210
Kodama, Y., 250
K-partite graphs, 142
Kratsch, D., 28
Kruskal, J., 232
Kruskal's algorithm, 232
`KSubsets`, 44, 284
K-subsets, 40
    compositions, 61
    constructing, 44
    odd graphs, 162
Kuratowski, K., 247
Kuratowski's theorem, 247
Kwan, M., 194

Labeled graphs, 103
Labeled trees, 151
`LabeledTreeToCode`, 151, 284
`Last`, 263
Lattice, 170
Lattice paths, 147
Lawler, E., 199
Leighton, T., 124
Lekkerkerer, C., 163
`Length`, 263
Lenstra, J., 199
Leonard, B., 48
`Less`, 15
Levin, L., 181
Lewis, D., 210
Lewis, G., 16
Lewis, P., 201
Lexicographic order, 4
    backtracking, 13
    compositions, 61
    distinct permutations, 14
    partitions, 51
    permutations, 4
    strings, 40
    subsets, 43, 45
    Young tableaux, 68
`LexicographicPermutations`, 4, 284
`LexicographicSubsets`, **43**, 47, **284**
LILA, 122
Line, 80
Linear algebra, 85
`LinearProgramming`, 252
`LineGraph`, 136, 284
Line graphs, 135
    coloring, 216
    complements, 167
    total graph, 166
Lin, S., 201
Lipman, M., 122
Lipton, R., 184
Lisp, 264
List construction operations, 267
List manipulation, 263
`ListPlot`, 8, 35, 36
Lists, 263
    arrays, 8
Liu, C., 53, 257
Lloyd, E., 193
`LongestIncreasingSubsequence`, 74, 284
Longest increasing subsequence, 73
Lovasz, L., 215, 219, 240
Lower bound, 201, 215
Lucas, E., 193
Lueker, G., 164, 257
Luks, E., 20, 181

`M`, 28, **81**, **284**
MacMahon, P., 29
Maeder, R., 259
Main diagonal, 226
`MakeGraph`, 28, 37, 52, 77, **161**, 166, 203, **285**
`MakeSimple`, 94, 285
`MakeUndirected`, 94, 285
Male-optimal marriage, 246
Manacher, G., 28
Manber, U., 258, 266
Manhattan distance, 227
Manhattan streets, 172
Mannila, H., 28, 30
Manvel, B., 215
`Map`, 265, 267
`MapAt`, 268
Marble, G., 215
Marriage problem, 47, 240

Matchings, 240
  maximal, 243
Mathematica, 2
  arrays, 3, 264
  calling a function, 260
  conditionals, 268
  conventions, 261
  data types, 262
  defining a function, 260
  dereferencing, 261
  efficiency, 7, 39, 84, 229, 266
  files, 119
  graphics, 54, 100
  iteration, 265
  linear programming, 252
  lists, 263
  manual, 259
  mapping functions, 265
  mathematical operations, 261
  numbers, 262
  number theoretic functions, 263
  recursion, 266
  scoping, 97, 261
  side effects, 264
  statements, 261
  timing, 5
Matrices, 263
Matrix multiplication, 263
Matrix operations, 263
`MatrixQ`, 268
Matroid, 232
Matula, D., 215
`Max`, 264
`MaximalMatching`, **244**, 252, **285**
`MaximumAntichain`, 242, 285
Maximum change order, 11
`MaximumClique`, **217**, 220, **285**
Maximum clique size, 219
Maximum degree, 157
`MaximumIndependentSet`, 219, 285
Maximum matchings, 240
`MaximumSpanningTree`, 233, 285
McAndrew, M., 166
McGeer, P., 9, 48
McKay, B., 122
Measures of order, 28, 30, 76
`MemberQ`, 268
Membership testing, 20
Menger, K., 178
Menger's theorem, 178
Meserve, B., 250
Metric function, 200
Middle binomial coefficients, 262
Miller, J., 57
`Min`, 264
`MinimumChainPartition`, 242, 286
Minimum change order, compositions, 62
  partitions, 53, 78
  permutations, 9
`MinimumChangePermutations`, 11, 286
Minimum degree, 157
`MinimumSpanningTree`, **233**, 252, **286**
Minimum spanning trees, 232
  Euclidean, 252
  path, 252
  traveling salesman bound, 201
  variants, 232
`MinimumVertexCover`, 218, 286
Minor, 235
Misleading embeddings, 137, 139
`Mod`, 262
Model of computation, 6, 9, 39
Modular arithmetic, 35
Molluzzo, J., 129
Monotone graph property, 154
Moon, J., 175
Moore, E., 225
Morpuro, R., 123
Moser, L., 175
Moses, L., 6
Mukhopadhyay, A., 253
Multigraphs, 81, 89
  degree sequences, 158
  Königsberg, 193
`Multinomial`, 14
Multinomial coefficients, 12
Multiplication, adjacency matrices, 131
Multiplication, matrix, 263
`MultiplicationTable`, 17, 286
Multiplication table, 17
Multiplicative inverse, 17, 32
Multisets, 10, 12
Munro, I., 71
Murty, U., 180, 221, 257

`N`, 262
Nash-Williams, C., 138
Nauty, 122
N-cube, 148
Necklaces, 25

Negative, 268
Negative cost cycles, 228
Nel, L., 236
Network, 225
NetworkFlow, 170, 220, **237**, **286**
Network flow, 237
  connectivity, 178
  linear programming, 252
  matching, 240
NetworkFlowEdges, 237, 286
NeXT, 122
NextComposition, 61, 287
NextKSubset, 45, 287
NextPartition, 53, 287
NextPermutation, 6
NextSubset, 41, 287
NextTableau, 68, 287
Ng, A., 9, 48
Nievergelt, J., 258
Nijenhuis, A., 9, 42, 45, 58, 68, 70, 72, 77, 153, 257
Node, 80
Non-planar graphs, 141, 143
NormalizeVertices, 113, 287
Normalizing embeddings, 121
North, S., 184
Not, 268
NP-completeness, 181, 258
  coloring, 211, 216, 219
  drawing trees, 111
  Hamiltonian cycles, 194, 196
  maximum clique, 217
  permutation graphs, 28
  spanning tree variants, 232
  traveling salesman, 199
NthPair, 155, 287
NthPermutation, 5, 287
NthSubset, 41, 288
Null graph, 141
NumberOfCompositions, 60, 288
NumberOfDerangements, 33, 288
NumberOfInvolutions, 32, 288
NumberOfPartitions, 57, 288
NumberOfPermutationsByCycles, 23, 288
NumberOfSpanningTrees, 236, 288
NumberOfTableaux, **70**, 72, **289**
Number theoretic functions, 263
Number theory, 51

Oakford, R., 6
Octahedron, 247, 250
Odd cycles, 213, 215
Odd graphs, 162
Odd partitions, 56, 77
Odd permutations, 24
OddQ, 260, 268
On-line help, 259
Opsut, R., 210
Or, 268
Ordered pairs, 81, 87
OrderedQ, 268
Order of a graph, 82
Ore graphs, 197
Ore, O., 197
Orientation, 174
OrientGraph, 94, **174**, **289**
Orlin, J., 239
Out-degree, 192, 206
Outerplanar graphs, 251

Pach, J., 251
Pages, 124
Pairwise flows, 239
Palmer, E., 153, 185
Pancake sorting, 48
Papadimitriou, C., 48, 232
Paradigms, 265
Parameterized graphs, 129
Parentheses, 63, 71
Parent relation, 189, 225
Parity, 187
Part, 263
Parter, S., 247
Partial order, 203
  Dilworth's theorem, 241
PartialOrderQ, 203, 289
Partial solution, 12
Partition, 264
PartitionQ, 51, 289
Partitions, 51, 289
Partitions, 50
  construction, 51
  counting, 57
  distinct parts, 77
  dominance, 52
  Ferrers diagram, 53
  predicate, 51
  random, 58
  representation, 263
  shape, 63, 70
PartitionsP, 57

PartitionsQ, 58
Pascal, 122
Patashnik, O., 24, 31, 34, 257, 258, 262
Path, 147, 289
Path, 152
    self-complementary graphs, 187
Path compression, 232
PathConditionGraph, 199, 226, **228**
Pentagram, 141
Perfect graphs, 219
Perfect matchings, 240
PerfectQ, 219, 289
Permutation generation, 4
    distinct permutations, 13
    lexicographic order, 4
    maximum change, 11
    minimum change, 10
    multisets, 12
    random, 6
    unranking, 5
Permutation graph, 28, 38
PermutationGroupQ, 19, 289
Permutation groups, 17, 47
PermutationQ, 3, 290
Permutations, 12
Permutations, 2
    encroaching lists, 76
    representation, 263
    scattered subsequences, 73
    special classes, 32
    tableaux, 65, 66, 78
    transposition, 162
Permute, **3**, 17, **290**
Perturbing embeddings, 114
Petersen graph, 128, 139, 162, 191
Peterson, J., 244
Pi, 262
Pigeonhole principle, 251
Pisanski, T., 167
Planar embeddings, dual, 147
Planar graphs, coloring, 210
    planarity testing, 247
    platonic solids, 247
    trees, 107
Planarity testing, 124
PlanarQ, **247**, 252, **290**
Plane partitions, 64
Platonic solids, 247
PlotRange, 101
Plummer, M., 240
Plus, 261, 266
Point, 80
PointsAndLines, 101
Poker, 221
Pollack, R., 251
Polya, 25, 290
Polya, G., 25
Polya's theory of counting, 25, 47
Polygons, drawing, 101
Polyhedra, convex, 247
Polynomials, 26
Polynomial time, 197
Ponder, C., 9, 48
Poset, 203
Position, 268
Positive, 268
Postman tour, 194
Postscript, 100
Power, 261
Power of a graph, 229
Precedence relation, 208
Predicates, 261
Pregel river, 192
Prepend, 263
PrependTo, 264
Presortedness, 28, 30, 76
Prim, R., 232
Prim's algorithm, 232, 235, 252
Priority queue, 38
Product, 263
Products, graphs, 133, 147, 148
Programming languages, 2
Properties of graphs, 170
Propp, J., 55
Prüfer codes, 151
Prüfer, H., 151
PseudographQ, 89, 290
Pseudographs, 89, 158
Pure functions, 265

Queuing system, 38

RadialEmbedding, 108, 290
Radial embeddings, 108
Radius, 108, 290
Radius, 107
RAM, 6
Random, 261
Random, compositions, 60
    Euclidean graphs, 252

k-subsets, 45
number generation, 196
partitions, 58, 77
permutations, 6, 37
subsets, 42
Young tableaux, 72
Random access machine, 6
`RandomComposition`, 60, 290
`RandomGraph`, **154**, **156**, 229, **291**
Random graphs, degree sequences, 158
directed, 156
exact, 155
labeled trees, 152
theory, 154
`RandomHeap`, 36, 291
`RandomKSubset`, 45, 291
`RandomPartition`, 50, 57, **58**, 77, **291**
`RandomPermutation`, **7**, 36, 67, 76, **291**
`RandomPermutation1`, 7
`RandomPermutation2`, 7
`RandomSubset`, **42**, 90, **291**
`RandomTableau`, 72, 291
`RandomTree`, **152**, 191, **291**
`RandomVertices`, 252, **292**
`Range`, 17, 264, 265
`RankedEmbedding`, 106, 292
Ranked embeddings, 106
bipartite graphs, 143
Hasse diagram, 206
`RankGraph`, 105, 292
Ranking, permutations, 5
spanning trees, 236
subsets, 42
`RankPermutation`, 5, 292
`RankSubset`, 41, 292
Rédei, L., 175
`ReadGraph`, **120**, 185, **292**
Reading from files, 119
Read, R., 123, 141, 221
`Real`, 262
`RealizeDegreeSequence`, **158**, 166, 215, **292**
Recurrence, 57
chromatic polynomial, 211
derangements, 33
Eulerian numbers, 31
heaps, 36
involutions, 32
parentheses, 71
partitions, 51, 58, 77
permutations, 4
Stirling numbers, 23
strings, 40
subsets, 43
Recursion, 266
Reddy, V., 147
Reductions, 217
Reference guide, 256
Reflexive relations, 18
`RegularGraph`, **159**, 221, **293**
Regular graphs, 159
cages, 191
complete graphs, 158
cubic, 177
cycle, 145
dodecahedron, 198
eigenvalue, 85
Harary graphs, 180
line graphs, 137
`RegularQ`, 159, 293
Regular subgraphs, 244
Reingold, E., 111, 258
`RemoveSelfLoops`, 89, 293
`Replace`, 264
`ReplaceAll`, 264
Research problems, 48, 78, 124, 167, 221, 253
`Rest`, 263, 264
`Return`, 266
`RevealCycles`, 22, 293
Reversals, 62
`Reverse`, 264
Reverse permutation, 29, 74
Richard, J., 122
Rinnooy Kan, A., 199
Roads, 225
Robbins, H., 174
Roberts, F., 19, 25, 175, 210, 257
Roman Empire, 34
Ronse, C., 196
Root, 107, 233
`RootedEmbedding`, **111**, 124, **293**
Rooted trees, 110, 151
Rorschach test, 227
Rosenberg, A., 124
Rosenkrantz, D., 201
`RotateLeft`, 263, 264
`RotateRight`, 140, 263
`RotateVertices`, **113**, 141, **293**
`Runs`, 30, 293
Runs, 30, 47, 75
Ryser, H., 157

Saaty, T., 210
Sabidussi, G., 133
Sachs, H., 85, 187
Safecracker, 195
Sahni, S., 258
`SamenessRelation`, 19, 293
Sameness relation, 19
`SameQ`, 268
Sandberg, J., 184
Satanic symbol, 80, 141
Savage, C., 53, 78
`Scan`, 265, 267
Scattered subsequences, 73
Scheduling, 210, 216
Schensted, C., 66, 73
Schmidt, D., 182
Scope rules, 97
Searching, 14
Secretary problem, 34, 47
Sedgewick, R., 3, 10, 37, 258
`Select`, 34, 268
`SelectionSort`, **15**, 39, **294**
Selection sort, 14
Self-complementary graphs, 187
`SelfComplementaryQ`, 187, 294
Self-conjugate partitions, 77
Self-help, 259
Self-loops, 82, 89
    degree sequences, 158
    representation, 165
    transitive reduction, 206
Semicolon, 103
Semirandom graphs, 158, 215
Set operations, 264
Set partitions, 24
Set union, 232
Sewerage, 237
`ShakeGraph`, 114, 294
Shannon, A., 111
Shannon, G., 122
Shape, 63
Shapely, L., 245
`ShapeOfTableau`, 63
Shapes, 51
Shift registers, 196, 221
Shmoys, D., 199
Shortest cycle, 191
`ShortestPath`, 226, 294
Shortest paths, 225
    conjecture, 125
    isomorphism, 182
    metric graph, 202
    minimum spanning trees, 232
    network flow, 237
`ShortestPathSpanningTree`, 226, 294
`ShowGraph`, 100, 294
`ShowLabeledGraph`, 103, 294
Shultz, H., 48
Side effects, 264
`Signature`, 24
`SignaturePermutation`, 24, 295
Sign, permutation, 24
Simple graphs, 89, 200
`SimpleQ`, 89, 295
`Simplify`, 212
Simulation, 38
`Sin`, 261
Singleton cycles, 21
Sink vertex, 237
Skiena, S., 28, 75, 99, 125, 147
Slot machines, 221
`Sort`, 14, 263
Sorting, analysis, 27
    complexity, 6
    heapsort, 38
    pancakes, 48
    selection sort, 14
Source vertex, 237
Soviet Union, Kalingrad, 193
Sparse directed graphs, 173
Sparse graphs, 86, 196
`Spectrum`, 85, 295
Spectrum, 85, 124
    sum, 125
Spelling, 53, 78
Spencer, J., 156
Spine, 124
Spokes, 146
SPREMB, 98, 119, 122
`SpringEmbedding`, **116**, 124, **295**
Spring embedding, 116
Square, adjacency matrix, 158
    cycle, 144
    geometric, 56
    power, 231
`StableMarriage`, 245, 295
Stable marriage problem, 252
Stanford Graphbase project, 123
Stanley, R., 53, 64, 145, 257
Stanton, D., 22, 52, 53, 56, 77, 162, 257

`Star`, 146, 295
Star, articulation vertex, 177
  automorphisms, 19
  construction, 145
  embedding, 104
  example, 83, 85, 87, 88
  girth, 191
  interval representation, 164, 166
  join, 132, 146
  matching, 241
  traversal, 96
  tree, 152
State space, 13
Stearns, R., 201
Steiglitz, K., 232
`StirlingFirst`, 23, 295
Stirling numbers, 23
`StirlingS1`, 24
`StirlingS2`, 24
`StirlingSecond`, 23, 296
Storage formats, 119
Straight-line drawings, 100, 165
`Strings`, **40**, 41, **296**
Strings, 40, 195, 196
Strong connectivity, 94, 172, 192
`StronglyConnectedComponents`, **172**, 220, **296**
Strongly connected graphs, 174
Subgraphs, clique, 215
  cliques, 143
  induced, 90
  perfect graphs, 219
`Subsets`, 43, 296
Subsets, 2
  boolean algebra, 170, 207
  combinations, 40
  Gray code, 43
  lexicographic order, 43
  representation, 263
`Subtract`, 261
`Sum`, 262, 267
Sum, graphs, 131
Supowit, K., 111
Survivor, execution, 35
Sussman, G., 259
Sussman, J., 259
Suwanda, H., 71
Symmetric graphs, 93
Symmetric group, 17
Symmetric relations, 18
Symmetries, 25, 184
Szekeres, G., 75, 167
Szele, T., 175

`Table`, 265, 267
`TableauClasses`, 74, 296
`TableauQ`, 63, 296
`TableauToYVector`, 69
`Tableaux`, 69, 296
`TableauxToPermutation`, 66, 296
`TableForm`, 46, 64, 68, 76
Tait, P., 198
`Take`, 263
Tallot, D., 122
Tamassia, R., 99, 104, 253
`Tan`, 261
Tarjan, R., 96, 172, 232, 235, 239, 247, 253
Taxi drivers, Manhattan, 172
Telephone book, 15
Tetrahedron, 247, 250
Thickness, 251
Thomassen, C., 175
Thompkins, C., 27
Tilford, J., 111
`Times`, 261
`Timing`, 5, 7, 39, 43, 266
Tindell, R., 175
`ToAdjacencyLists`, 86, 297
`ToCycles`, **20**, 33, 47, **297**
`ToInversionVector`, 27, 297
`ToOrderedPairs`, 87, 297
`TopologicalSort`, **208**, 220, **297**
Topological sort, 208
Torus, 192
Total graphs, 166
Total order, 14
Touchdown, 262
`ToUnorderedPairs`, 87, 297
Tournaments, 175
Transformation rules, 264
`TransitiveClosure`, **203**, 220, **297**
Transitive closure, 203
  inversion poset, 162
  shortest paths, 228
  strongly connected components, 174
`TransitiveQ`, 18, 203, 297
`TransitiveReduction`, 204, 298
Transitive reduction, 203
  equivalence relations, 18
  Hasse diagram, 206
Transitivity, 203

`TranslateVertices`, 113, 298
Transmission errors, 42
`Transpose`, 263, 264
Transpose, 55, 93
`TransposeGraph`, 193
`TransposePartition`, 55, 298
`TransposeTableau`, 63, 298
Transpositions, 9
  cycles, 21
  graph, 2, 10
  involutions, 32
  permutations, 162
  signatures, 24
`TravelingSalesman`, 199, 298
`TravelingSalesmanBounds`, 201, 298
Traveling salesman problem, 199
  Euclidean, 252
  lower bounds, 201
Tree, 236
`TreeQ`, 189, 298
Trees, 151
  bipartite, 213
  chromatic polynomial, 211
  girth, 191
  number of edges, 190
  spectra, 124
  star, 145
Triangle, 144, 190, 191
Triangle inequality, 200
`TriangleInequalityQ`, 200, 299
Triangular matrices, 81
Triconnected graphs, 179
Trigonometry, 101, 104
Tripartite graphs, 80
Trotter, H., 11
`True`, 265
Tucker, A., 25, 257
`Turan`, 144, 299
Turan graph, 144, 218
Turan, P., 143
Tutte, W., 179, 198, 221
Two-colorable graphs, 142, 213
`TwoColoring`, 213, 299
Two-dimensional lists, 264

Ullman, J., 176, 203, 258
Unary predicates, 268
Undirected cycles, 188
Undirected edges, 102
Undirected graphs, 93
`UndirectedQ`, 94, 299
`Union`, 12, 264
Union-find, 232
Union, graphs, 129
`UnionSet`, 232
Unordered pairs, 88
Unranking, permutations, 5
  spanning trees, 236
  subsets, 42
Unweighted graphs, 81, 226
`UnweightedQ`, 89, 299
Urrutia, J., 28

`V`, 81, 299
Valency, 157
van Rooij, A., 138
Variable dereferencing, 261
Vaucher, J., 111
Vertex, 9, 80
  coloring, 210, 220
  connectivity, 177, 179, 237
  cover, 217, 218
  degrees, 192
  transformations, 113
`VertexColoring`, 214, 299
`VertexConnectivity`, **178**, 220, **299**
`VertexCoverQ`, 218, 300
`Vertices`, 81, 300
Vizing, V., 216
VLSI layout, 124

Wachs, M., 145
War and Peace, 123
Weak connectivity, 172, 192
`WeaklyConnectedComponents`, 172, 300
Weighted, graphs, 81, 87
  independent sets, 232
  matchings, 240
Wetherell, C., 111
`Wheel`, 146, 300
Wheel, chromatic number, 212
  connectivity, 179
  construction, 146
  eccentricity, 108
  embedding, 104
  example, 91
  join, 132
  properties, 166
`While`, 265, 266
White, D., 22, 52, 53, 56, 77, 162, 257

Whiting, P., 225
Whitney, H., 138, 178
Wildcard, 260
Wilf, H., 9, 11, 42, 45, 58, 68, 70, 72, 77, 85, 124, 138, 153, 214, 220, 257
Wilson, R., 153, 193
Windows, 101
Winners and losers, 175
Wolfram Research, Inc., 119
Wolfram, S., 259
World War II, 193
Wormald, N., 159
`WriteGraph`, 121, 300
Writing to files, 121

Young's lattice, 77
Young tableaux, 50
  definition, 63
  insertion algorithm, 64
  involutions, 32
  lexicographic order, 68
  pairs of tableaux, 66
  random, 72
  representation, 263
  shape, 51
Y-vector, 69
`YVectorToTableau`, 69

Zeilberger, D., 70
Zeno's paradox, 262

Software Information

Available Versions of Mathematica

Macintosh Versions

- Standard Version
- Enhanced Version (requires 68020/30 and numeric coprocessor)

5 megabytes recommended. (Less memory is required if a virtual memory system is used.)

Microsoft Windows Version

Requires Windows 3.0 or later, MS-DOS 3.0 or later, and a 386 or 386SX-based computer with 4 megabytes RAM. Supports Notebook user interface, and supports all printers and graphics cards supported by Windows 3.0.

386-based MS-DOS Versions

- 386 Version (does not require numeric coprocessor)
- 386/7 Version (requires 287 or 387 coprocessor or 486 CPU)
- 386/Weitek Version (requires Weitek 1167, 3167, or 4167 coprocessor)

640K memory and 3 megabyte extended memory required. Supports CGA, EGA, VGA, MCGA, 8514, and Hercules graphics standards, PostScript, LaserJet, Epson FX, and Toshiba P3 compatible printers.

Other Systems

386 Unix–386/ix and V/386

CONVEX–C1 and C2 Series

Data General–AViiON

Digital Equipment–Ultrix (VAX and RISC), and VAX/VMS

Hewlett-Packard Apollo–HP 9000/300 and HP Apollo 9000/400, HP 9000/800, DN 2500–4500, DN 10000

IBM–RISC System/6000

MIPS–RISComputer and RISCstation

NeXT–NeXTstation, NeXTcolor, NeXTcube and NeXTdimension

Silicon Graphics–Personal IRIS, Professional, Power Series

Sony–NEWS

Sun–3, 4, and SPARC-based systems

Now shipping *Mathematica* 2.0 on most platforms. Single-machine and Floating License options are available on workstation versions. Educational and student discounts are available. All specifications are subject to change without notice.

For information on these and other versions of *Mathematica*, visit your local software dealer, or contact Wolfram Research at 1-800-441-MATH.

Software Packages from This Book

Programs and data files used in *Implementing Discrete Mathematics* are available on Macintosh and MS-DOS diskettes from Wolfram Research for $15.00. To order a diskette, fill out the following information and return this card with payment in an envelope or by business reply mail. The programs are also distributed with *Mathematica* Version 2.0 in `Packages`DiscreteMath`Combinatorica`

Please type or print legibly

Name:

Organization:

Street:

City, suburb, or ward: State, province, or prefecture:

Country: Postal code:

Telephone: Email:

Disk format ☐ Macintosh ☐ MS-DOS ☐ 3.5" ☐ 5.25"

Include $15.00

☐ Enclose check drawn on U.S. bank

☐ VISA/Master Card # Expiration date / /

Signature

BUSINESS REPLY MAIL

FIRST CLASS MAIL PERMIT NO. 1212 CHAMPAIGN, ILLINOIS

POSTAGE WILL BE PAID BY ADDRESSEE

Discrete Mathematics Disk
Wolfram Research, Inc.
P. O. Box 6059
Champaign, IL 61826-9905